THEORY AND PRACTICE OF BIOLOGICAL WASTEWATER TREATMENT

NATO ADVANCED STUDY INSTITUTES SERIES

Proceedings of the Advanced Study Institute Programme, which aims at the dissemination of advanced knowledge and the formation of contacts among scientists from different countries.

The series is published by an international board of publishers in conjunction with NATO Scientific Affairs Division

A Life Sciences	Plenum Publishing Corporation
B Physics	London and New York
C Mathematical and Physical Sciences	D. Reidel Publishing Company Dordrecht and Boston
D Behavioural and Social Sciences	Sijthoff & Noordhoff International Publishers B.V.
E Applied Science	Alphen aan den Rijn, The Netherlands and Germantown, Maryland, USA

Series E: Applied Science - No. 35

THEORY AND PRACTICE OF BIOLOGICAL WASTEWATER TREATMENT

edited by

KRITON CURI

Assistant Professor
Civil Engineering Department
Bogazici University, Istanbul

and

W. WESLEY ECKENFELDER, Jr.

Distinguished Professor
Vanderbilt University
Nashville, Tennessee

SIJTHOFF & NOORDHOFF   1980
Alphen aan den Rijn, The Netherlands
Germantown, Maryland, U.S.A.

Proceedings of the NATO Advanced Study Institute on
Theory and Practice of Biological Wastewater Treatment
Istanbul, Turkey
July 1976

ISBN-13: 978-94-009-9138-5   e-ISBN-13: 978-94-009-9136-1
DOI: 10. 1007/978-94-009-9136-1

FOREWORD

Among the challenges to mankind, few are more critical than the need to protect the environment. The rapid increase in population, coupled with the enormous rate on industrialization had a negative effect on the realization of this goal. As a result, the environment (water, air and earth) has been deteriorating more and more every day. It is only with the proper treatment of the wastes which are produced by man and his activities, that this deterioration can be stopped.

Wastewater, is a major polluter of the environment. Although in many areas, science and technology has reached a level capable of preventing pollution, the reduction has not been realized for two reasons:

(a) Lack of communication and transfer of knowledge to the desired extend between engineers and scientists,

(b) Economic reasons.

Good knowledge of the *Biological Wastewater Treatment* processes is essential to overcome the economic handicaps. Because of that the improvement and dissemination of knowledge in this field was selected as the goal of the NATO - Advanced Study Institute held in Istanbul in July 1976. The lectures presented at this meeting have been compiled in the present volume.

The proceedings which follow are not arranged in order of presentation, rather they are grouped in a logical order. Problems solved during this ASI are given with their solutions in the last chapter. It is recommended that this chapter be referred to after reading each of the previous chapters.

VI

Thanks are expressed to all the lecturers. This volume is the result of their contributions. Special tribute must go to *W.W. Eckenfelder, Jr.,* Scientific Advisor of the ASI, for his invaluable assistance in helping to organize the seminar. Gratitude is owed to the Scientific Affairs Division of the North Atlantic Treaty Organization. The realization of this Advanced Study Institute became possible only with their financial support.

Appreciations must also be expressed to *Miss Meral Akyol,* for her skillful typing of the manuscripts and to *Mrs. Cana Balay,* secretary to the Director of the ASI for her assistance in general.

Bogazici University
October 1979

KRITON CURI
Director of A.S.I.

TABLE OF CONTENTS

VIII

CHARACTERIZATION OF WASTEWATERS

Kriton Curi

Civil Engineering Department,
Boğazici University, Istanbul, Turkey

A thorough determination and understanding of the characteristics of wastewater is essential in (a) selecting the best method of treatment, (b) designing the treatment facilities, (c) determining the efficiency of treatment facilities and (d) supervising them adequately.

A dependable characterization of wastewater is based on proper sampling, analyses and interpretation of results.

SAMPLING

There is no universal procedure for sampling. The sampling procedure should vary in accordance with the purpose of the particular investigation and the local conditions. There are, however, some requirements which are always valid. These are the following:

(a)   The samples should be *representative,* i.e. its composition should be identical with that of the material being sampled.

(b)   The sample should be preserved adequately, so that no significant change takes place in its characteristics between the time of collection and the performance of the tests. Detailed procedures for preventing biochemical and physicochemical change of the samples are given elsewhere (1). A brief summary of these procedures is presented in *Table 1.*

According to the character of the wastewater sampled and the purpose of the conducted study, *grab, composite* or *integrated* samples can be used.

TABLE 1 : Preservation of Wastewater Samples

Test to be Conducted	Maximum Storage time	Preservative	Temperature of Storage	Minimum Sample Size mℓ	Type of Container	Other Details
Acidity	24 hrs	-	4°C	100	Polyethylene, Borosilicate glass	1.Avoid sample agitation 2.Avoid prolonged exposure to air 3.Fill sample bottles completely and cap them tightly
Alkalinity	24 hrs	-	4°C	200	Polyethylene, Borosilicate glass	same as acidity
BOD	24 hrs (it is preferred not to exceed 6 hrs)	-	4°C	1000	Polyethylene or glass	
COD	As soon as possible	H_2SO_4 to pH=2	-	100	Polyethylene, or glass	
Chlorine	Immediately	-	-	500	Polyethylene, or glass	Avoid excessive light & agitation

TABLE 1 : Preservation of Wastewater Samples (continued)

Test to be Conducted	Maximum Storage time	Preservative	Temperature of Storage	Type of Container	Other Details
pH	Immediately			Polyethylene or Borosilicate glass	
Phenols	4 hrs or preserve for 24 hrs after adding preservative	H_3PO_4 to pH=4 and 1g $CuSO_4 \cdot 5H_2O/\ell$ of sample	4°C	Glass	If H_2S or SO_2 is known to be present, briefly aerate or stir the sample
Phosphate	Filter immediately after collection	40 mg $HgC\ell/\ell$ of sample	-10°C	Glass rinsed with 1 + HNO_3	Do not use detergent containing phosphate for glassware cleaning
Total Organic Carbon	As soon as possible	$HC\ell$ to pH=2	0°C (when $HC\ell$ is not used as preserver)	Glass (dark)	Minimize exposure to light

A *grab sample* represents only the composition of the wastewater at the particular moment of sampling, at that particular point from where the sample is collected. Another sample taken earlier or latter may show different characteristics. Grab samples are used when it is expected that the composition of the waste remains fairly constant with time or when it is desired to study the variations in the characteristics of the wastewater. In the latter case, a series of samples are collected at given time intervals and each of them is analyzed separately.

Composite Samples: The term *composite* refers to a series of grab samples collected from the same source at constant time intervals and then mixed in proportion to their rate of flow. This type of sampling is useful for determining the mean values of the characteristics of the wastes, but it cannot provide information about the extreme values. Usually a 24-hour period of sampling is considered satisfactory; sometimes, however, this time may be chosen equal to the time required to complete one cycle of a periodic operation.

Integrated Samples: are a mixture of grab samples collected at the same time from different points. This type of sampling may be used when it is expected that the characteristics of the sampled water vary from point to point and it is desired to determine the mean characteristics.

ANALYSIS for MAJOR
WASTEWATER CHARACTERISTICS

The procedures for the determination of the characteristics of wastewaters are given in the Standard Methods for the Examination of Water and Wastewater (1). In this paper, it is considered useless to repeat these procedures, instead the definition, significance and application of each characteristic will be examined.

PHYSICAL CHARACTERISTICS

(a)   Residue is the solid matter which may be suspended or dissolved in wastewater. This solid matter may be derived from domestic and industrial usage as well as from ground water infiltration, or storm water. Residue can be classified as *total residue* which can be separated to nonfiltrable (suspended) and filtrable (dissolved), *volatile residue, fixed residue, settleable matter*, etc. The residues as absolute values are of limited significance, they are, however, very useful as a control of inplant operations. The volatile residue reflects the biological stability, while the settleable residue approximates the quantity of sludge that may be removed by sedimentation. The filtrable residue (dissolved so-

lids), on the other hand, may affect many wastewater treatment processes. For effective biological treatment, the dissolved salt concentration should not exceed 16,000 mg/ℓ. Similarly, chloride concentrations of 8000 - 15,000 mg/ℓ have also adverse effect on biological systems (2).

(b)   Temperature is a very important parameter, because it
         effects the rates of chemical and biochemical reactions
as well as the adequate life. Furthermore, the saturation value of
oxygen in water decreases as temperature increases.

(c)   Color and Odor are two characteristics indicating the age
         of the sewage. Fresh sewage usually has a light yellowish-
grey color and an unnoticeable odor. As the oxygen is used up and
the sewage is aged, the color darkens and when finally anaerobic
conditions are reached, its color changes to black and its odor becomes unpleasant.

CHEMICAL CHARACTERISTICS

Organic Parameters

The amount of organic matter present in the wastewater is
one of the most important parameters effecting the design of biological wastewater treatment. It is a common practice in environmental engineering instead of trying to determine directly the amount
of each individual organic compound, to estimate their total amount
indirectly through some other parameters. Among these the Biochemical Oxygen Demand, the Chemical Oxygen Demand and the Total Organic
Carbon are the most important.

Biochemical Oxygen Demand (BOD) is the amount of oxygen required by microorganisms for the biological
oxidation or stabilization of the biodegrable organic matter.Thus,
BOD can be considered as an indirect method for the determination
of the gross concentration of organic matter present in the wastewater, or in other terms, for the determination of the strength of
the sewage, BOD is one of the most widely used tests. For   the
first time, it was proposed by the British Royal Commission on Sewage Disposal in the year 1913. In their report, it was stated
that the amount of dissolved oxygen required for the biochemical
oxidation of organic matter during 5 days at a temperature of 18.3°C
(65°F) be taken as a measure of the pollutant load of samples (3).
This method after twenty-three years was accepted as standard
method and in 1936 was incorporated in the Eighth Edition of the
"Standard Methods" accepting 20°C as the temperature of incubation.
From that time on, several attempts have been made in order to mo-

dify this test and/or to decrease the time of incubation (4, 5, 6, 7, 8, 9, 10, 11), but none of them has been accepted as "standard" method. It is important however, to note that manometric systems like the Warburg respirometer, as well as electrolysis systems, although not classified as "standards", are available (12). BOD being the most important parameter in relation to biological treatment of wastes will be examined in this section in details.

The BOD is usually separated as "first stage" or "carbonaceous" and "second stage" or "nitrogenous" BOD.

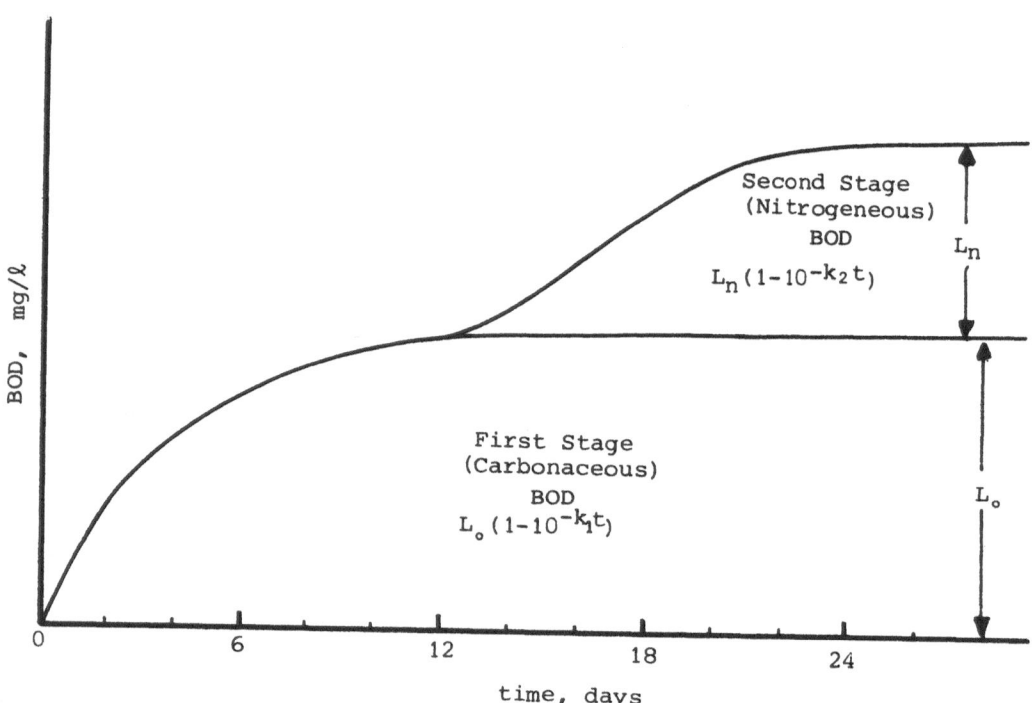

Fig. 1 : BOD Variation with Time

The first stage BOD is generally formulated as a first-order reaction, and it is expressed by

$$y_1 = L_0 (1 - 10^{-k_1 t}) \qquad \qquad \dots \quad (1)$$

where

y_1 = the amount of first stage BOD that has been exerted at any time t

L_0 = ultimate carbonaceous BOD

k_1 = rate coefficient of carbonaceous oxidation

t = time

On the other hand, the second stage BOD is expressed similarly as

$$y_2 = L_n (1 - 10^{-k_n t}) \qquad \dots \ (2)$$

where

y_2 = the amount of second stage BOD that has been exerted at any time t

L_n = ultimate nitrogeneous BOD

k_n = rate coefficient of nitrogeneous oxidation

t = time

The total amount of BOD exerted (y), at any time t is

$$y = y_1 + y_2 \qquad \dots \ (3)$$

In the first stage, the bacteria responsible for oxidation of the carbonaceous material are aerobic heterotrophs which multiply very rapidly. The bacteria, however, which are responsible for the second stage (*nitrosomonos, nitrobacter*, etc.) multiply rather slowly and their maximum population is reached after a relatively long period of time. Because of that the rates of nitrogeneous oxidation are initially much lower than those of carbonaceous oxidation. The oxygen demand due to nitrification becomes significant 8-10 days after the reaction has been started and a considerable part of the carbonaceous demand has been satisfied. For this reason, the standard 5-day BOD test cannot provide information for the extent of oxygen demand due to nitrification. A recent study conducted by Esen has shown that the results of the BOD_5 are sometimes misleading, especially when the oxygen deficiency of water bodies like estuaries and lakes is to be estimated (13, 14).

The rate coefficients k_1 and k_n as well as the ultimate BOD values L_o and L_n depend on the characteristics of the sewage as well as on the environmental conditions. As it is stated above, in most BOD studies usually only the first stage demand, is examined, because of that only the methods used for the determination of k_1 and L_o will be considered in this paper.

Determination of k_1 and L_0

For many years, k_1 was assumed to be equal to 0.1 day^{-1}, recent studies however have shown that k_1 rarely has this value and that it may vary from less than one half to more than twice of this (1).

Values which are recommended by Eckenfelder (25) as typical values of the mean rate constant k_1 are given in Table 2.

Table 2 : Typical k_1 Rate Constants at 20^0C

SUBSTANCE	k_1 (day^{-1})
Untreated Wastewater	0.15 - 0.28
High-Rate Filters & Anaerobic Contact	0.12 - 0.22
High-Degree Biotreatment Effluent	0.06 - 0.10
Rivers with Low Pollution	0.04 - 0.08

(b) Chemical Oxygen Demand (COD) is another indirect method used for the determination of the organic matter. According to Murray (16), COD test was first proposed in 1850 at Copenhagen by Professor Forschamer and from that time on was widely used. In this test, the oxygen equivalent of the organic matter that can be oxidized by a strong chemical oxidant is measured. As oxidant, potassium dichromate and as catalyst, silver sulphate is used. Because the amount of oxidizable organic matter is proportional to the amount of potassium dichromate consumed when the sample is refluxed, the amount of organic matter can be estimated easily. The reaction which takes place can be written as

$$C_xH_yO_2+Cr_2O_7^{=} + H^+ \xrightarrow[\text{catalyst}]{\text{heat}} Cr^{+++} + CO_2 + H_2O$$

Although COD test takes approximately three hours, attempts have been made to decrease this time further. The "rapid" COD test is one example (18). The precision of the COD test is 10.8% (1). The disadvantages of this method is that it does not give any information on the amount of the waste that can be decomposed by bacteria and the rate in which this decomposition can take place.

There are many different methods by which k_1 and L_o can be determined, after first a series of BOD measurements, at certain time intervals (say 1, 2, 3..... x days) are conducted. Three of them are summarized below.

i. Log-Difference Method

If the first stage BOD equation (Eq. 1) is differentiated with respect to t, it yields

$$\frac{dy}{dt} = r = 2.303\ L_o k_1 \cdot 10^{-k_1 t} \qquad \ldots\ (4)$$

In this equation, r is the rate of oxygen utilization with time. Taking logarithms of both sides

$$\log r = \log (2.303\ L_o k_1) - k_1 t \qquad \ldots\ (5)$$

is obtained. This is a semilog plot of r vs t, from which L_o and k_1 can be calculated. In other terms if daily differences $\Delta y/\Delta t$ are plotted vs time on a semilog graph paper, the slope of the best-fit straight line will be equal to $-k_1$ and the intercept value to $2.303\ L_o k_1$.

ii. Method of Moments (Moore, et al.,(26))

$\frac{\Sigma y}{L_o}$ vs k_1 and $\frac{\Sigma y}{\Sigma ky}$ vs k_1 curves are plotted

for n-day (that is 1, 2, 3 ..... n days) sequence of BOD measurements. Making use of the following equations

$$\Sigma y = L_o(1-10^{-1k_1}) + L_o(1-10^{-2k_1}) + \ldots + (1-10^{-nk_1})$$
$$= L_o\ (n-(10^{-k_1} + 10^{-2k_1} + \ldots 10^{-nk_1}))\ \ldots (6)$$

and

$$\frac{\Sigma y}{L_o} = n - \frac{10^{-k}(10^{-nk_1} - 1)}{10^{-k} - 1} \quad\quad \dots\dots (7)$$

Also

$$\Sigma ty = 1 L_o(1-10^{-1k_1})+2L_o(1-10^{-2k_1})+$$

$$\dots\dots+ nL_o(1-10^{-nk_1})$$

$$= L_o(\sum_{i=1}^{i=n} t_i - \sum_{i=1}^{i=n} t_i \cdot 10^{-k_1 t}) \quad\quad \dots\dots (8)$$

From Equations 7 and 8,

$$\frac{\Sigma y}{\Sigma ty} = \frac{n - (10^{-k}(10^{-nk_1}-1)/(10^{-k_1}-1))}{\displaystyle\sum_{i=1}^{i=n} t_i - \sum_{i=1}^{i=n} (t_i \cdot 10^{-k_1 t_i})} \quad\quad \dots\dots (9)$$

can be obtained.

From Equations 7 and 9 it can be easily seen that both $\Sigma y/L_o$ and $\Sigma y/\Sigma ty$ are only functions of k_1. Thus, for a given n one can assume values of k and plot $\Sigma y/L$ vs k_1 and $\Sigma y/\Sigma ty$ vs k_1 (See *Fig. 2*).

Using these curves and the results of a sequence of daily BOD measurements k_1 and L_o can be determined.

EXAMPLE: Given the following BOD determinations at 20°C, find k_1 and L_o

Time, days	1	2	3	4	5	6	7
BOD exerted, mg/ℓ	72	100	141	152	165	176	184

Solution: $\Sigma y=990$ $\Sigma ty= 4472$ $\Sigma y/\Sigma ty= 0.221$

From *Fig. 2* $k_1=0.178$ day^{-1} and $\Sigma y/L_o=5.15$ from which $L_o = \Sigma y/5.15 = 192$ mg/ℓ.

iii. Thomas' Graphical Method (26)

This method is based in the similarity of the following two functions:

$$(1-10^{-k_1 t}) = 2.3\ k_1 t \left[1- \frac{2.3k_1 t}{2} + \frac{(2.3k_1 t)^2}{6} - \frac{(2.3k_1 t)^3}{24} + \dots \right] \quad \dots\ (10)$$

and

$$2.3k_1 t \left[1+ \frac{2.3k_1 t}{6}\right]^{-3} = 2.3k_1 t \left[1 - \frac{2.3k_1 t}{2} + \frac{(2.3k_1 t)^2}{6} - \frac{(2.3k_1 t)^3}{21.6} + \dots\right] \quad \dots\ (11)$$

Equation 1 may therefore be approximated by relation

$$y = L_o (2.3k_1 t) \left[1+ \frac{2.3k_1 t}{6}\right]^{-3} \quad \dots\ (12)$$

which taking the inverse and rearranging takes the straight-line form

$$\left(\frac{t}{y}\right)^{1/3} = (2.3k_1 L_o)^{-1/3} + \frac{(2.3k_1)^{2/3}}{6\,L_o^{1/3}}\,t \quad \dots\ (13)$$

From Equation 13 is understood that if the results of a sequence of BOD measurements are plotted having $(t/y)^{1/3}$ as the ordinate vs t as the abcissa

$$k_1 = \frac{6b}{2.3a} = 2.61\,\frac{b}{a} \quad \dots\ (14)$$

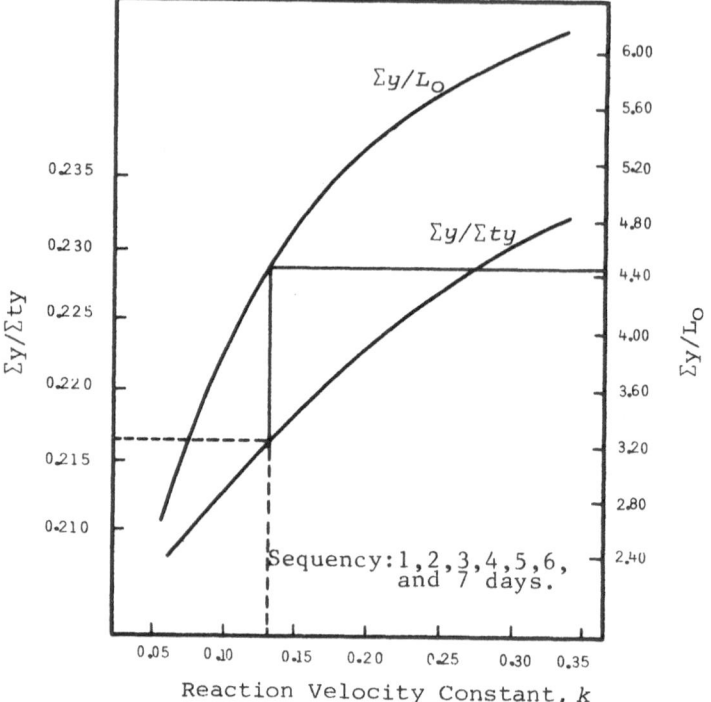

Fig. 2 :   Calculation of k_1 and L_o
from the Method of Moments
(Sequence of 7 days)

$$L_o = \frac{1}{2.3k_1 a} \qquad \qquad \dots (15)$$

where b = the slope of the line
a^3 = the intercept

iv. The Least-Square Method

This method involves fitting the best curve through
a set of data points. This is achieved using the
following equations;

$$na + b\Sigma y - \Sigma y' = 0 \qquad \dots (16)$$
$$a\Sigma y + b\Sigma y^2 - \Sigma yy' = 0 \qquad \dots (17)$$

where

  n = number of data points

$$y' = \frac{dy}{dt} = \frac{y_{n+1} - y_{n-1}}{24t}$$

From Equation 16 and 17 k_1 and L_o can be calculated by

$$k_1 = -\frac{b}{2.3} \qquad \dots (18)$$

$$L_o = \frac{a}{b} \qquad \dots (19)$$

Given the results of a sequence of BOD measurements, Σy, $\Sigma y'$, Σy^2 and $\Sigma yy'$ are calculated. Solving Equations 16 and 17, simultaneously a and b are determined and using these results and Equations 18 and 19, k_1 and L_o can be calculated.

Relationship Between k_1 and BOD_5/L_o

Equation 1 can be written for a incubation period of 5-days as

$$BOD_5 = L_o(1-10^{-5k_1}) \qquad \dots (20)$$

From this

$$\frac{BOD_5}{L_o} = (1-10^{-5k_1}) \qquad \dots (21)$$

can be obtained. Using Equation 21 *Fig. 3* can be prepared. As can be seen from this *Figure* BOD_5/L_o approaches the unity as the k_1 value increases.

Temperature Effects on the BOD

The BOD reaction rate coefficient k_1 is directly affected by temperature. According to Van't Hoff-Arrhenius Law the rate coefficient at any temperature T can be estimated by the following

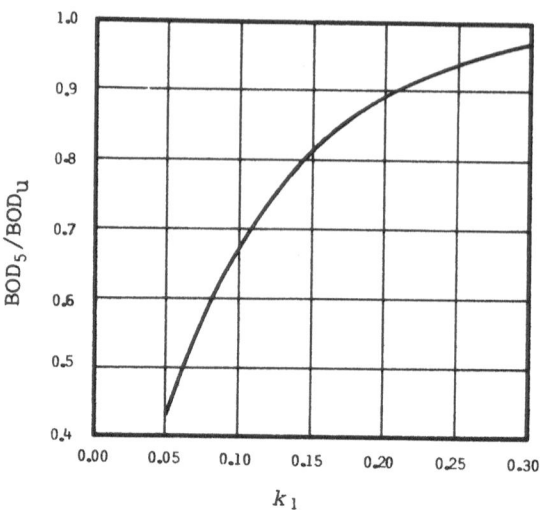

Fig. 3 :   Relationship Between k_1
           and BOD_5/BOD_u

equation:

$$k_T = k_{20} \theta^{(T-20)} \qquad \qquad \dots \text{(22)}$$

where

k_T = the rate coefficient at T°C

k_{20} = the rate coefficient at 20°C

The constant θ according to Phelps is equal to 1.047, Schroepfer
however recommends 1.056 for temperatures between 20-30°C, and 1.135
for temperatures between 4-20°C (25).

The BOD test, although widely used has two important
disadvantages: (a) the length of time required to run the test makes
the results sometimes practically useless, and (b) the precision of
the test (±15%) is poor (1).

Chemical Oxygen Demand (COD) is another indirect method
used for the determination of the organic matter. According to Murray (16), COD test was first proposed in 1850 at Copenhagen by Professor Forschamer and from that time on was widely used.
In this test, the oxygen equivalent of the organic matter that can
be oxidized by a strong chemical oxidant is measured. As oxidant,
potassium dichromate and as catalyst, silver sulphate is used. Because the amount of oxidizable organic matter is proportional to

the amount of potassium dichromate consumed when the sample is refluxed, the amount of organic matter can be estimated easily. The
reaction which takes place can be written as

$$C_xH_yO_2 + Cr_2O_7^= + H^+ \xrightarrow[\text{catalyst}]{\text{heat}} Cr^{+++} + CO_2 + H_2O$$

Although COD test takes approximately three hours, attempts
have been made to decrease this time further. The "rapid" COD test
is one example (18). The precision of the COD test is 10.8% (1).
The disadvantages of this method is that it does not give any information on the amount of the waste that can be decomposed by bacteria and the rate in which this decomposition can take place.

Total Organic Carbon (TOC) measures the carbon dioxide
produced upon complete oxidation of organic matter in the
waste. As can be seen from the reaction below

$$C_xH_yO_2 + O_2 \longrightarrow CO_2 + H_2O$$

the amount of CO_2 produced is directly proportional to the amount of
carbon in the organic material (15). In practice the test is performed by injecting a known quantity of sample in a special high-
temperature furnace, where the carbon turns to CO_2, in the presence
of a catalyst. The amount of CO_2 produced is measured automatically
by an infrared analyzer.

Due to the short time required to perform TOC test and its
precision (5-10%) (1), it is believed that soon it will replace
the other methods.

Other Methods which can be used for the determination of
the organic parameters, but which are not included in the
"Standard Methods" yet, are the Total Oxygen Demand (TOD) and the
Oxygen Demand Index (ODI) (19). Finally, the Theoretical Oxygen
Demand (ThOD), which corresponds to the stochiometric amount of oxygen required for the complete oxidation of a given sample is an
other method which can be used.

RELATIONSHIP BETWEEN THE RESULTS of THE DIFFERENT
METHODS USED FOR THE DETERMINATION
of THE ORGANIC PARAMETERS

In order to be able to correlate the results of the different
methods used for the determination of organic pollutants present in
a wastewater, it is necessary to state clearly what each of them is
measuring. COD and TOC test are measuring indirectly the strength
of the organic pollutants by means of a chemical reaction, while
BOD determines the same parameter by means of a biological activity.
Results of BOD tests depend on variables like toxicity of the sample,
acclimation of the seed, pH, etc. The same parameters however do
not effect the results of COD and TOC tests. Non-biodagrable matter can not be determined by BOD test but it is easily detectable
by COD and TOC tests. Similarly some organic compounds may not be
oxidizable by dichromate and thus not included in the results of
COD test, but the same compounds can be determined by the TOC test.

When the nature and source of waste is constant, the development
of a relationship between the different methods is possible. Usually TOC, COD and TOD can be correlated to each other, but their correlation to BOD is difficult, because of the problems involved in
the biochemical reaction. According to Rhame (17) a dependable correlation between the first stage BOD and COD exists. It is reported
that in U.S.A., the ratio BOD/COD for typical domestic waste varies
from 0.4 to 0.8 (12).   This result is verified also by values reported by Eckenfelder (2).   Studies conducted in Turkey, however,
indicated that the ratio BOD/COD for domestic wastes may vary between 0.2-0.6 with an average of 0.5. Ratios between BOD/TOC in domestic wastes are reported to vary from 0.8-1.0 (12) or 0.35-2.62
(21) or to be equal to 1.87 (22). Values obtained in Turkey give
this ratio as 0.6-1.2 (23).   Finally the ratio between COD/TOC from
a stoichiometric point should be equal to 2.66, however, experiments
have proven that this varies from 1.75 to 6.65 (2). The large differences between the values reported above indicates that it is not
possible to convert with an acceptable accuracy results of one form
to another if the constituents of the sewage do not remain relatively constant. However, it is apparent that the following relationship always exist.

$$ThOD > COD > BOD$$

INORGANIC PARAMETERS

The inorganic parameters which are used for the characterization of a wastewater are many. The most important among them are:

(a)   pH   is a parameter which has a direct effect on the treat-
      ability of wastewaters by biological means. Usually a pH
varying between 6.5 - 8.0 is required for a proper biological
treatment.

(b)   Nitrogen: Determination of nitrogen compounds was a test
      which was used for many years for examining the sanitary
quality of waters. This was based on the fact that the form of
nitrogen is changing with time in the following way.

       Organic Nitrogen → Ammonia → Nitrite → Nitrate

     Thus, waters containing organic or ammonia nitrogen can
be considered as recently polluted, while those containing nitrate are polluted long time ago.

     The significance of nitrogen in biological wastewater
treatment arises from the fact that it is an essential nutrient
required for the growth of microorganisms which are responsible
for the biological decomposition. Because of that, addition of
nitrogen compounds to the wastewater is sometimes required in
order to make the wastes biologically treatable.

(c)   Phosphorus is another nutrient which like nitrogen is es-
      sential for the biologically treatability of wastes. In
natural waters phosphorus occurs usually in the form of ortho-
phosphate, condensed phosphate (pyrophosphate, metaphosphate
and polyphosphate) and organic phosphate.

(d)   Toxicity: Wastewater, especially those originating from
      industrial sources contain certain elements, like arsenic,
boron, copper cyanide, lead etc., which are toxic to microorganisms responsible for biological treatment. Toxic threshold for
copper, zinc, cadmium, etc., has been reported as approximately
1 mg/ℓ (2).

COMPOSITION of WASTEWATERS

     The composition of wastewaters is an important parameter used
in the design of wastewater treatment plants. The character of sewage differs from place to place according to habits and living
standards of population. *Table 3* gives the characteristic values
for different countries, while *Table 4* gives the daily per capita
wastewater and BOD production.

TABLE 3 :   Wastewater Characteristics

Country/City	BOD mg/ℓ	COD mg/ℓ	Total Solids mg/ℓ	Total N mg/ℓ	Total P mg/ℓ	Source
Brazil						
Belem	110	204	227			(27)
Sao Paulo	250	445				(28)
Rio de Janeiro						
Jacarepogua	200		650	15	7.0	(29)
Downtown	407		670	33		(30)
Penha	382		784	25		(31)
S.Cristovao	200	710				(31)
Alegria	200	630				(31)
China						
Chung-Hsin	73					(32)
Keelung	200					(32)
Germany	300		900	8-15		(33)
Ghana						
Acra-Tema	280		219			(32)
Greece						
Athens	350	910	950			(34)
India						
Ahmadabad-Guayat	216		1408	18.7	7.2	(35)
Kodungaiyur	282			30		(36)
Iraq						
Baghdad	342-432					(32)
Israel						
Haifa	340	870	500	40		(37)
Italy						
Milano	156	221	561	38	23	(38)
Japan						
Kyoto	93-114					(32)
Tokyo	88-158					(32)
Kenya						
Nairobi	448			67		
Korea						
Seoul	312					(32)
Peru						
Lima	175					(39)
Philippines						
Manila	120	93-725				(32)
Turkey	150	350	1100	30	4.5	(20,23,24)
U.S.A.	200	500	700	40	10	(12)

TABLE 4 :   Amount of Wastewater and Pollutants
Produced Per Capita Per Day

COUNTRY	Q ℓ/cap-day	BOD g/cap-day	Source
Brazil	250-350	76-120	(32)
Denmark	310	57	(40)
Finland	610	100	(41)
France	200	24-34	
Germany	153-280	54-71	(43)
Ghana	140	44	(32)
Greece	110-138	45	
India		85	(44)
Kenya		23	
Korea	51-192	60	(32)
Philippines	144	45-52	(32)
Swiss	200-350	75	(42)
Turkey	130	32	(20)
U.K.	180	54	(45)
U.S.A.	610	69	(45)
Zambia		36	

Differences in composition of industrial wastewaters are larger and most probably caused from differences in the production processes (20). Experience has shown also that the composition of wastewater in developing countries is changing drastically with their development. This is one reason why the interpretation of the results obtained from different analyses should be done carefully and the necessary predictions for future characteristics should be made when these values are to be used as data for the design of treatment facilities.

REFERENCES

1. *Standard Methods for the Examination of Water and Wastewater,* 14th Edition, American Public Health Association, Washington, 1976.

2. Eckenfelder, W. W. and Ford, D.L., *Water Pollution Control,* The Pemberton Press Jenkins Book Publishing Company, New York, 1970.

3. McGowan, G., et. al., Determination of Dissolved Oxygen Absorbtion in 5-days, Royal Commission on Sewage Disposal, 8th Report, H.M.S.O., London, 1913.

4. Orford, H.E. and Matusky, F.E., Comparison of Short-Term with 5-day BOD Determination of Raw Sewage, Sewage and Industrial Wastes, 31, 3, 1959.

5. Tidwell, W. and Sorrels, J.H., Comparative Evaluation of Special Two-Day and Standard Five-Day BOD Tests, Sewage and Industrial Wastes, 28, 488, 1956.

6. LeBlanc, P.J., Review of Rapid BOD Test Methods, Journal WPCF, 46, 2202, 1974.

7. Domaniç, R., The Reaeration BOD Technique, M.S. Thesis, Boğaziçi University, Istanbul, 1974.

8. Cadwell, D.H. and Langelier, W.F., Manometric Measurement of the Biochemical Oxygen Demand of Sewage, Sewage Works Journal, 20, 202, 1948.

9. Jenkins, D., The Use of Manometric Methods in the Study of Sewage and Trade Wastes, Waste Treatment, Pergamon Press, New York, 1960.

10. Arthur, R.M., Let's Upgrade the BOD Test, Water and Sewage Works, 121, 6, 1974.

11. McGhee, T.J. and Walsh, R.D., The Effect of Mixing on the BOD Determination, Water and Sewage Works, 121, 3, 1974.

12. Metcalf and Eddy Inc., *Wastewater Engineering,* McGraw Hill Book Co., New York, 1972.

13. Esen, İ.İ. and Alpay, C., Göl Kirlenmesinde İkinci Kademe Biyo-kimyasal Oksijen İhtiyacının Etkisi (A Stochastic Birth-and-Death Model for Predicting the Effect of Second Stage BOD in Lakes), TBTAK, Ankara, 1975, (in Turkish).

14. Esen, İ.İ., Second Stage Biochemical Oxygen Demand, Boğaziçi University Journal, Engineering, 2, 117, 1974.

15. Clark, D.W., BOD: A Re-Evaluation, Water and Sewage Works, 121, 5, 1974.

16. Murray, K.A., The Oxygen Consumed Test: Historical Review, Pro-ceedings of Institute of Sewage Purification, 9, 5, 1956.

17. Rhame, G.A., Rationalization of the COD-BOD Relationship, Water and Sewage Works, 121, 11, 1974.

18. Jeris, J.S., A Rapid COD Test, Water and Wastes Engineering, 1967.

19. Reynolds, J.F. and Goelliner, K.A., Statistical Evaluation of BOD5 versus ODI, Water and Sewage Works, 121, 1, 1974.

20. Ward, R.F., Curi, K. and Dakkak, J., Domestic and Industrial Waste to Discharges to Izmit Bay, Boğaziçi University Journal, 1976.

21. Mohlman, F.W. and Edwards, G.R., Ind. Eng. Chem., Anal., Ed., 119, 1931.

22. Wuhrmann, K., Hauptwirkungen und Wechsel Wirkkengen Einiger Bet-riebs-parameter im Belebschlammsystem Ergebnisse Mehrjahriger Grossversuche Vortag, Zurich (1964).

23. Curi, K., et. al., Gemlik Deniz Deşarjı Araştırma ve Tatbikat Projesi, Mart-Nisan 1976 Faaliyet Raporu, Istanbul, 1976 (un-published report).

24. CAMP. TEK-SER, Istanbul Sewage Project, Master Plan Revision, Istanbul, 1975.

25. Eckenfelder, W.W., Jr., Water Quality Engineering for Practicing Engineers, Barnes and Noble, New York, 1970.

26. Moore E.W., Thomas H.A. Jr., and Snow W.B., Simplified Method for Analysis of BOD Data, Sewage and Industrial Wastes, 22, 1343, 1950.

27. ESB-CENSA, Preliminary Report for the Subaquatic Outfall for the City of Belem, 1976.

28. GEGRAN, Master Plan for Sewage Collection, Treatment and Dispo-sal, 1974.

29. ENCIBRA S/A, Master Plan for the Sewage System for Jacarepagua and Barra da Tijuca Regions of Rio, 1975.

30. ENCIBRA, S/A and Engineering Science, Inc., A Master Plan for Waste Disposal for the City of Rio de Janeiro, 1965.

31. ENCIBRA S/A and Engineering Science Inc., Submarine Outfall Studies and Design for the City of Rio de Janeiro.

32. Schmidt, O.J., Master Sewage System Plan for Metropolitan Manila, ASCE, JSED, 98, SAI, 1972, 125-152.

33. Mayer, V., Personal Communication.

34. Environmental Pollution Control Project-Athens, Interim Technical Report, Ministry of Social Services, Athens, 1976.

35. Mehta, M.D., Personal Communication.

36. Raman, A. et. al., Low Cost Waste Treatment, CPHERI, Nagpur, 1972.

37. Rebhum, M. and Streit, S. "Physico-Chemical Treatment of Strong Municipal Wastewater", *Water Research,* 8, 1974, 195-24.

38. Frangipane, E.F., Personal Communication.

39. Valdez-Zamudio, F., Science of the Total Environment, 2, 406, 1974.

40. Løholtt, J., Raspildevans Indhold af BIS, N og P, Stads-og Havneingeniøren, 64, 7, 1973 (in Danish).

41. Drifts Problem vid Avloppsreningsverk, Nordforsk, Milfövardssekretariatet, Publication 1975: 9, Helsingfors 1975 (in Danish).

42. Swiss Federal Office for Environmental Protection, Personal Communications.

43. Pöpel, F., Lehrbuch für Abwassertechnik und Gewässerschutz, Deutcher Fachschriften-Verlag, Mainz-Wiesbaden.

44. Sehgal, J.R. and Siddigi, R.H., Characterization of Wastewater for Kampur City, Environmental Health (Nagpur), 11, 1969, 95-107.

45. Painter, H.A., Chemical, Physical and Biological Characteristics of Wastes and Waste Effluents, in Ciaccio L.L. (editor) Water and Water Pollution Handbook, V. 11, Marcel Dekker, Inc., New York.

MICROBIOLOGY OF WASTEWATER TREATMENT

Wolf-Dietrich Linke

Institut fur Siedlungswasserbau und
Wassergutewirtschaft
der Universitat Stuttgart

Classical microbiology investigates the morphological attributes, nutritional requirements and metabolism of bacteria, actinomycetes and fungi. Microbiology as the science of microorganisms is not determined by the size of the organisms, but by the fact that these organisms are 1) unicellular and 2) that these single cells have their own physiology without the need to live with other cells in a community. This means that one singular cell has an independent metabolism with nearly all and the same metabolic pathways that higher organisms have. Pure microbiologists exclude the unicellular protozoae and sometimes even algae because some of them do not live as single cells, forming long chains even though they are able to live as single cells.

One area of research in microbiology having great economic value is the investigation of the transformation of substances. Some of these transformations such as baking bread and making cheese and wine, were dealt with long before microorganisms were detected as their cause. A less pleasing aspect of this conversion ability is the great variety of illness and disease in man, animals, and plants caused by bacteria, fungi, or viruses. Modern microbiology has economic value in the production of metabolites as ascorbic, glutaminic, lactic or citric acid, polysaccharides, alcohols (isopropanol, butanol and ethanol), and the wide spectrum in antibiotics.

Fermentation is defined as microbial metabolism in which energy is derived from the use of organic compounds as both electron donors and electron acceptors. Fermentations are done with a defined substrate under sterile conditions and optimal pH, temperature and aeration, and with a pure culture of one well-known

strain of bacterium, actinomycete or fungus. Because none of
the criteria just listed are fulfilled in biological wastewater
treatment, it is not considered to be fermentation in the classical
sense.

APPLICATION TO WASTEWATER TREATMENT

A microbial physiologist or biochemist is allowed to simplify
complex systems by working with synthetic media, optimal substrate
concentrations, cell extracts, etc., in order to obtain reprodu-
cible results from experiments under well-defined conditions.
However, these results are not very helpful as design criteria
for biological waste treatment because of the total disregard for
the process' complexity. Steady state systems, though not repro-
ductions of natural conditions, do offer the possibility of studying the influence of one or a few parameters on the system.
Continuous culture systems are of great interest, especially with
respect to the behavior of mixed bacterial populations and the
mode of conversion of complex substrates. Investigations in
continuous culture have been made by Jannasch and Mateles [1],
Jones [2], Tempest [3] and others in order to make an approach
describing activated sludge processes mathematically.

All sorts of microorganisms are present on principle   in biological wastewater. In order to specify we may begin in the region
of a hundred nanometer in size, that is 10^{-7} m, where the virus
may be seen in an electron microscope. These are the smallest
biological units but they need a host cell to augment life. Nearly
the same size are ricketsiae which are similar to viruses but have
their own enzyme systems. These microorganisms are responsible
for spotted fever and similar deseases. The group of bacteria
shows an enormous variety in size, form, possible substrate, metabolic end products and tolerance against environmental conditions
such as acidity, temperature and dryness. Yeasts, actinomycetes
fungi and protozoa shall only be listed though their activity in
biological wastewater treatment is not negligable.

Bacteria are found everywhere in nature-in water, soil, and in
air. Their distribution and concentration depends on the presence
of several factors, and because of the range of environmental
conditions, they have developed into a great number of species.

Bacteria are classified by their morphology and their physiological behavior while other characteristics, such as motility,
necessity for oxygen (aerobic or anaerobic), and pigments are
means of further definition. Bacteria form three groups morpholo-
gically:rods, called "bacillus", spheres, called "coccus", and
spirals, called "spirillum". The rod, as the most common form,

can be distinguished in three configurations: individual cells, diplo cells, and chains of cells. These types can also be observed in the coccus forms, but they extend beyond these aggregations to cubes (Sarcina) and clusters (Staphylococcus). Sizes vary rather widely within the range of 1 to 15 μm.

According to their physiological behaviour, microorganisms which gain their energy from oxidation and reduction processes are described as "chemotrophs". Where light is the donor of energy (plants, blue algae, thiobacteriaceae), they are called "phototrophs". In addition, all phototrophs and nitrifying bacteria are called "lithotrophs" in that they have all inorganic proton-donors: H_2, NH_3, Fe^{2+}, H_2S. By comparison, organisms consuming organic compounds-animals and most microorganisms-are called "organotrophs".

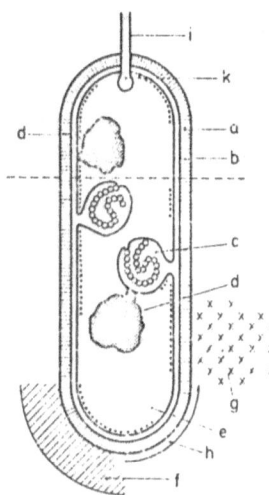

a. Cellwall
b. Cytoplasmatic Membrane with assoc associated Ribosomes
c. Mesosomes
d. Nucleus Material
e. Cytoplasma
f. Macro Capsule
g. Slime
h. Micro Capsule
i. Flagellum
k. Basal Corpus

Below: Association of nucleus material with mesosomes (*bacillaceae*).
Above: Association of nucleus material with peripher cytoplasmatic membrane (*entero-bacteriaceae*).

Fig. 1: Procaryotic Cell Type (bacterial cell)

One method for distinguishing bacteria is their behaviour against coloring after Gram, where the mass of bacteria are divided into "Gram-positive" and "Gram-negative" bacteria. Whereas the Gram-positive cells hold back the iodine color-complex and become blue, the Gram-negative cells become red when colored with carbon-fuchsin. This differing behaviour is explained by the construction of the cell wall, which in Gram-negative bacteria contains lipo-proteins and lipopolysaccharides, where the others do not.

The metabolism of a cell or organism may be defined as the totality of the chemical processes that it can perform. This implies that the metabolism of even a simple unicellular organism may not be specifically expressed at a given instant. Control of metabolism is performed by an intricate set of interrelated checks and balances having both intrinsic and extrinsic components. Intrinsic factors are understood as genetic whereas extrinsic factors are environmental or physiological. These two effects rule the metabolism as a whole.

The metabolic pathways, that is the network of connected enzyme-catalyzed chemical reactions, design the transformations of certain organic compounds vital to the organism. These compounds are referred to as metabolites.

Metabolic pathways are classically divided into two types: catabolic and anabolic. Catabolic routes define degradative processes. Large organic molecules, supplied as food for the organism are broken down and with the aid of oxidation type reactions transformed into simpler cellular constituents with the consequent release of chemical energy. This energy is then utilized by the organism for its maintenance, growth and replication, and for transformation into other forms of energy. Anabolic routes, on the other hand, define synthetic processes, in which complex organic cellular constituents are produced from simpler precursors.

CELL

The cell is a self-regulating system controlled by feedback mechanisms. The regulating mechanisms enable the cell to coordinate the metabolic functions of the single enzyme systems in an economical manner. In this way, an excess of end-product production is avoided. On the other hand, it is guaranteed that substances necessary for life are present at the right time and in the required quantity. A change in environmental conditions is followed by an adjustment of the cell metabolism. The adjustment of metabolism is possible by

1. Regulation of the enzyme quantity
2. Regulation of the enzyme activity
3. Competition for substrate, phosphate, co-enzyme.

The quantity of an enzyme may be controlled either by the rate of synthesis or by rate of reduction. Microbes can take up foods with low molecular weight, like oligopeptides, nucleotides and small organic phosphates. To use macromolecular food, -the predominant form of dead plants or animals returning to the soil-, preliminary extracellular hydrolysis is necessary. This is achieved by the secretion by various gram-positive bacteria and fungi, exoenzymes into the medium (extracellular enzymes). Others

remain attached to the cell (surface enzymes). Extracellular
enzymes include proteases and peptidases, polysaccaridases,
nucleases, hipases, etc.. Neither the mechanism of secretion
of the enzymes nor the reason why only gram-positive but not
gram-negative bacteria produce enzymes is yet understood.

EFFECT OF OXYGEN

Bacteria are classified into several groups with respect to
the effect of oxygen on their growth. The "aerobes" grow exclusively in the presence of oxygen while the "anaerobes" grow
only in the absence of oxygen. "Facultative" organisms are able
to grow both under aerobic and anaerobic conditions; the "micro-
aerophile" forms need for growth a reduced oxygen tension.

The aerobic organisms have a complete respiratory chain in
which the subtrate serves as electron donor and the oxygen as
electron acceptor. The electrons flow from DPNH (Diphospor
Phyridine-Nucleotide) to a flavoprotein and via cytochromes to
oxygen. In the accompanying process of oxidative phosphorylation,
three molecules of ATP (Adenin-Tri-Phosphate) are generated per
two [H] oxidized. Calculated with an energy content of -12000 cal
(ΔG) of one molecule ATP the total energy release of $\Delta G = -52000$
cal corresponds to an efficiency of 70%. This efficiency is
normally given as the p/o ratio which in bacteria is only 1 or
less. The rest appears in the medium as heat.

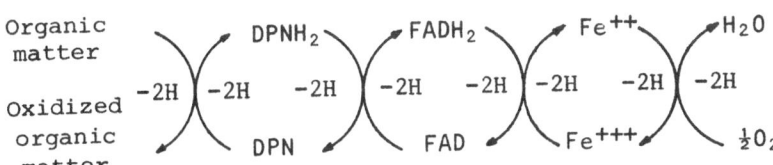

Fig. 2 : Scheme of aerobe hydrogen transfer (respiratory chain)

For aerobic bacteria oxygen is an absolute requirement in the
same way as it is toxic for anaerobes. This means that in the
presence of oxygen anaerobes are prevented from continuing to
ferment and grow. The probable mechanism is that certain enzymes
are maintained in an oxidized state which prevents them from
carrying out an essential reductive reaction. Some anaerobes such
as Clostridia are only inhibited but are killed by the presence
of oxygen.

Oxygen supply is essential for successful aerobic biological wastewater treatment. The difficulty with this process is that the maximal oxygen concentration that aeration provides, decreases with increasing temperature as well as when microbial activity increases and more oxygen is consumed. Neglecting the source and the manner in which oxygen supply is done, oxygen has to be delivered subsequently because its consumption is often as great as the input to the system and sometimes requirements are even greater. Whether or not oxygen becomes a limiting factor in an activated sludge system depends on the equipment for aeration and/or the sludge loading. If oxygen concentration decreases to a critical value, efficiency will decrease as well. This can lead to several factors such as reduced substrate oxidation, death and decay leading either to new substrate or to toxic substances for still living bacteria. The content of living and substrate reducing bacteria in activated sludge plants may often be limited by insufficient oxygen supply. Raising the oxygen supply up to saturation concentration, neglecting costs, will undoubtedly have a positive effect. Whether this is done by hyperbaric aeration or by use of pure oxygen or hydrogen-peroxide, the microbiology will not be affected. ZoBell and Hittle [4] reported no negative effect of dissolved oxygen concentration up to 35 mg/ℓ to Bacillus subtilis, Bacillus mega-terium, E.coli and others. In contrast to this, Gottlieb [5] is of the opinion that hyperbaric oxygen could be toxic to micro-organisms. Chin et al [6] report on the successful use of hydrogen-peroxide for activated sludge treating domestic and other wastewaters. Investigations of Cole et al [7] with H_2O_2 showed success in the control of bulking sludge because free growing filaments were eliminated. Hydrogen-peroxide, said to be toxic to bacteria, is really not when being applied as an additional oxygen source in activated sludge treatment because it is immediately disintegrated

$$2H_2O_2 \rightarrow 2H_2O + O_2 \qquad \qquad \text{.... (1)}$$

due to heavy metal ions and the enzyme catalase being present in activated sludge.

EFFECT OF TEMPERATURE

Temperature is a main factor influencing reaction rates in chemical as well as in biological systems. Microorganisms possess as a specific characteristic a temperature optimum which is defined as the range of maximum growth; however, fermentation and respiration rates may have a different optimum. Most micro-organisms show optimum growth between 25^0 and 37^0C, a temperature

not usual for activated sludge treatment. The psychrophiles, some
fungi and bacteria, especially the Pseudomonas species, reach
optimal growth beyond 20°C. Thermophile algae, fungi, actinomycetes and bacteria show optimal growth at temperatures of 65°C,
and, in a few cases, even 98°C. Microorganisms are adaptive and
can grow in a wide range of approximately thirty degrees, but
maximum growth is in a narrow range around their optimum. Benedict
and Carlson [8] reported that acclimatization of an activated sludge
at low temperatures was essentially complete within two weeks
after shock temperature change of 10° to 15°C and that it took
months to acclimate at above 30°C.

EFFECT OF pH

In biological wastewater treatment pH becomes an important
parameter when it does not remain constant. Variations in pH
happen in systems with insufficient buffer capacity. As a result
of a decrease in pH value, the rate of metabolism falls until
acids have been metabolized and the pH rises again. With a change
of pH other parameters such as solubility of salts, ionization of
substances and the carbonic acid buffer system change as well.
pH-values observed normally in activated sludge are between 6.5
and 9.0. The maintenance of a definite pH level is a function of
buffer capacity. This is demonstrated by intensive nitrification
in connection with a low buffer capacity when pH will drop to
values beyond 6.0.

The growth or multiplication of a pure culture of microorganisms
is a succession of phases with different growth rates. A unit
volume of growing culture containing x_1 cells at time t_1 after n
divisions it will contain

$$x = x_1 \cdot 2^n \qquad \qquad \dots (2)$$

If r is the division rate per unit time, then at time t_2, the
number of cells will be

$$x_2 = x_1 2^{r(t_2 - t_1)} \qquad \qquad \dots (3)$$

The classical pattern of the different phases of a growing
bacterial culture are given by Monod [9] by means of the variation
of the division rates μ:

1. lag phase $\mu = 0$
2. acceleration phase μ increases
3. exponential phase $\mu = $ constant

4. retardation phase μ decreases
5. stationary phase $\mu = 0$
6. phase of decline μ negative

This is a rather generalized picture of the growth of a bacterial culture. In many cases, one or several of these phases may be absent. Lag and acceleration phases may often be suppressed, possibly due to enzymatic adaptation. The retardation phase is frequently too short to remark what may be true to the stationary phase in batch culture.

In activated sludge systems only the exponential phase, steady state phase and phase of decline are observed. The different steps of growth, in these systems, depend on sludge loading. High loading (0.8 - 3.0 kg BOD/kg MLSS d) will bring the biomass to exponential growth. The stationary phase is characteristic for low loading (< 0.3 kg BOD/kg MLSS d) whereas the decline phase indicates sludge stabilization.

Diauxie is the division in the development of a pure culture containing a mixture of substrates into cycles of growth separated by a lag phase. Interactions in the synthesis of adaptive enzymes were detected by this phenomenon. The course of two cycles corresponding to two substrates is given in *Fig. 3* [Monod, 9].

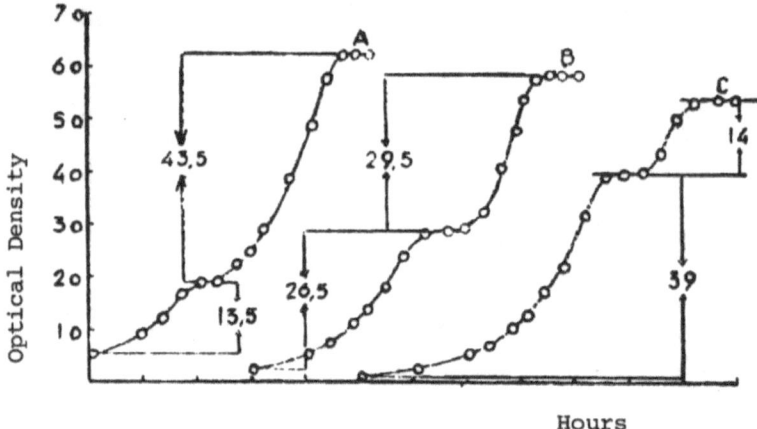

Fig. 3: Diauxie. Growth of E.coli in synthetic medium with glucose and sorbitol as carbon source.

The evidence indicates that each of these cycles corresponds to the exclusive utilization of one of the constituents of the mixture. This effects of enzymatic adaptation shows an inhibitory effect of the compound being consumed on the formation of the enzyme attacking the other. At an activated sludge plant with two major components this may lead to a reduction of the one compound without remarkable decrease of the other's concentration.

The microorganisms in a trickling filter reflect the faculta- tive nature of the filter. The predominant microorganisms are aerobic, facultative and anaerobic bacteria. The aerobic spore- formers Bacillis are easily found in the upper aerobic surfaces of the filter. The great majority of bacteria in the filter are facultative, living aerobically as long as dissolved oxygen is present and anaerobically when oxygen is already consumed. This group is composed of various species of the genera Pseudomonas, Alcaligenes, Flavobacterium, Micrococcus as well as Enterobacteri- aceae. In the regions where oxygen is completely absent, the anaerobic Desulfovibrio exist.

Desulfovibrio is able to utilize sulfate as H-acceptor and convert it to hydrogen sulphide.

$$8[H] + SO_4^{2-} \rightarrow H_2S + 2H_2O + 2OH^- \qquad \text{.... (4)}$$

Organic acids and alcohols which are converted to acetic acid are mainly used as hydrogen donors.

Desulfovibrio is to be found in anaerobic digesters and is the organism responsible for the corrosion of iron pipelines:

$$4Fe + SO_4^{2-} + 4H_2O \rightarrow FeS + 3Fe(OH)_2 + 2OH^- \qquad \text{.... (5)}$$

The same effect may be caused by Thiobacillus aerobically. This group of bacteria is able to oxidize various sulfur compounds according to the following reactions:

$$S^{2-} + 2O_2 \rightarrow SO_4^{2-} \qquad \text{.... (6)}$$

$$2S + 2H_2O + 3O_2 \rightarrow 2SO_4^{2-} + 4H^+ \qquad \text{.... (7)}$$

$$S_2O_3^{2-} + H_2O + 2O_2 \rightarrow 2SO_4^{2-} + 2H^+ \qquad \text{.... (8)}$$

Thiobacillus thiooxydans is an example of toleration of extreme pH-value. It can survive in values as low as 1.0.

The nitrifying bacteria Nitrosomonas and Nitrobacter belong to the autotrophic chemolithotrophic microorganisms because they have inorganic H-donors and are using carbon dioxide as carbon source. In activated sludge treatment the presence and cooperative work of these bacteria are desired because they oxidize ammonia which is in its unionized form is toxic to fish, even in concentrations as low as 0.2 mg NH_3/ℓ.

$$2NH_4^+ + 3O_2 \rightarrow 2NO_2^- + 4H^+ + 2H_2O \qquad\qquad \cdots\cdots (9)$$

By a simultaneous reaction the buffer capacity of water is lowered.

$$4H^+ + 4HCO_3^- \rightarrow 4H_2O + 4CO_2 \qquad\qquad \cdots\cdots (10)$$

Nitrobacter oxidizes in the second step of nitrification, the nitrite to nitrate

$$2NO_2^- + O_2 \rightarrow 2NO_3^- \qquad\qquad \cdots\cdots (11)$$

and the overall equation for nitrification is:

$$NH_4^+ + 2O_2 + 2HCO_3^- \rightarrow NO_3^- + 3H_2O + 2CO_2 \qquad\qquad \cdots\cdots (12)$$

From the above equation it can be estimated that a loss of 7.14 mg/ℓ $CaCO_3$-alkalinity is taking place at the nitrification of 1 mg N/ℓ. As a result in water with insufficient buffer capacity a decrease of pH will follow and affect the efficiency of other microorganisms. For completing the biological nitrogen elimination this aerobic stage has to be followed by the anaerobic denitrification, where the nitrate is reduced to molecular nitrogen. The nitrate respiration occurs only in absence of oxygen, because, the nitrate oxygen serves as hydrogen acceptor in this case. Various organic compounds such as carbohydrates, organic acid, and methanol may be hydrogen donors.

Activated sludge is an accumulation of microorganisms which are classified as bacteria, fungi, protozoa, rotifers. On principle all sorts of bacteria are present in activated sludge, but the incoming wastewater with its various concentrations and composition of organic and inorganic matter determines which bacterial genera will predominate. For example, Prakasam and Dondero [10], Benedict and Carlson [11] sampled and identified the aerobic heterotrophic bacteria from activated sludge. Seiler [12] observed the variation in their concentration over a certain period. He observed approximately thirty organisms. Sphaerotilus which is said to cause bulking sludge is also present in many activated sludge plants. Hawkes (14) states that fungi are rarely present in activated sludge but he admits that this rarity may be due to lack of reports on this subject.

33

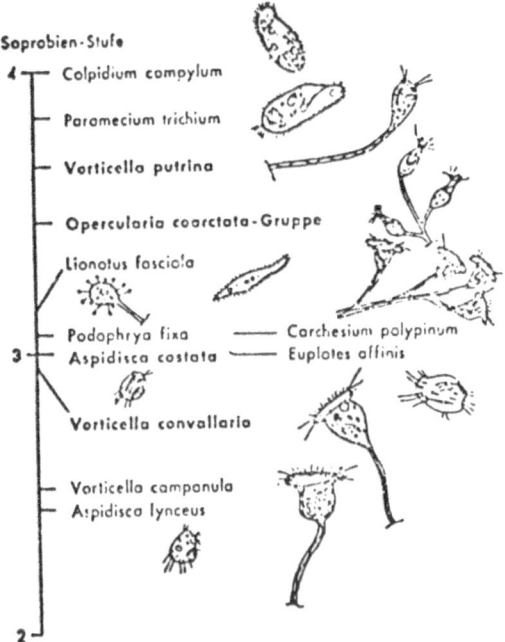

Fig. 4 : Distribution of Ciliates in the LIEMBANN
Saprobic System [Buck, 16].

Yeasts from the genera Candida, Rhodotorula, Torulopsis and
Trichosporon were said to be common in activated sludge. These
are an interesting group of facultative aerobes able to degrade
hydrocarbons.

Baines et al [15] declared protozoa as indicator organisms in
activated sludge and concluded that the types of protozoa present
could be used as qualitative measure of sludge condition. According to Hawkes [14] a plant operator could predict when trouble
will develop by following the changes of species distribution in
activated sludge. A sudden disappearance of protozoa may be
caused by the depletion to dissolved oxygen, or by toxic substance.

The ciliates are the most important group for sludge analysis
because the presence of various representatives is a simple and
rather good indicator of the quality of activated sludge.

A highly loadad sludge in a wastewater treatment plant shows
regularly certain representatives of the protozoa group. Masti-
gophora and Rhizopoda are to be found in a greater extent only
in new plants during the first phase of operation. The presence of

suctoria and rotatoria is relatively rare, but indicates a sufficient oxygen supply. The frequent occurrence of Colpidium campylum indicates an insufficient oxygen supply whereas Opercularia, Vorticella convallaria, Euplotes affinis and Aspidisca costata announce favorable oxygen conditions with concentrations around 2 mg O_2/ℓ. The sudden appearance of Aspidisca costata in connection with nitrification is mentioned by Hamm [17] and explained by the favorable conditions created for this organism by increasing nitrification.

A population of stalked ciliated protozoa such as Vorticella or the free swimming ciliate Stylonichia are often the causes of the absence of dispersed bacteria and consequently also a 5-day BOD of the settled effluent reduced to 5-10 mg/ℓ.

Baines et al [15] performed a detailed study of the utility of protozoa as quality indicators in activated sludge. Pirt and Bazin [18] reported that protozoa that preyed on bacteria reduced significantly the efficiency of wastewater treatment. This could be prevented by constructing a two stage process and regulating the dilution rates to avoid predation in the first and promote it in the second unit, thus achieving clarification of the effluent.

Curds [19, 20] developed model systems based on the activated sludge process to illustrate the relationship between bacterial predators and dispersed wastewater bacteria.

Predator-prey kinetics were studied by Sudo and Aiba [21], and Dive (22) who found a preference of gram-negative bacteria serving as nutrient for Colpidium campylum. Microbiology in wastewater treatment is sensitive to a lot of inhibitory and toxic compounds being present in the influent.

Inhibition of microbial growth may be caused by antibiotics, as reported by Csanady and Deak [23] for hospital effluent's effect on a trickling filter, but normal causes will be different. Lamband and Tollefson [24] found cupric ions (Cu^{2+}) more toxic than chromate ($CrO_3{}^{2-}$) or chromic (Cr^{3-}) ions to laboratory sludges. Aluminum used for phosphorus precipitation was found to have an adverse effect on both protozoa and rotifer population in excess of 15 mg/ℓ in natural and synthetic wastewater [Anderson and Hammer, 25]. An apparent negative effect of aluminum on nitrification was observed by Long et al [26]. Tomlinson et al [27] examined and found a lot of substances having a negative effect on nitrification.

REFERENCES

1. J.W. Jannasch and R.I. Mateles, Experimental Bacterial Ecology Studied in Continuous Culture, *Adv. Microb. Physiol.* 11, 165, 1974.

2. G.L. Jones, Bacterial Growth Kinetics: Measurement and Significance in the Activated Sludge Process, *Water Research*, 7, 1475, 1973.

3. D.W. Tempest, The Continuous Cultivation of Microorganisms I. Theory of the Chemostat. *Meth. Microbiol.*, 2, 260, 1970.

4. C.E. ZoBell and L.L. Hittle, Some Effects of Hyperbaric Oxygenation on Bacteria at Increased Hydrostatic Pressures, *Can. J. Microb.*, 13, 1311, 1967.

5. S.F. Gottlieb, Effect of Hyperbaric Oxygen on Microorganisms, *Ann. Rev. Microbiol.*, 25, 111, 1971.

6. D. Chin, G. Hicks, C.A. Geisler, Bio-oxygen Stabilization Using Hydrogen Peroxide, *JWPCF*, 45, 283, 1973.

7. C.A. Cole, L.D. Ochs, F.C. Funnell, Hydrogen Peroxide as a Supplemental Oxygen Source, *JWPCF*, 46, 2759, 1974.

8. R.G. Benedict, D.A. Carlson, Temperature Acclimation in Aerated Biooxidating Systems, *JWPCF*, 45, 10, 1973.

9. J. Monod, The Growth of Bacterial Cultures, *Ann. Rev. Microb.* 3, 371, 1949.

10. T.B.S. Prakasam, N.C. Dondero, Aerobic Heterotrophie Bacterial Populations of Sewage and Activated Sludge, *Appl. Microbiol.*, 15, 461, 1967.

11. R.G. Benedict, D.A. Carlson, Aerobic Heterotraphie Bacteria in Activated Sludge, *Water Res.*, 5, 1023, 1971.

12. H. Seiler, Beurteilung von Klüranlagen durch Bakterienanalytik, 3, Symp. Techn., Mikrobiologie Berlin, 1973.

13. D.H. Eikelboom, Filamentons Organisms Observed in Activated Sludge, *Water Research*, 365, 1975.

14. H.A. Hawkes, The Ecology of Wastewater Treatment, *MacMillan*, New York, 1963.

36

15. S. Baines, H.A. Hawkes, C.H. Heritt and S.H. Jenkins, Protozoa as Indicators in Activated Sludge, *Sew. Ind. Wastes*, 25, 1023, 1953.

16. H. Buck, Die Ciliaten des Belebtschlammes in ihrer Abhängigkeit vom Klärvergahren, Mühchner Beiträge zur Abwasser-, *Fischerei- und Flussbiologie*, 5, 206, 1968.

17. A. Hamm, Untersuchungen über die Ökologie und Variabilität von Aspidisca costata (hypotricha) in Belebtschlamm. *Arch. Hydrobiol.*, 60, 286, 1964.

18. S.J. Pirt and M.J. Bazin, Possible Adverse of Protozoa on Effluent Purification Systems, *Nature*, 239, 290, 1972.

19. C.R. Curds, A Theoretical Study of Factors Influencing the Microbial Population Dynamics of the Activated Sludge Process I, *Water Res.*, 7, 1269-1284, 1973.

20. C.R. Curds. A Theoretical Study of Factors Influencing the Microbial Population Dynamics of the Activated Sludge Process II. A Computer Simulation Study to Compare Two Methods of Plant Operation, *Water Res.*, 7, 1439-1452, 1973.

21. R. Sudo, S. Aiba, Mass + Monoxenic Culture of Vorticella Microstroma Isolated from Activated Sludge, *Water Res.*, 7, 615, 1973.

22. D. Dive, Nutrition Holozoique de Colpidium Campylum Phenomenes de Selectio.

23. M. Csanady, Z. Deak, Trickling Filter Experiment for Purification of Antibiotic Containing Hospital Sewage, *Water Res.*, 6, 1541, 1972.

24. A. Lamb, E.L. Tollefson, Toxic Effects of Cupric Chromate and Chromic Ions on Biological Oxidations, *Water Res.*, 7, 599, 1973.

25. D.T. Anderson, M.J. Hammer, Effects of Alum Addition on Activated Sludge Biota, *Water Sew. Works*, 120, 63, 1973.

26. D.A. Long, J.B. Nesbitt, R.R. Koutz, Soluble Phosphate Removal in Activated Sludge Process, *Water Res.*, 7, 321,1973.

27. T.G. Tomlinson, A.G. Boon, C.N.A. Trotman, Inhibition of Nitrification in the Activated Sludge Process of Sewage Disposal, *J. Bact.*, 29, 266-291, 1966.

EFFLUENT VARIABILITY AND ITS CONTROL

A. J. Englande, Jr.

Associate Professor, Department of Environmental
Health Sciences, Tulane University, New Orleans, La.

FACTORS AFFECTING EFFLUENT VARIABILITY

System design and management geared at minimizing effluent variation can be effected once the factors influencing variability
are recognized. Methods of control will become apparent when the
"cause and effect" relationships between system operation and effluent quality are understood. Consequently, it is necessary to
review the nature and quantity of performance variation associated
with wastewater treatment facilities. It should be noted that effluent variability can be reduced but not eliminated, due to "inherent" variations characteristic of the treatment system.

The overall variability of effluent quality is affected by:

1- the variability of the wastewater (raw waste load) before
   waste treatment resulting from the manufacturing operations

2- variability inherent in the operation of waste treatment
   technology.

RAW WASTE LOAD VARIABILITY

Manufacturing operations vary from a relatively simple single
product process to a complex multi-product operation for any one
of which the nature and timing of production sequences may change
at any time. Significant factors which contribute to fluctuations
in the quantity and quality of wastewater discharged from manufacturing processes are summarized in *Table 1*. These factors result
in an input to the waste treatment facility of variable flow, organic concentration, and substrate composition. Although the impact

of raw waste load variation can be reduced by in-plant changes or by equalization, surge, and "off-spec" basins prior to treatment, transitory loads cannot be completely sequestered. Raw waste load variability, then, is inherent in the manufacturing operation.This influent variability may be magnified by the treatment process.

TABLE 1 :   Causes of Variability in Raw Waste Load

Variation in Plant Product Mix

Production Rate Changes

Nature of the Manufacturing Process (batch vs. continuous)

Changes in Raw Materials

Variations in Production Unit Efficiency

Non-Specification Product Discharge

Maintenance   (Equipment Shut-down and Clean-out)

Miscellaneous Leaks and Spills

Contamination Drainage from Rainstorms

Waste Treatment Variability

Effluent quality from a properly designed and operated treatment system will exhibit an "inherent variability" which is attributable to the basic nature of the treatment process, the characteristics of the raw waste load, and geographical and climatological conditions. Factors contributing to secondary effluent variability are summarized in *Table 2*. Several factors for biological system effluent variation are discussed as follows:

Influent Wastewater Strength: At constant organic loading the soluble effluent substrate concentration is directly proportional to the influent substrate concentration. Increased organic concentrations result in increased microbial growth and activity. Biomass growth is often too slow to respond to these variations to prevent effluent lost of substrate. Increased activity may create an oxygen demand in excess of transfer capabilities which result in increased effluent organic levels. Other adverse effects, such as from toxicants, salts, and temperatures can be heightened at marginal D.O. levels (approximately 1 to 2 mg/ℓ). Other possible reasons for an apparent dependence of achievable effluent levels on influent concentration include the short-circuiting of feed which is inherent in non-plug-flow systems and the presence of metabolic intermediates which may degrade more slowly than the original materials. Increasing influent organic concentrations can also result in increased effluent solids concentrations due to the stimulated growth of non-flocculative bacteria.

TABLE 2 :   Factors Contributing to Secondary
            Effluent Variability

Influent Concentration

Influent Variability

Inherent Variability of the Biological Process

Ambient Temperature Changes

Operating Techniques and System Design

Variations in Total Dissolved Solids

Presence of Inhibitory Substances

Changes in influent flow rate will obviously affect retention time within the various unit operations. This affects both aeration basin efficiency and clarifier performance where turbulence and overflow rate must be controlled. Work at Stevenage has indicated that compliance to effluent standards specifying a range variation can be achieved at a constant flow plant at higher loadings than in one of the same capacity with fluctuating flow. However, the solution, equalization vs. extra aeration tank capacity, must be cost effective.

Influent Wastewater Composition: Changes in wastewater composition will affect biological degradability and hence the substrate removal rate coefficient, k. Qualitative wastewater variability may require population shifts and/or the formation of inducible enzymes ·for degradation of the varying components. The time delay involved can result in "leakage" of substrate to the effluent. Population shifts or changes in the physiological state of the sludge resulting from the influent variability also can substantially affect sludge settling properties. Phenomena such as predator-prey relationships between protozoa and bacteria which depend on inherently variable physiological mechanisms cannot be influenced by control. The periodic presence of one substrate may prevent the metabolism of another, due to repression of enzyme synthesis or inhibition of enzyme activity. It is very difficult to control or monitor the degradability of a wastewater and the design k must be determined statistically.

The Influence of Temperature: Temperature will have a significant effect in biological wastewater treatment plant efficiency. The temperature effect on the reaction coefficient, k, can be defined by the relationship:

$$k_t = k_{20^\circ C} \, \theta^{(T-20)} \qquad \dots \dots (1)$$

The coefficient θ will depend on the nature and complexity of the wastewater. θ has been shown to vary from 1.055 to 1.10 for several soluble industrial wastewaters. Less temperature sensitivity is displayed when most of the organics are in the colloidal or suspended form due to "biosorption". Seasonal performance variations are generally related to temperature variations.

Variation in Dissolved Solids: Biological systems normally function more efficiency when treating low TDS wastewaters. Abrupt changes in dissolved solids will produce a pronounced deleterious effect on system efficiency. Changes in osmotic pressure will disrupt biochemical mechanisms reducing organic removal capacity until the system can readjust. Effluent suspended solids will increase due to the proliferation of non-flocculating microorganisms, biological die-off, and an increase in water density.

DESCRIBING EFFLUENT QUALITY VARIABILITY

The variability of raw wastewaters and effluents from treatment facilities can be examined statistically to evaluate the relationship between the magnitude of values and their frequency of occurrence which is necessary for control purposes. The nature and extent of control required will be dictated by effluent standards as set by regulatory guidelines.

Because of the stochastic properties of the data, effluent standards should be formulated based on the results of an evaluation of a frequency distribution to generalize data. Application of standards should be influenced by an evaluation of the inhomogeneities of the data, particularly with respect to sampling procedures. Due to the stochastic properties of effluent parameters, a fixed standard will theoretically always be exceeded. Consequently, a realistic effluent standard should consist of a mean and a permissible probability of exceeding it. Perhaps an average value from composite samples over a 30-day period might be more a realistic guideline. If the substance is not acutely toxic, some degree of infringement tolerance should exist depending on the nature of the receiving water. The degree of tolerance of infringement will have an extremely important effect on system design.

To facilitate the application of the results of the frequency distribution, the analysis selected for the evaluation of data should be as simple as possible. The Gaussian "normal distribution", log-normal distribution, and Pearson's Type 111 skew distribution have adequately described collected data. Generally, data is positively skewed, which can be eliminated by the log-normal distribution (1-5).

Time series analysis can be employed to gain insight into the trends and frequencies existing in effluent data (6). Most process changes occurred too fast for a daily composite sample to reflect sensitive time-series trends. However, certain treatment systems have extensive detention and time series analysis can be used to establish system delay as well as cross-correlate effluent load with the effect of process change on raw waste load. Auto-covariant functions and power spectra have been shown capable of describing:

a. variability related to raw waste load
b. variability generated by the treatment process
c. relationships between effluent parameters
d. relationship between raw waste load and final effluent
e. relationship between production parameters and raw waste load.

EFFLUENT VARIABILITY CONTROL

APPROACHES to MINIMIZE EFFLUENT VARIATION

Control measures to reduce variability of the quantity and quality of wastewaters should be instituted both at their source and within the wastetreatment complex. As noted by Chalmers [1], effluent control measures applied within the manufacturing process can reduce treatment cost expenditure by a factor of 10. Prior to design, therefore, a critical examination of individual wastestreams should be undertaken to determine the volume and strength, variability of pertinent parameters, why and how they are generated, and what modifications are possible to reduce their impact on the wastetreatment facility. Approaches to minimize effluent variability by control at the manufacturing plant are summarized in *Table 3*.

TABLE 3 : Approaches to Minimize Effluent Variability

CONTROL at the MANUFACTURING PLANT

Production Scheduling
Improved Process Control
Improved Process Technology
Pre-treatment of Problem Streams
Monitoring of Process Discharges
Use of Holding or Blending Tanks
Reuse and Recycle
General Good Housekeeping

CONTROL at the WASTETREATMENT FACILITY

Basic Process Selection and Design
Equalization
Emergency Holding Basin
Automated Monitoring of Influent and Effluent
Improved Operating Techniques
Adequate Oxygen Supply for Biological Systems
Control of Biomass Recycle Concentration,
   MLSS Level, and Sludge Blanket Level
Variation of Substrate and Sludge Recycle Feed Locations
Automated Control of Chemical Additions
Stand-by Capacity to Reduce Peak Effluent Impurities

Table 3 also indicates methods of control possible at the waste-treatment facility. Proper process selection and design will insure process stability, adequate mixing, proper flow configurations, sufficient oxygen supply, etc. Complete-mixing, for example, has the advantage of damping and minimizing surge loadings, influent toxic materials concentrations, and equalizing other variations of influent wastewater characteristics. As noted by Downing [2], two stage processes can be advantageous in reducing variability arising from the presence of inhibitors as with the nitrification process. Andrews, et. al. [3], simulated the step feed activated sludge process possesses the ability to vary the location of substrate and recycled sludge input and therefore offers a valuable control action. This modification enables the rapid transfer of sludge to and from the reactor and solids–liquid separator and can be used to reduce effluent substrate variability and reduce bulking.

One important factor which affects the variability of effluent quality is the operating techniques of treatment plant operators. Consistent process efficiency will ultimately depend on the human factor. The detection of and proper response to a chemical spill and the control of the clarifier sludge blanket, for example, require personnel attention.

SYSTEM MONITORING for VARIABILITY CONTROL

Effluent variability can be reduced by increasing the physical size of the process, providing an equalization basin, or by modern control systems. The use of control systems offer the potential of decreased effluent variation at lower capital cost. The degree of treatment can be quickly matched with the applied load yielding a more consistent performance. Downing[2] indicates that the kinds of action that might be initiated by automatic control systems include:

a-   diversion of excess flows to balancing tanks or to storage for separate treatment,

b-   by-passing certain units so as to avoid unnecessary treatment

c-   adjusting the extent of treatment given, for example, by altering;

   1- rate of addition of chemicals where these are used

   2- air supply in activated sludge plants

   3- rate of recycling

d-   using stand-by capacity either to reduce loadings on existing units or to provide treatment to effluent peaks.

A schematic, illustrating a monitoring and control diagram for a typical biological wastewater facility as presented by Adams [4], is shown as *Fig. 1*. A holding basin is provided for infrequent high organic or toxic shocks to avoid overdesign of the equalization facility. If the organic strength as measured by total organic carbon (TOC) or total oxygen demand (TOD) exceeds a preset level, the wastestream is automatically diverted to the spill pond for storage and an alarm system alerts the operator. The operator can analyze the contents of the spill basin and determine a proper discharge rate with which to feed the ponds contents to the equalization basin.

The presence of toxic constituents may require a separate basin. Continuous monitoring for toxicity is difficult but may be effected by a bench-scale reactor. Substrate is fed to the flow-through

44

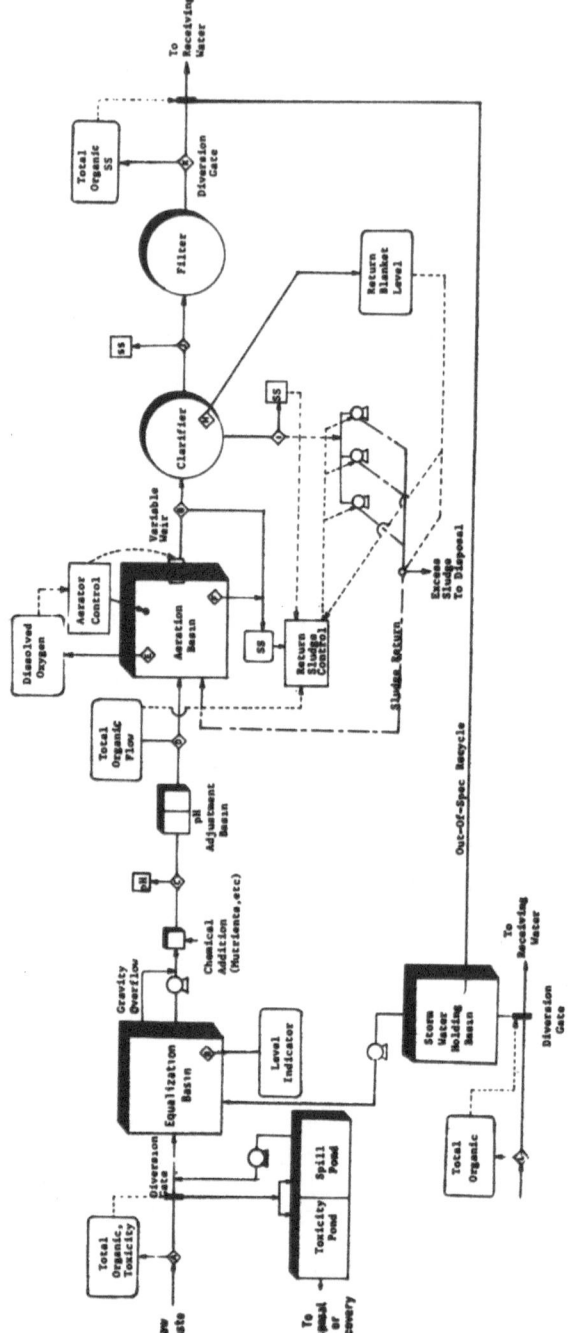

Fig. 1 : Monitoring and Control Diagram for Wastewater Treatment Facility

reactor after neutralization, nutrient addition, and possible supplemental substrate addition. Toxicity can be determined by continuously monitoring the oxygen uptake, dehydrogenase activity, or
the production of ATP.

The equalization basin is designed to reduce variation in the
biological system. The complete-mix variable volume equalization
basin has been observed more effective in reducing variability than
a constant volume reactor. Pumpage to the biological reactor is
at a constant rate so that the only variation is in concentration.

The design of the neutralization system will depend on the nature of the waste. If it is either acid or alkaline "feed-back"
control will suffice. If the influent is highly variable a "feed-
forward" control combination is required.

Influent to the aeration basin is usually monitored with respect to total organics (TOC, TOD) and flow. In order to maintain
a constant organic loading within the aeration basin, it is necessary that the MLVSS be increased with increased organic loading.
This increase in MLVSS can be effected by an increase in the rate
of sludge return. To control recycle it is necessary to continuosly monitor aeration basin solids and the underflow concentration
of suspended solids from the clarifier. The clarifier sludge blanket level should also be monitored to indicate accumulation of
sludge in the clarifier or a poor settling sludge. A photoelectric
probe is usually employed with an alarm set at an upper limit. The
operator then can determine the cause and either recycle or waste
sludge. It is desirable to monitor effluent suspended solids from
the clarifier prior to filtration.

Dissolved oxygen probes can be located in the aeration basin to
indicate the need for additional aeration. Standby blowers or a
variable speed compressor can be used for diffused air systems.
Two-speed aerators can be installed with the surface aeration system. Depending on the economics, excess horsepower might be continuously provided rather than a control system.

Andrews et al.[3], have proposed the specific oxygen utilization rate (SOUR) as a control signal for sludge return since it is
a direct measure of the activity of the sludge and shows a rapid
response to both substrate loading changes and the input of toxic
materials. SOUR would be especially appropriate to the high purity
oxygen process since these reactors are covered and can be used as
on-line respirometers. It could also be employed to calculate on-
line reaction rates.

Because of regulatory requirements, it may be necessary to divert the initial portion of storm flows to the biological system.
The use of organic monitoring can divert the contaminated water to

a holding basin for subsequent treatment. When the organic concentration falls below a predetermined level, the remaining storm flow can discharge directly to the receiving waters.

The treated effluent can be monitored continuously by a TOC, TOD, or ultra-violet instrumentation and if the effluent quality deviates from the required levels, the flow can be diverted to a spill basin for subsequent treatment.

The most important parameters used to control variability are suspended solids concentration, flow, organic concentration, and dissolved oxygen measurements. The various techniques for monitoring these parameters are presented in Table 4. Adams [4] has rated these methods on a scale from 1 to 10 according to their present state of development, proven application in the field, and maintenance tendencies.

TABLE 4 : Rating of Methodology for Continuous Monitoring

Parameter	Monitoring Method	Rating
Suspended Solids Density (on-off control only)	Ultrasonic	7
	Nucleonic	5
Concentration	Electrical Conductivity	2
	Specific Gravity	3
	Turbidity (optical)	6
Organics	BOD	1
	COD	3
	TOC	8
	TOD	8
	UV	8
Salts	Conductivity	9
	Ion Selective Probes	7
Enzymes	Automated Wet Chemistry	3
Metals	Total: Mass Spectrophotometry	2
	Specific: Atomic Adsorption	5
Specific Ions	Ion-Selective Probes	5-7
	Other	3-5
Flow Sludge	Magnetic Flow Meters	8
Water	Closed Pipe Measurement	6
	Standing Flume Measurements	10

REFERENCES

1. Chalmers, R.K., "Manufacturing Operations and Effluent Quality," presented at International Conference on Effluent Variability from Wastewater Treatment Processes and Its Control, New Orleans, Louisiana, December 2-4, 1974.

2. Downing, A. L., "Variability of the Quality of Effluent from Wastewater Treatment Processes and Its Control," presented at International Conference on Effluent Variability from Wastewater Treatment Processes and Its Control, New Orleans, Louisiana, December 2-4, 1974.

3. Andrews, J. F., Buhr, H.O., and Stenstrom, M. K., "Control Systems for the Reduction of Effluent Variability from the Activated Sludge Process," presented at International Conference on Effluent Variability from Wastewater Treatment Processes and Its Control, New Orleans, Louisiana, December 2-4, 1974.

4. Adams, C.E., "Techniques of Monitoring for Variability Control," presented at International Conference on Effluent Variability from Wastewater Treatment Processes and Its Control, New Orleans, Louisiana, December 2-4, 1974.

PRINCIPLES OF BIOLOGICAL TREATMENT

W. Wesley Eckenfelder, Jr.

Distinguished Professor,
Vanderbilt University,
Nashville, Tennessee

INTRODUCTION

Many types of micro-organisms are active in the breakdown of organic matter and the resulting stabilization of organic wastes. These micro-organisms may be broadly classified as aerobic, facultative or anaerobic. Aerobic organisms require molecular oxygen for their metabolic processes. Anaerobic organisms function in the absence of oxygen and obtain their energy from organic compounds. Facultative organisms can function aerobically in the presence of oxygen or anaerobically in the absence of oxygen. A majority of the organisms found in biological wastewater treatment processes are of the facultative type. Most biological systems treating organic wastes depend upon heterotropic organisms which utilize organic carbon as an energy source and as a carbon source for cell synthesis.

Autotropic organisms, on the other hand do not require an organic carbon source, but rather use an inorganic carbon source such as CO_2 or bicarbonate. Chemosynthetic autotrophs obtain energy from the oxidation of inorganic compounds such as nitrogen or sulfur. Photosynthetic autotrophs utilize solar energy for the synthesis of carbon dioxide to cellular protoplasm and produce molecular oxygen as a by-product.

AEROBIC SYSTEMS

In aerobic biological treatment systems the reactions occurring are:

$$\text{organics} + (a')O_2 + N + P \xrightarrow{k} (a)\text{cells} + CO_2 + H_2O +$$

(BOD,COD,TOC) nondegradable soluble residue ....(1)

$$\text{cells} + O_2 \xrightarrow{b} CO_2 + H_2O + N + P + \text{nondegradable}$$

cellular residue...(2)

(It should be noted that these reactions also occur in streams and natural waters and in the BOD bottle). Biological treatment design involves the balancing of these equations for the waste in question.

The pertinent process parameters a, a", b and k are shown in equations (1) and (2). For a known single compound such as glucose or phenol it is possible from available information to balance their equations. For the complex mixed wastewaters usually found in practice,however, it is not possible to balance these equations and a laboratory as pilot-plant study is needed to define the parameters. They may be defined:

a – fraction of substrate removed converted to cells as VSS
a – fraction of substrate oxidized for energy
b – fraction per day of degradable cell mass as VSS
k – substrate removal rate coefficient.

In the stabilization of an organic substrate, a portion of the energy obtained from the reaction is used for biological synthesis and the remainder to satisfy the energy requirements for growth. A small portion of the energy is used for cellular maintenance (Figure 1). Equations (1) and (2) are schematically shown in Figure 2. The symbols and parameters used in the development of design relationships are also shown in Figure 2. As can be seen from Figure 1 and Equation (1) all but a small portion of the substrate removed is either converted to cell mass as VSS or oxidized for energy. Therefore on a COD basis, $a + a" \sim 1$. It should be noted as shown in Example 1 that this does not apply when using BOD_5 since the BOD_5 is a fraction of the COD or ultimate BOD. In order to illustrate the breakdown of organic matter by aerobic oxidation, a hypothetic example using 1 Kg COD is shown in Figure 3. It is assumed in this example that $a' = 0.5$.

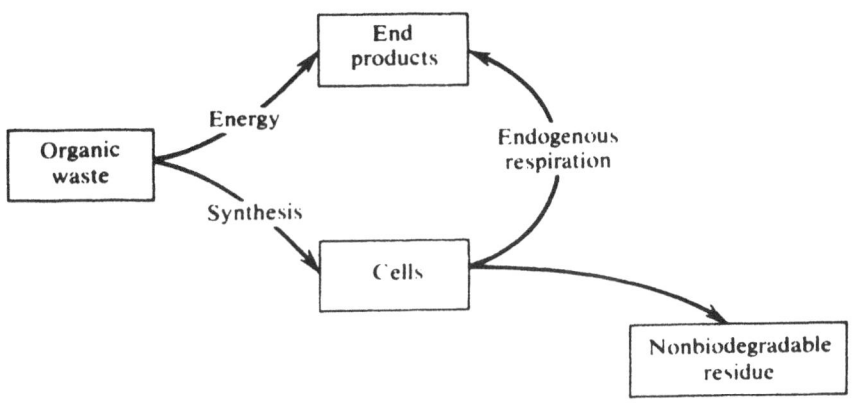

Fig. 1: The Mechanism of Aerobic Biological Oxidation

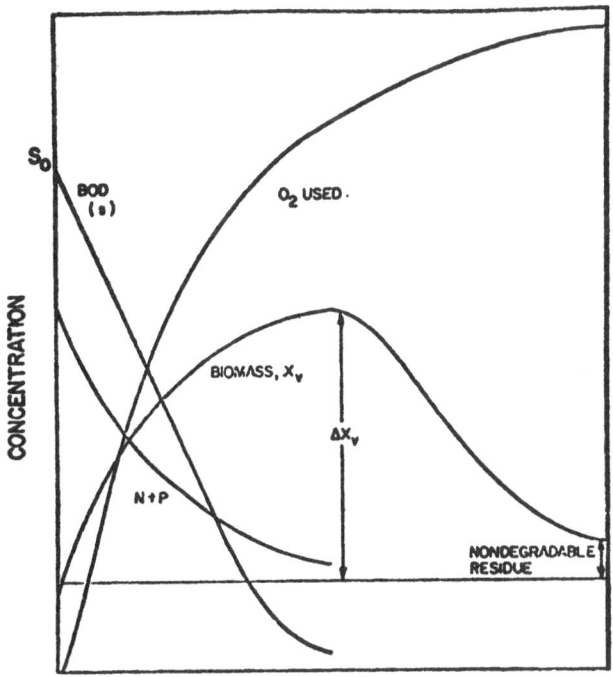

Fig. 2: Schematic of Aerobic Bio-Oxidation Process

It has been shown by McCarty [1] that the similarity in the biochemistry of synthesis of all micro-organisms under a wide variation in environment permits calculation of cellular yields from thermodynamic considerations when the composition of the substrate is known.

CELL YIELD

In aerobic growth, energy is released from the conversion of organic carbon, resulting in considerable energy being available for synthesis, and hence a relatively high yield coefficient, a. McCarty [1] obtained values for a, varying from 0.30 to 0.51 for glucose, analine, lactate, and acetate. Servizi and Bogan [2] showed a value of 0.39 for a variety of compounds. Eckenfelder and O'Connor [3] showed yield coefficients varying from 0.37 to 0.46 for several readily degradable organic wastes.

In anaerobic systems, less energy is obtained from the organic conversion, and hence the growth yield is much less than for aerobic systems. Yield coefficients varying from 0.032 to 0.27 were found depending on the substrate. A detailed study by Andrews et.al., [4] on a synthetic soluble substrate showned a yield coefficient of 0.14. It should be realized that most of these growth yields include both the acid formers and the methane organisms.

In autotropic growth, considerable energy must be expended to convert CO_2 to an intermediate for cell synthesis. As a result, yield coefficients of less than 0.1 are usually obtained. Several sources reported a values for *Nitrosomonas* as 0.015 to 0.03 and 0.02 for *Nitrobacter*.

The measured yield coefficients for various systems are summarized in Table 1 and include the influence of volatile suspended solids originally present in the waste and hence denoted as a.

TABLE 1: Aerobic Biological Waste-Treatment Parameters[a]

Waste	a (BOD_5 basis)	a´ (BOD_5 basis)	b (ℓ/day)
Domestic	0.73	0.52	0.075
Refinery	0.49-0.62	0.40-0.77	0.10-0.16
Chemical and Petrochemical	0.31-0.72	0.31-0.76	0.05-0.18
Brewery	0.56	0.48	0.10
Pharmaceutical	0.72-0.77	0.46	−
Kraft Pulping and Bleaching	0.5	0.65-0.8	0.08

[a] All parameters include the effect of influent suspended solids.

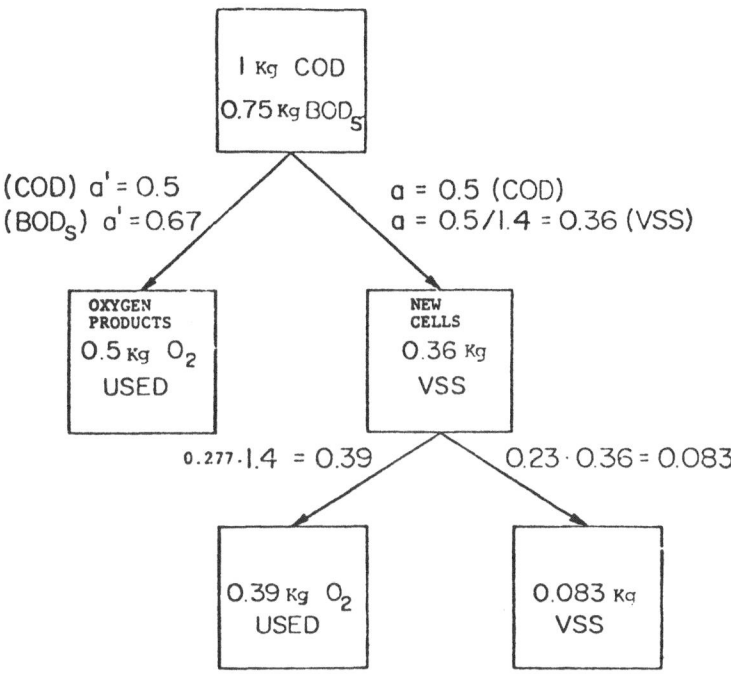

Fig. 3: The Breakdown of Organic Matter

Porges [5] showed that activated sludge from the treatment of
dairy wastewaters has an average composition of $C_5H_7NO_2$. Other
investigators have shown a similar composition for sludge treating
other wastewaters. The degradable portion of the biomass (Equation 1) has been reported as 77 percent with 23 percent of the
volatile suspended solids as non-degradable residue.

It can be seen from Equation (1) and (2) that biological volatile solids will be generated from organic removal (Equation 1)
and oxidized from endogenous metabolism (Equation 2). In addition
if the wastewater contains volatile suspended solids in the influent wastewater (as might be expected in domestic wastewater
or pulp and fiber in pulp and paper mill wastewaters) the non-
degradable portion of these solids will contribute to the accumulated volatile suspended solids in the system.

The sludge yield for a biological system can be estimated from
the relationship:

$$\Delta x_v = fs_i + as_r - bxX_v$$

54

where:

a = cell yield coefficient

b = cell endogenous rate

x = biodegradable fraction of the mixed-liquor suspended solids

X_v = mixed-liquor volatile suspended solids

s_i = influent volatile suspended solids

s_r = BOD removed

f = fraction of influent volatile suspended solids not degraded.

The coefficient f can be related to sludge age since the longer the solids remain in the aeration system the greater will be the rate of degradation. Typical data is shown in Figure 4.

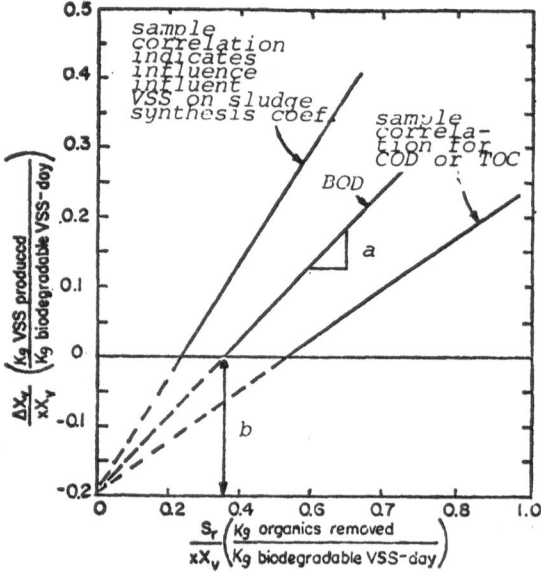

Fig. 4: Determination of Sludge Production Coefficients

SLUDGE AGE

Sludge age is defined as the average length of time the biomass is under aeration. In a flow through system, sludge age is the dilution rate (1/Q). For growth to occur and to effect BOD removal, the growth rate becomes

$$\mu = \frac{1}{G} = R_\mu$$

in which G is the sludge age and R_μ the dilution rate.

In a system with sludge recycle and a wastage of excess sludge, sludge age is defined as:

$$G = \frac{X_v}{\Delta X_v} = \frac{X_v}{as_r - bxX_v}$$

where X_v and ΔX_v are expressed in kg and kg/day (lb and lb/day), respectively.

This equation also applies relative to the limiting growth rate for specific organisms.

For a soluble wastewater the degradable fraction, x, will be related to the sludge age or the length of time the biomass is under aeration since as the aeration time is increased the percentage of non-degradables also increases. This is shown in Figure 5. For a soluble wastewater x can be computed from the relationships:

$$x = \frac{as_r + bX_v - \sqrt{(as_r - bX_v)^2 - (4bX_v)(0.77as_r)}}{2bX_v} \quad \ldots (4)$$

When the wastewater contains volatile suspended solids Equation (4) may be modified to:

$$x = \frac{as_r + bX_v + fs_i - \sqrt{(as_r - bX_v + fs_i)^2 - (4bX_v)(0.77as_r)}}{2bX_v} \quad \ldots (5)$$

In the case of soluble wastewaters the active mass of the biomass has been considered to be the degradable fraction, x, divided by 0.77 as calculated by Equation (4). A number of other procedures have been employed to determine the active fraction of the biomass.

These are adenosine tri-phosphate (ATP), dehydrogenase enzyme (TTC) plate counts and oxygen uptake rate. While these procedures are useful for experimental investigation and for plant operational control they are less applicable for process design and evaluation. The computed degradable fraction, x, reasonably correlates with these other parameters as shown in Figure 7 and should be applicable to design procedures. In the case of activated sludge where the loadings are high and a major portion of the sludge

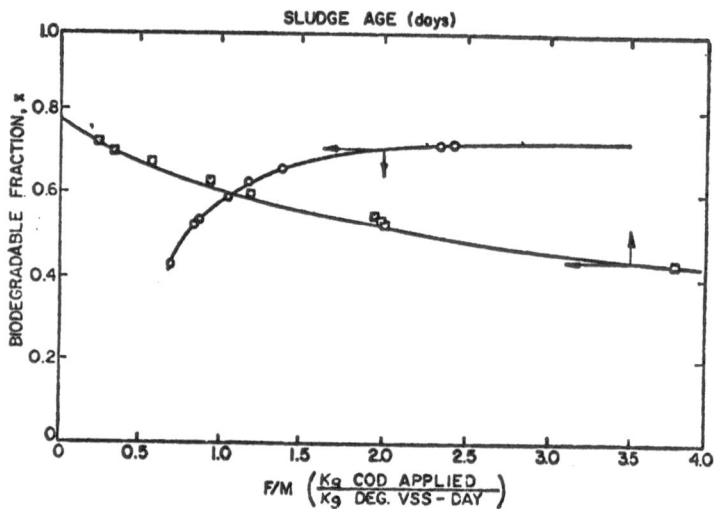

Fig. 5: Variation of Degradable Fraction with Sludge Age and with F/M for Synthetic Sewage

is active biomass Equation (3) can be expressed:

$$\Delta X_v = as_i + fs_i - bX_v$$

It should be emphasized that in many cases it is not feasible to divide the sludge yield ΔX_v into the contribution by microbial synthesis and the accumulation of volatile suspended solids originally present in the waste. The experimental coefficient a is then used for engineering design (Table 1), and Equation (3) becomes

$$\Delta X_v = as_r - bxX_v \qquad \qquad \dots (6)$$

It should be noted that for a soluble waste, $\bar{a}$ is approximately equal to a. Sludge production from the treatment of domestic sewage by the activated sludge process is shown in Figure 6.

In a flowthrough system, without recycle, the concentration of solids in the effluent will be equal to the concentration in the reactor, and the sludge yield is equal to the solids lost in the effluent. Equation (3) can be re-expressed for the flowthrough system as

$$X_v Q = S_o Q + as_r Q - bxX_v V$$

Dividing by Q, and rearranging,

$$X_v = \frac{S_o + as_r}{1 + bxt} \qquad \dots (7)$$

in which X_v is the concentration of volatile suspended solids maintained in the reactor. Equation (7) applied to aerobic lagoons.

ENDOGENOUS METABOLISM

Endogenous metabolism occurs in all cells in which energy is utilized for cellular maintenance. Endogenous metabolism may be defined by the coefficient, b, which has the units of reciprocal time; that is, there is a fractional decrease in cell mass per day. It should be noted that the coefficient, b, applies to the oxidation-type systems [3]. Endogenous values reported for several systems are summarized in Table 1.

When high aeration solids are carried, the accumulation of non-biodegradable mass reduces the percentage of active organisms present in the system based on total volatile suspended solids. For example, the data of Wuhrmann [6] would indicate 59 percent active organisms at 6,000 mg/ℓ relative to substantially total activity at less than 500 mg/ℓ aeration solids at the same applied loading.

OXYGEN UTILIZATION IN AEROBIC SYSTEMS

In aerobic systems, the portion of the substrate not utilized for cellular synthesis uses oxygen for energy. In addition, oxygen is used for cellular maintenance (endogenous respiration) as shown in Equations (1) and (2). The resulting relationship is:

$$O_2 = a's_r + b'xX_v \qquad \dots (8)$$

and is shown in Figure 7.

58

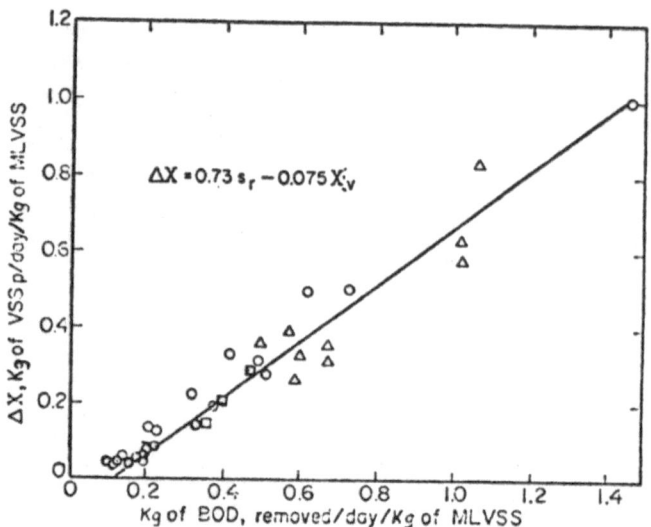

$$\Delta X = 0.73\,s_f - 0.075\,X'_v$$

Fig. 6: Sludge Production in the Activated Sludge
Process Treating Domestic Sewage (After Wuhrmann)

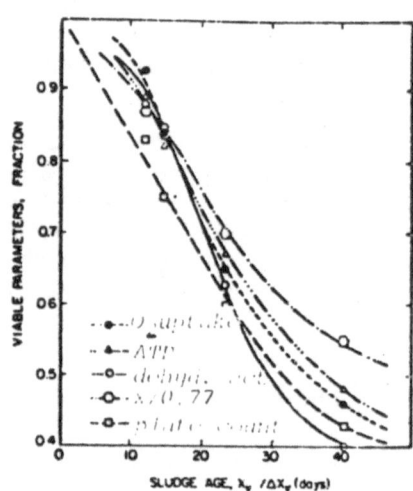

Fig. 7: Variation of Viable Cell Parameters (fraction)
with Sludge Age

When volatile suspended solids are undergoing slow degradation in the aeration system, a' will reflect this oxygen usage. Also, nitrification will vastly increase the value of a'. The coefficient b' reflects the oxygen used for endogenous respiration. Assuming a mean cellular composition of $C_5H_7NO_2$ [5] the oxygen requirements can be computed:

$$C_5H_7NO_2 + 5O_2 \rightarrow 5CO_2 + 2H_2O + NH_3$$

and

$$\frac{5O_2}{C_5H_7NO_2} = \frac{116}{113} = 1.42$$

When a material balance is based on COD or BOD_u, $1.4a + a = 1$, because the organic substrate carbon results in the production of CO_2 or the formation of biological cells. When based on BOD_5, the following conversion must be made:

$$a_5 = -\frac{1}{BOD_5/BOD_u} = 1.42a_5$$

The oxygen uptake rate in a biological process can be determined in several ways, namely by oxygen uptake rate, by off-gas analysis or by a COD balance. Oxygen uptake rate is measured by withdrawing a sample of sludge, aerating to raise the dissolved oxygen and measuring by means of a dissolved oxygen probe the decrease in dissolved oxygen. Caution must be exercised in interpreting the results of this test. Since a sample is withdrawn from an aeration basin receiving wastewater, and no wastewater is added during the course of the test, results will usually read lower than the actual uptake rate. The error will increase as the oxygen uptake rate increases. It is possible to correct for this error by adding wastewater to the test cell.

At the same rate as addition to the aeration tank, off-gas analysis has been applied to high purity oxygen systems which employ covered aeration basins, but is not feasible for mechanical air processes. In cases where strippable volatile constituents are not present in the wastewater, a COD balance should yield the most accurate measure of total oxygen requirements. In this case, the oxygen utilized is:

$$O_2 = COD_{INF} - COD_{EFF} - COD_{VSS\ generated}$$

The oxygen usage can then be computed by a COD balance over a reasonable time period.

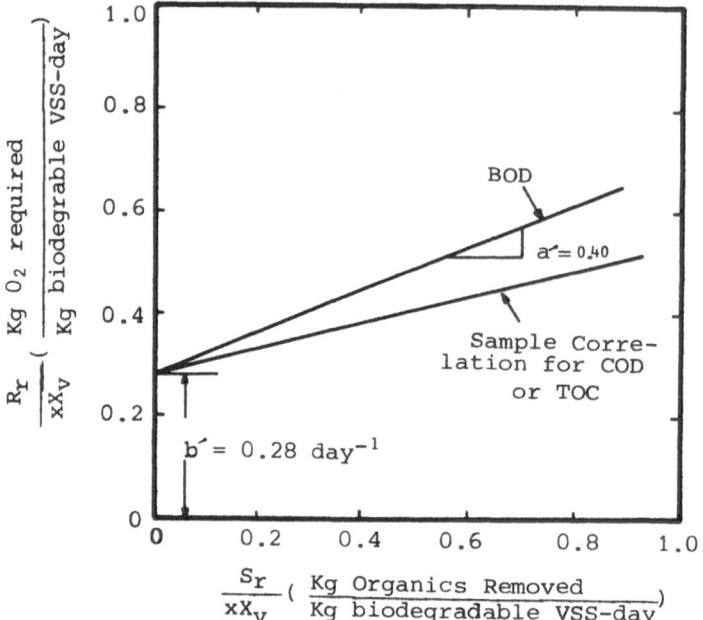

Fig. 8: Determination of Oxygen Utilization Coefficients

The mechanism of organic removal from soluble mixed substrates is a complex phenomena. Grau et. al., [7] has indicated that the mechanism can generally be described as a sequence of three complex processes, namely, contact of a cell with a molecular substrate, transport of the molecule into the cell and intermediate metabolism of the substrate. Large molecules must first be broken down before they can be classified into three main groups:

a) Single component substrates which are directly transportable into the cell.

b) Multi-component substrates which are represented by a mixture of several single substrates.

c) Complex substrates which have to be changed externally prior to transport into the cell.

The removal of single component substrates has been described by linear removal kinetics or a zero order reaction by Wuhrmann [8], Ticshler and Eckenfelder [9]. The linear removal concept states that substrate removal will follow zero order kinetics to very low concentration levels. Data reported by Tischler and Eckenfelder are shown in Figure 9.

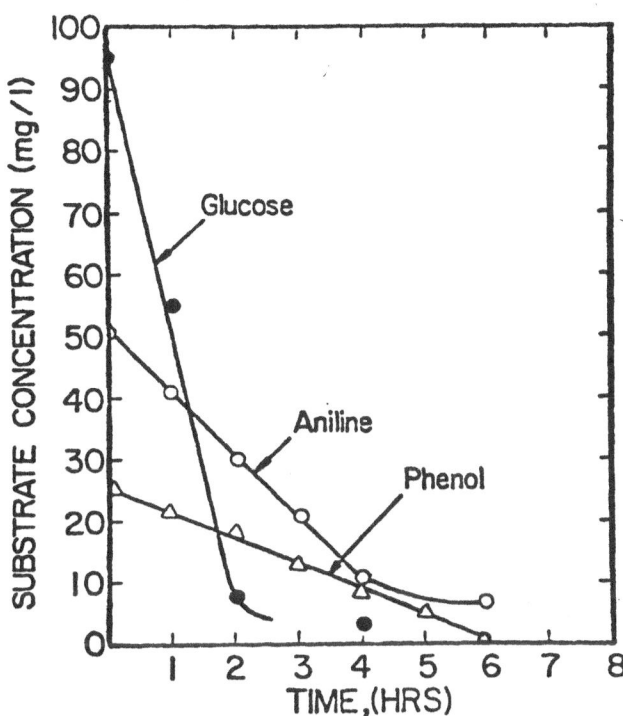

Fig. 9: Linear Substrate Removal

Single substrate removal has also been defined by the Monod equation:

$$-\frac{ds}{dt} = (\frac{\mu_m X}{a}) (\frac{s}{k_s + s})$$

.... (9)

in which:

s =   substrate concentration

μ_m =   maximum growth rate of organisms

a =   biomass yield coefficient

$$x = \text{biomass concentration}$$

$$k_s = \text{Monod's constant}$$

and

$$\frac{dX}{dt} = a\frac{ds}{dt} - bX \qquad \qquad \dots\text{(10)}$$

At high concentrations of s when substrate concentration is not limiting microbial growth, $s > k_s$ and Equation (9) reduces to

$$\frac{ds}{dt} = \frac{\mu_m}{a} \cdot X \qquad \qquad \dots\text{(11)}$$

When the increase in biological solids, ΔX, is less than $2X_a$, Equation (11) can be expressed

$$\frac{as_r}{X_v t} = \mu_m \qquad \qquad \dots\text{(12)}$$

At very low concentrations of s, $k_s < s$ and Equation (9) reduces to

$$\frac{ds}{dt} = -\frac{\mu_m}{k_s} sX_v \qquad \qquad \dots\text{(13)}$$

The linear removal concept and the Monod kinetic relationship are compatible since at low k_s values, the Monod equation closely approximates a zero order model. Monod's studies indicated that in fact, k_s values are often very small and that growth rates observed are commonly independent of substrate concentration of very low levels. Monod reported a k_s value of 4 mg/ℓ for a pure culture of E.coli growing on glucose and Wuhrmann has reported a value of 0.2 mg/ℓ for a mixed culture activated sludge utilizing the same substrate.

It is rare in treating industrial wastewaters that a single component will be considered, rather, many components will exist in the wastewater and undergo biological removal. Tischler and Eckenfelder developed a mathematical model for multi-component removal of organics. This model was based on the fact that when considering an acclimated mixed culture, organics would be removed simultaneously all at a zero order rate of very low concentrations. This is shown in Figure 10. Assume a wastewater contains three components, A, B, and C. With an acclimated culture, all components are removed simultaneously at different rates as shown in Figure 10 (a). When an overall organic parameter such as COD is used,

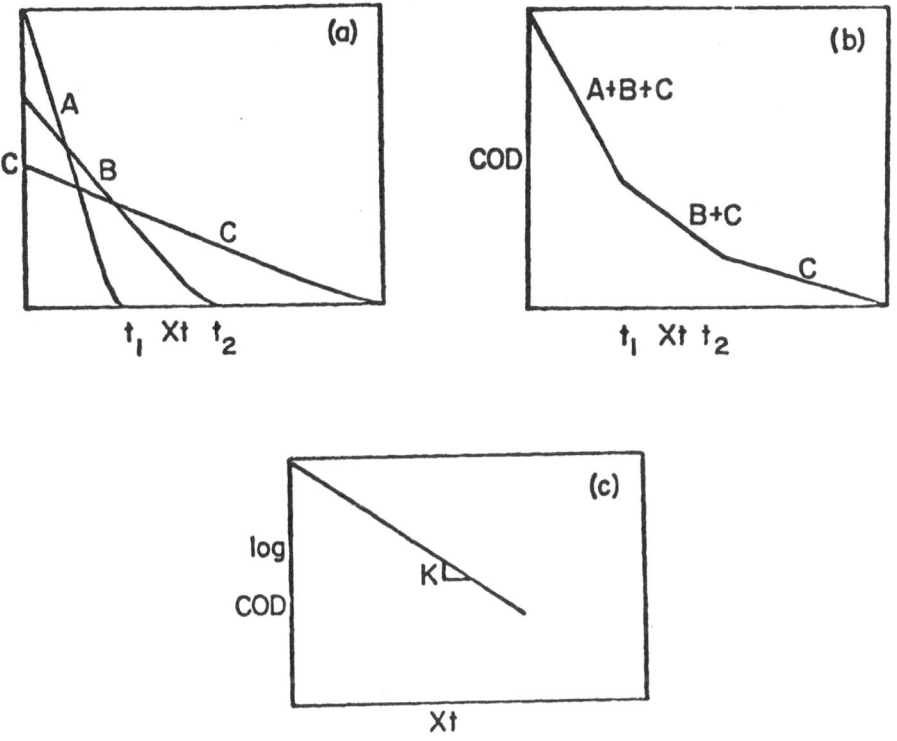

Fig. 10:   Schematic Representation of Organic
Removal

the total COD is the sum of all components remaining. The removal
rate will be constant until time t_1 when component A is substantially removed. The rate will then reduce reflecting components
B and C until time t_2 when component B is substantially removed.
The rate will then reflect only component C. In most wastewaters
there are many components and Figure 10 (b) appears as a smooth
curve. The curve can usually be linearized on a semi-logarithmic
plot as shown in Figure 10 (c).

Sequential growth and removal have also been described in some
cases for mixed cultures [10]. It was further shown by Tischler
as BOD, COD, or TOC, that the overall removal relationship for a
batch reaction can be described by a first-order type equation or
second-order type equation. A second-order reaction will apply
when one or more components have a very low removal rate.

$$\frac{s}{s_o} = e^{-k_1 xt} \qquad \ldots (14)$$

or

$$\frac{s}{s_o} = \frac{1}{1 + k_2 xt} \qquad \ldots (15)$$

The effect of initial substrate concentration in batch oxidation was studied by Tischler [11], Grau et. al. [7]. Through a mathematical analysis, Tischler showed that the removal rate coefficient is inversely proportional to the initial substrate concentration and that Equation (14) and (15) can be modified:

$$\frac{s}{s_o} = e^{-K_1 xt/s_o} \qquad \ldots (16)$$

and

$$\frac{s}{s_o} = \frac{1}{1 + K_2 xt/s_o} \qquad \ldots (17)$$

Through similar reasoning, Grau et. al. arrived at the same conclusion. When considering continuous completely mixed reactors, a similar analysis can be made. Adams and Eckenfelder [11] through a series of continuous activated sludge studies on peptone and an organic chemical wastewater showed that the kinetics of organic removal considering initial concentration will follow the relationship as shown in Figure 11 and defined by Equation (18).

$$\frac{s_o - s}{xt} = K_1 \frac{s}{s_o} \qquad \ldots (18)$$

or

$$s \doteq \frac{s_o}{\dfrac{K_1}{F/M + 1}} \qquad \ldots (19)$$

where:

$$F/M \quad = \quad \text{organic loading} = s_o/xt$$

The implication of Equation (19) is that at a constant organic loading, F/M, expressed as kg BOD applied/day-kg VSS, the effluent soluble BOD is directly proportional to the influent BOD. For example, the influent BOD is reduced from 400 mg/ℓ to 20 mg/ℓ in 0.5 days at a F/M of 0.27. If the influent BOD is increased to 800 mg/ℓ, F/M must be reduced to 0.13 and the retention time increased to two days in order to maintain the same effluent quality of 20 mg/ℓ. If the same percent removal is desired, the F/M remains the same and increasing the influent BOD from 400 to 800 mg/ℓ requires an increase in retention time from 0.5 to 1.0 days. The relationship between influent and effluent BOD for various organic loading levels is shown in Figure 12.

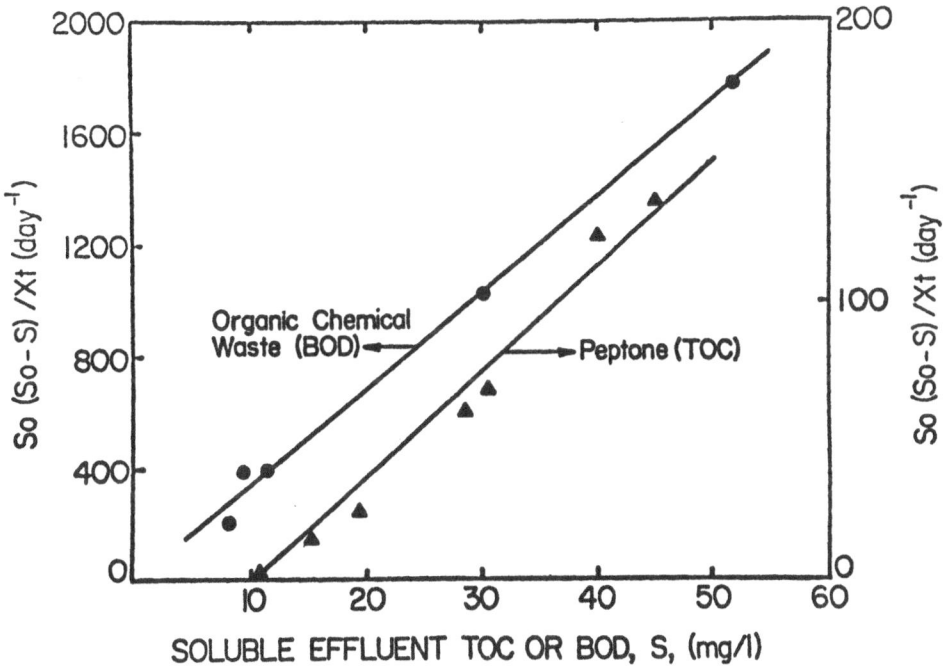

Fig. 11: Kinetic Correlation for Completely Mixed Process

Consideration of Equation (18) leads to the conclusion that the reaction rate coefficient K_1 is dependent on the composition of the wastewater. Values computed for several industrial waste-waters are shown in Table 2. While K_1 might be expected to be relatively constant in wastewaters of constant composition such as a dairy waste, a variable wastewater composition resulting from changes in produce mix as is encountered in the organic chemicals

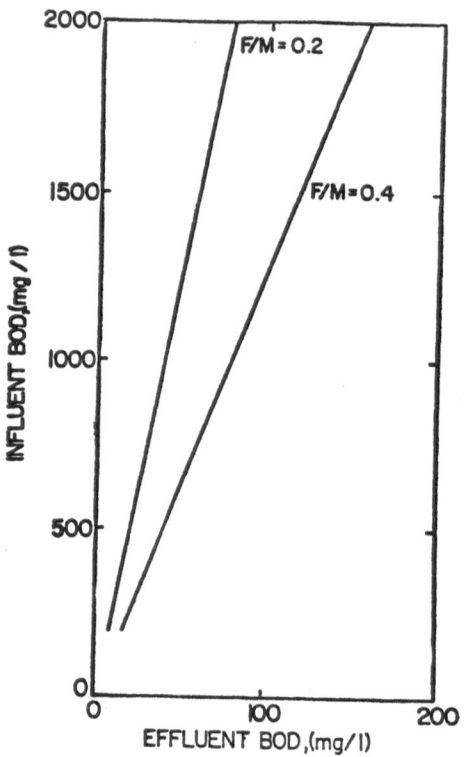

Fig. 12: Relationship Between Influent and Effluent BOD at
Various Loading Levels

industry will result in a variable K. In order to define an effluent quality and its variability from a biological wastewater
treatment facility for this type of wastewater, the reaction rate
K must be treated as a statistical variable.

BOD removal from domestic sewage presents a unique case. In a
domestic sewage the major part (75-90 percent) of the BOD is
present either in colloidal or suspended form. When treated by
the activated sludge process these organics are rapidly removed
by adsorption and enmeshment in the biological floc. It is for
this reason that contact periods of less than 15 minutes result
in BOD removals in excess of 80 percent. This phenomena provides

TABLE 2: Reaction Rate Coefficients for Organic Wastewaters

Wastewater	K Days^{-1}
Potato Processing	36.00
Peptone	4.03
Sulfite Paper Mill	5.00
Vinyl Acetate Monomer	5.30
Polyester Fibre	14.00
Formaldehyde, Propanol, Methanol	19.00
Cellulose Acetate	2.60
AZO dyes, Epoxy, Optical Brigtheners	2.20

the basis for the contact stabilization process and the physical-chemical process in which suspended and colloidal BOD is removed by coagulation followed by soluble BOD removal on activated carbon. Figure 13 illustrates this phenomena. The mechanisms of BOD removal from domestic wastewater can therefore be considered as a two phase reaction, an initial rapid removal of BOD followed by a slow removal of soluble BOD. It has been shown that this removal process can be modeled as a second-order type reaction or as a composite exponential which considers two reactions. These are shown in Figure 14. McCauley [13] concentrated the soluble fraction of domestic sewage and ran BOD removal studies as shown in Figure 15.

The kinetic relationships assume that all the organics removed in the process undergo biological oxidation and synthesis. Some wastewaters contain volatile organics which will undergo simultaneous biological oxidation and air stripping in the aeration process. In these cases it is necessary to operate a sterile test unit under the same aeration conditions to determine the air stripping rate coefficient. The rate coefficient from biological unit which includes both bio-oxidation and stripping is then corrected for stripping to reflect only biological oxidation. When scaling to full-scale design, the stripping effect should be estimated for the equipment and turbulence to be used.

The kinetic relationships previously developed relate to soluble effluent. The BOD contributed by the effluent suspended solids in a process must therefore be added to the soluble BOD as a terminal step in a process design calculation.

$$BOD_{total} = BOD_{soluble} + (f)_{suspended\ solids}$$

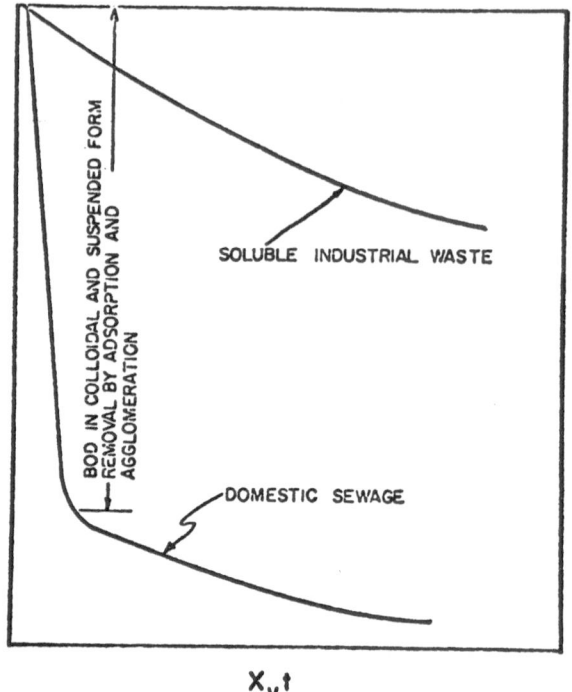

Fig. 13: BOD Removal from Sewage and Industrial Waste

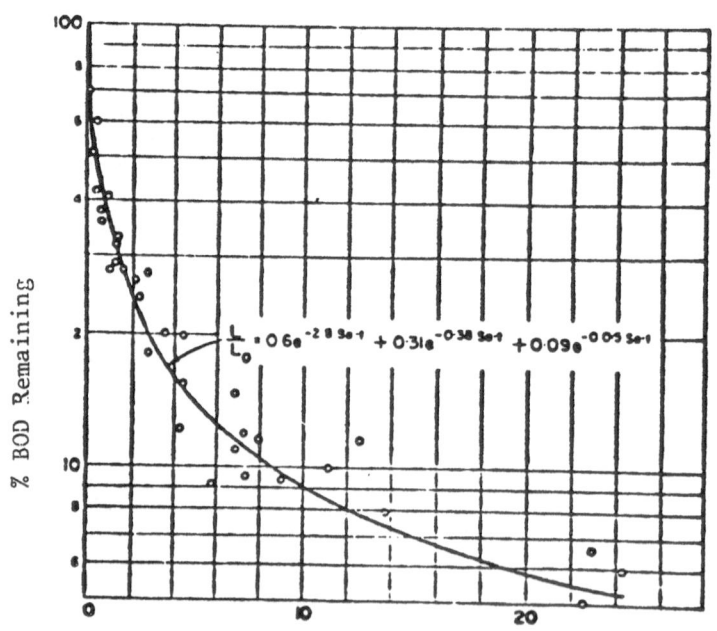

Fig. 14: BOD Removal Characteristics from Domestic Sewage

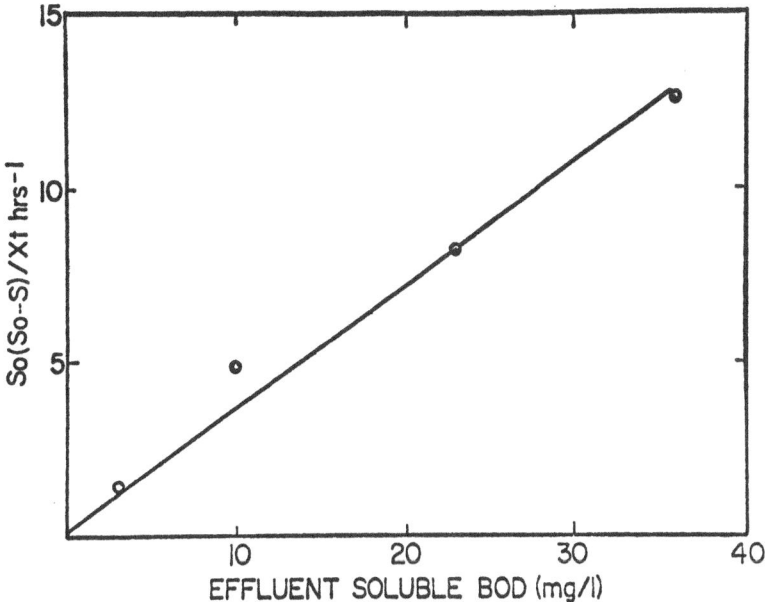

Fig. 15:   Soluble BOD Removal Kinetics from Domestic Sewage[8]

The coefficient, f, is related to the influent waste charac-
eristics (primarily the nature of its suspended solids content,
f any) and to the sludge age, since increasing sludge age inc-
eases the inert content of the sludge. A typical correlation
s shown in Figure 16.

It should also be noted that small residual BOD or COD, s, will
emain even after long periods of aeration, because auto-oxidation
f the sludge results in resolubilization of cellular material
hich is subsequently used for synthesis as shown in Figure 17.

It should be recognized that in any one plant as the composi-
ion of the waste changes, the overall removal rate may also change,
wing to variation in removal rate of particular constituents and
heir initial concentration. This results in a variation in the
oefficient, k. As seen in Figure 18, a wastewater with a constant
omposition such as a dairy waste yields a near constant k, while
 varying composition chemical wastewater shows a high variability
n k.

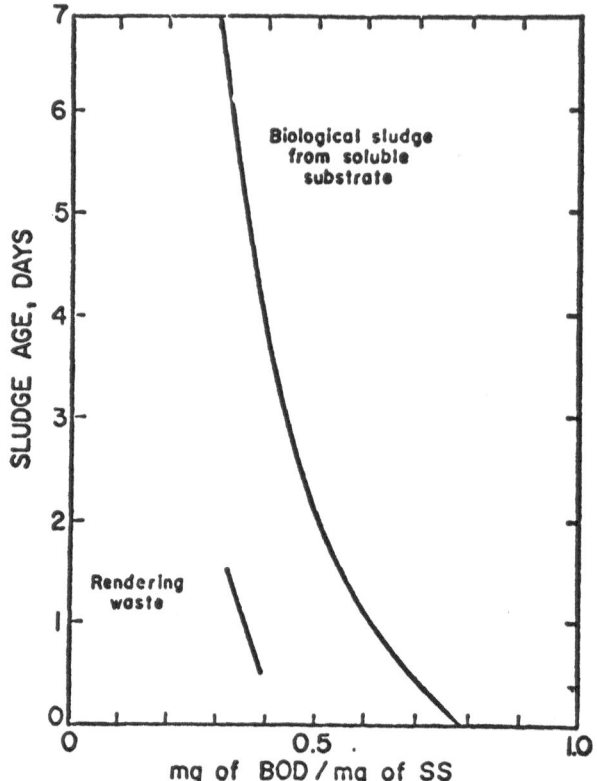

Fig. 16: BOD Characteristics of Biological Sludges

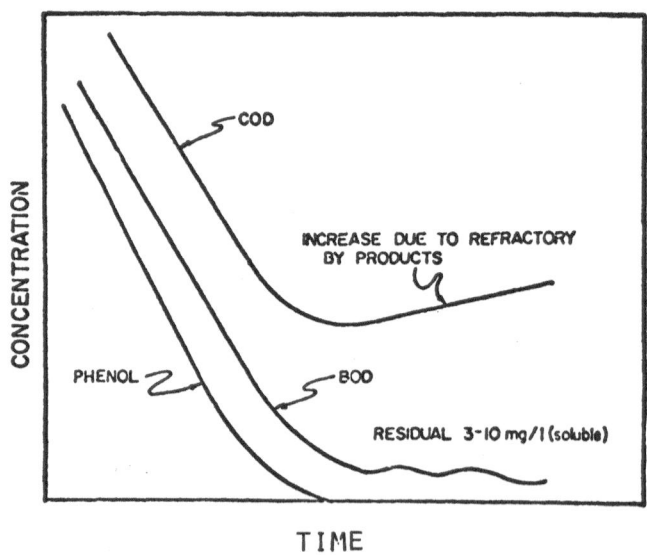

Fig. 17: Effluent Levels During Biological Oxidation

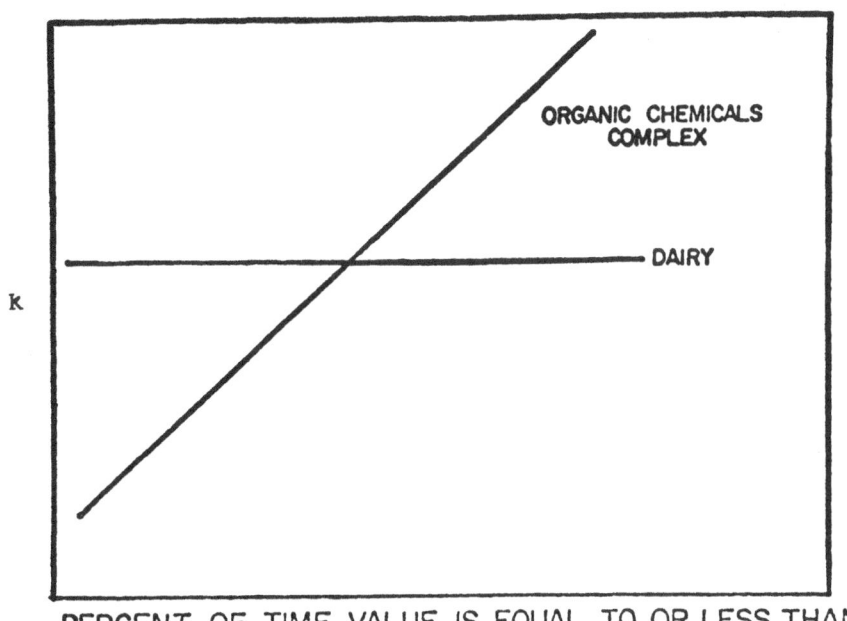

Fig. 18: Variability in k as Related to Wastewater Composition

OXYGEN PENETRATION INTO BIOLOGICAL FLOCS

Numerous investigators have shown that under the mixing and aeration conditions in the activated sludge process as conventionally employed, the biological floc may be composed of aerobic surface layers with an anaerobic center. The activity of the floc results from diffusion of oxygen and nutrients into its mass. If we assume that only the aerobic portion of the floc is effective in organic removal, then on a weight basis, increasing the percentage of the floc which is aerobic will result in an increase in overall organic removal efficiency. As can be seen from Equation (20),

$$c_i - c_c = \frac{kd^2}{24D} \qquad\qquad \dots (20)$$

in which

C_i =   oxygen concentration at the floc surface

C_c =   critical oxygen concentration to support a
          maximum uptake rate, assumed as 0.1 mg/ℓ

k =   oxygen utilization rate, mg O_2/cm^3-sec

d =   floc diameter, cm

D =   oxygen diffusivity in floc material, cm^2/sec,
      assumed as 5×10^{-6} cm^2/sec at 15°C,

the aerobic portion of the floc will depend upon the oxygen gradient across the surface film, ΔC, the floc size, d, and the oxygen
uptake rate k. In any system with a defined organic loading, the
aerobic portion of the floc can be increased either by increasing
the power level (Kw/1000 m^3)   and thereby reducing the floc size
or by increasing the dissolved oxygen driving force, ΔC, which is
accomplished by increasing the dissolved level in the process. As
the loading to the process is reduced and the oxygen uptake rate
decreases it will take less power and/or oxygen driving force to
maintain the floc aerobic since the oxygen uptake rate has decreased.

The endogenous rate coefficient, b, related to the auto-oxidation of the biomass, that is the Kg VSS oxidized/day-Kg VSS under
aeration. As more of the floc becomes aerobic it is reasonable
to expect that b will increase since more aerobic activity will
result.

Rickard and Gaudy [14] showed   the relationship between
power level, organic removal, and oxygen uptake rate and the
endogenous rate, b. Busch [15] has recomputed their data and
showed that increasing the hydraulic shear intensity from 300 to
1000 resulted in an increase of organic removal rate from 0.3 mg
COD/g. MLSS-hr to 0.7 mg COD/g MLSS-hr and an increase in the
endogenous coefficient, b, of 0.14 to 0.20. It is not possible
to relate, however, experimental results to common field operating
conditions, since the hydraulic shear rate does not readily relate
to the power level in aeration basins expressed as Kw/1000 m^3 .

These observations are consistent with the results of Pasveer
[16]. Mueller, et.al. [17], who developed relationships between
floc size and the dissolved oxygen concentration necessary to
render the floc completely aerobic, assuming a floc size of 115μ.
Their results showed that at an oxygen uptake rate of 80 mg/ℓ-hr,
4.0 mg/ℓ-hr required a dissolved oxygen level of 0.6 mg/ℓ. Unfortunately, Mueller, et al, did not relate their results to power

level, so it is not possible to define floc size in the aeration basins under various turbulence levels. Figure 18 shows calculated results from Equation (2) using the data of Mueller, et.al., relating floc size to maintain fully aerobic conditions to dissolved oxygen concentrations.

Decreasing the floc size by increased power levels will decrease the required dissolved oxygen concentration to maintain aerobic conditions. Zahardka [18] conducted pilot plant studies on Prague sewage in which he varied the organic loading and the power level. He showed that activated sludge under conditions of conventional loading and mixing intensity has considerable removal potential. Increasing the mixing intensity resulted in increased removal as related to loading.

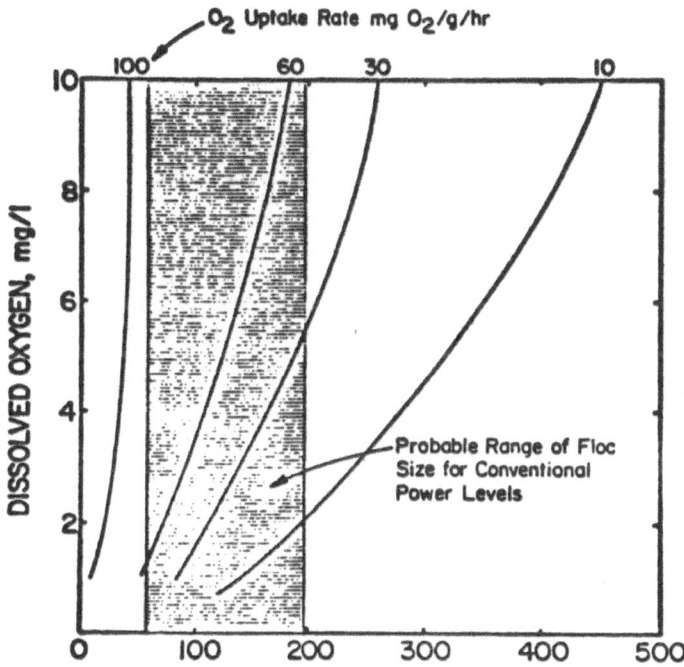

Fig. 19 : Oxygen Uptake

Figure 19 would indicate that at endogenous levels of 10 mg O_2/g-hr, 1.0 to 2.0 mg/ℓ of O_2 are required for fully aerobic conditions.

SLUDGE SETTLING

Sludge settling and compaction characteristics are a primary requisite to successful operation of the activated sludge process. With a poor settling sludge, solids carry-over in the effluent will contribute to the BOD (due to endogenous respiration of the activated sludge solids in the BOD bottle). The poor compaction will result in a low concentration of return sludge solids, which in turn will limit the mixed-liquor suspended solids level.

A poor settling of bulking sludge can be the result of the propagation of filamentous organisms (i.e..Sphaerotilus) or from diffuse bacterial growth due to nutrient limitations.

Many filamentous growths are obligatory aerobes which flourish in the presence of an available carbon source such as glucose. At low concentrations of dissolved oxygen in the mixed liquor (<0.5 mg/ℓ) there is little oxygen penetration into the biological floc and only a small fraction of the bacterial mass will exhibit aerobic growth [16,19]. The filaments, on the other hand, have a very high surface area/volume ratio and will quickly outgrow the bacteral population [15]. High oxygen tension will favor the growth of floc forming bacteria. When the carbon source is exhausted, the filaments will tend to disappear from the system.

Many filamentous organisms are aerobic and can be destroyed by prolonged periods of anaerobiasis. Most of the bacteria, on the other hand, are facultative and can exist for extended periods without oxygen. Although available data are somewhat contradictory, it would appear that at least 12 hours under anaerobic conditions are necessary to eliminate the growth of these filaments.

In practice, culture control can frequently be achieved by holding the mixed liquor in the final settling tank for a period sufficient to eliminate the filaments. Filamentous bulking can also be controlled by:

a)   the addition of chlorine or hydrogen peroxide [20] to the return sludge. Dosages are in the order of 20-50 mg/ℓ. The high surface area to volume ratio of the filamentous growths relative to the flocculated biomass favors a selective kill.

b)   the addition of polyelectrolytes to flocculate and settle the filamentous growths.

Filamentous growths are suffused at high oxygen levels such as is carried in the high purity oxygen processes.

Endogenous sludges can be maintained under anaerobic conditions for long periods because the exogenous food supply has been exhausted. At high loadings, however, gasification due to anaerobic activity may occur, resulting in rising sludge. At very low loadings, denitrification may also cause rising sludge.

Various investigators have related sludge bulking to organic loading. Indications have shown that bulking becomes progressively more severe as the loading exceeds 0.5 kg of BOD/day-kg of MLSS (.5 lb of BOD/day-lb of MLSS) [21]. Von der Emde [22], however, reported excellent settling characteristics with loading in excess of 2.0 kg of BOD/day-kg of MLSS (2.0 lb of BOD/day-lb MLSS). This apparent contradiction can possibly be explained by considering the cause of bulking and its relationship to the operation of the process.

Sludge bulking can be related to the growth rate or metabolic activity of the sludge, which in turn is related to the food/micro-organisms ratio, F/M. At very high F/M ratios, the organisms have a maximum growth rate (log growth phase) and flocculation does not occur [21], and filamentous organisms can develop. At very low loadings, unoxidized fragments of floc remain in suspension, resulting in poorer settling. These regimes are schematically shown in Figure 20. Over an optimal F/M range, the organisms flocculate.

Since the significant parameters is the concentration of food in contact with the organisms, the geometry of the system and the mode of introduction of the waste deserve consideration. For example, in a long, rectangular aeration tank or in a batch-treatment system, the sludge is initially in contact with the sewage entering the system and the F/M ratio is high at the head end the tank (or the start of aeration in the case of a batch-treatment system). A diffuse floc developed under these conditions could persist throughout the aeration period. By contrast, when a complete mixing system is used, and the sewage distributed throughout the aeration tank, the F/M ratio at anytime is low (i.e., the sludge is always in contact with a BOD concentration approximately equal to the effluent). As a result, high loading levels expressed as kg of BOD/day-kg (lb of BOD/day-lb) of MLVSS, can still yield a low F/M ratio and a dense floc.

Recently Grau et.al. [7] indicated that batch or multiple completely mixed reactors produced a denser sludge than a parallel completely mixed system. The rationale indicated that a greater organic driving force in the initial reactors selectively favored the growth of flocculated bacteria. Further work is needed to deliniate these effects.

When considering the effect of the F/M ratio on metabolic activity, the availability of the substrate must be considered. Soluble sugars, for example, are immediately available to the organisms and consequently yield an immediate growth response. Suspended organics, on the other hand, must undergo sequential breakdown to simpler substrates before being available for synthesis. The growth response is therefore much slower, even at high F/M ratios.

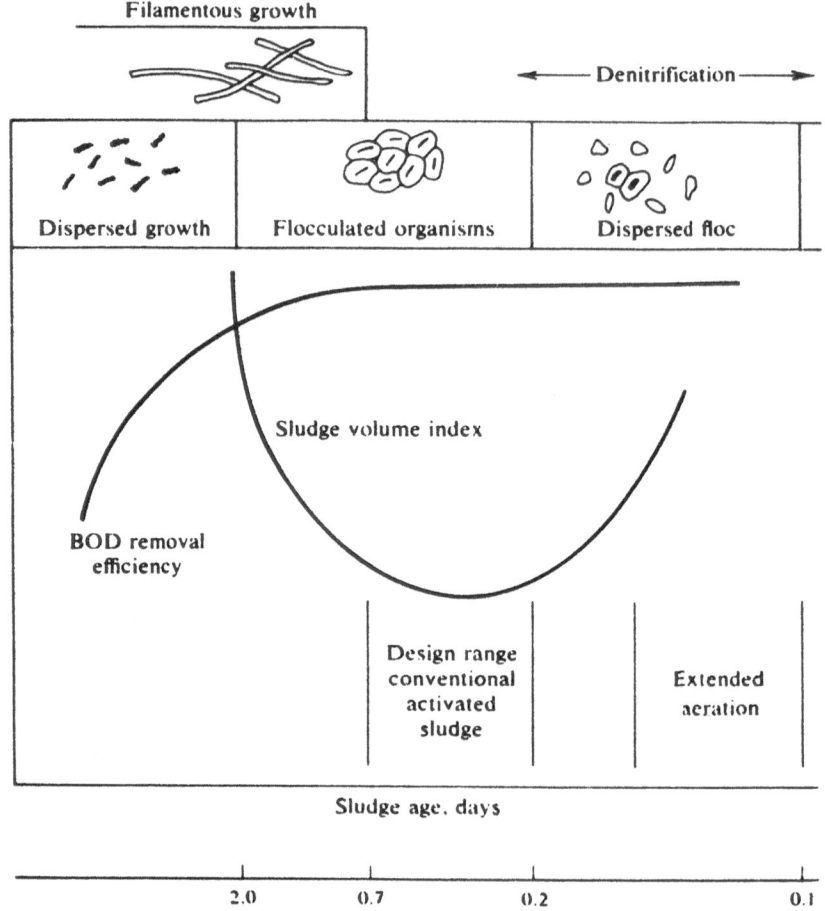

Fig. 20: Flocculation and Settling Characteristics of
Activated Sludge as Related to Organic Loading

In summary, floc characteristics as related to growth rate
will be influenced both by the availability of the substrate,
the mode of introduction of the waste to the system, and the
dissolved oxygen and turbulence in the system. Typical results
for an industrial waste are shown in Figure 21.

NUTRIENT REQUIREMENTS

Aerobic organisms require minimum quantities of nitrogen and
phosphorus and other trace elements for optimal activity. The

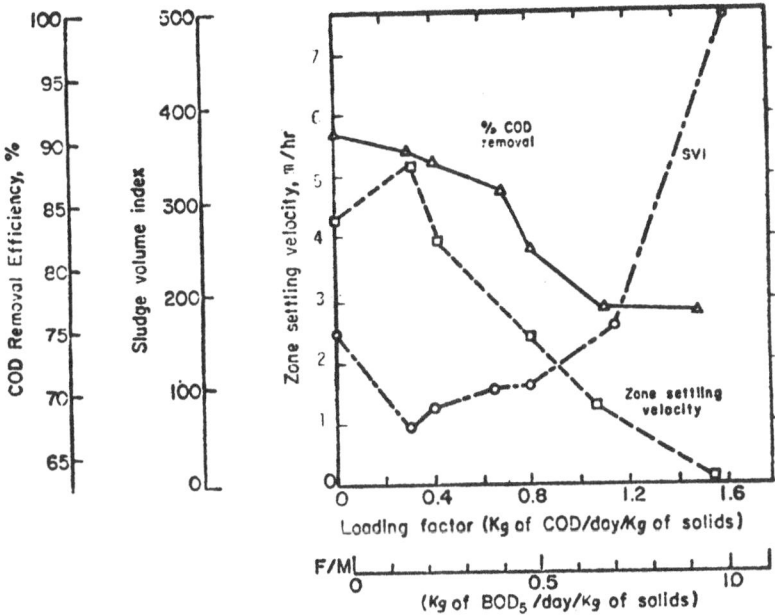

Fig. 21: Parameter Response to Organic Loading-Brewery Waste
(after Ford)

trace elements required are usually present in sufficient quantity
in the carrier water (one exception is when the wastewater emanates
from distilled or deionized water). The trace elements required
are shown in Table 3. Nitrogen and phosphorus are frequently
deficient in industrial wastewaters and must be added as a nutrient
supplement. (In combined treatment of domestic and industrial
wastewaters, the excess nitrogen and phosphorus in the sewage may
supply the requirements, for the industrial waste as shown in
Figure 21). The quantity of nitrogen and phosphorus required
relates to the composition of the biomass. Active biomass contains approximately 12.3 percent nitrogen and 2.6 percent phosphorus. The cellular residue after endogenous metabolism has been
reported to contain 7 percent nitrogen and 1 percent phosphorus.
Nitrogen and phosphorus will be lost from the process in the
excess sludge.

TABLE 3: Trace Nutrient Requirements for Biological Oxidation

	(mg/mg BOD)
Mn	10×10^{-5}
Cu	14.6×10^{-5}
Zn	16×10^{-5}
Mo	43×10^{-5}
Se	14×10^{-10}
Mg	30×10^{-4}
Co	13×10^{-5}
Ca	62×10^{-4}
Na	5×10^{-5}
K	45×10^{-4}
Fe	12×10^{-3}
CO_3	27×10^{-4}

The nitrogen lost will be that present in the active mass plus that present in the non-degradable residue. The minimum quantities of nitrogen can then be computed:

$$N(kg/day) = 0.123 \frac{x}{0.77} \Delta X_v + 0.07 \frac{(0.77 - x)}{0.77} \Delta X_v$$

In like manner the minimum phosphorus can be computed:

$$P(kg/day) = 0.026 \frac{x}{0.77} \Delta X_v + 0.01 \frac{(0.77 - x)}{0.77} \Delta X_v$$

EFFECT OF TEMPERATURE

One of the significant variables in the selection of type of process is the effect of temperature on process performance. The

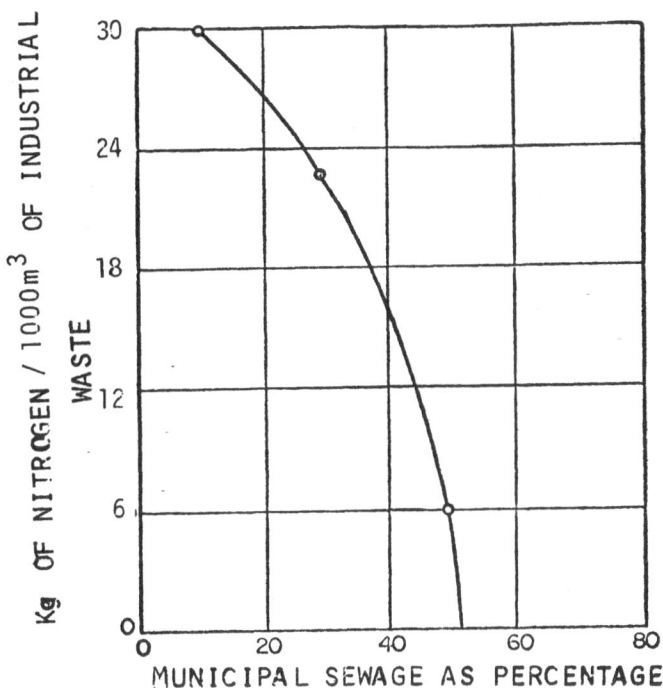

Fig. 22: Nitrogen Requirements for Combined Industrial-
Municipal Biological Treatment Plant

effect of temperature on biological activity is shown in Figure
23. In order for nitrogen to be consumed in the process it must
be available for assimilation by the organisms. Ammonia nitrogen
and nitrate nitrogen are available. Organic nitrogen must first
be hydrolyzed in the process in order to be available to the biomass. Depending on the nature of the wastewater a variable portion of the organic nitrogen may become available for synthesis.

In many cases, in aerated lagoons treating pulp and paper mill
wastewaters, nitrogen and phosphorus has not been added, but rather the retention time increased. Calculated reaction rate coefficients are shown in Table 4. High BOD reductions at minimal
nutrients can be achieved if sufficient retention time is provided.

The temperature effect on the reaction rate can be expressed
by the relationship

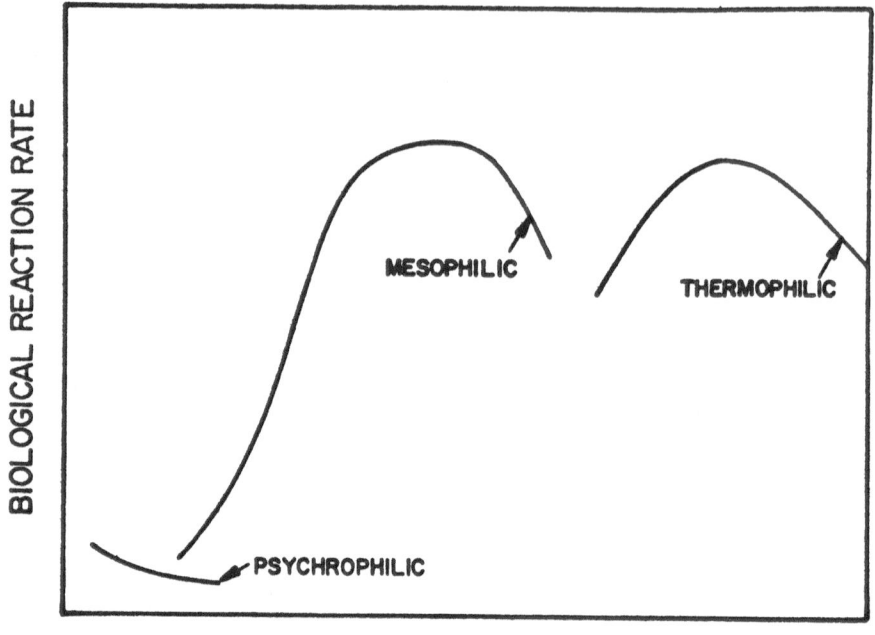

Fig. 23: Effect of Temperature on the Biological Reaction
Rate

TABLE 4 :   Reaction Rate Coefficients

WASTE	Rate	
	Without Nutrients	With Nutrients
Kraftpaper	0.35	1.33
Boardmill	0.70	3.20
Hardboard	0.34	1.66

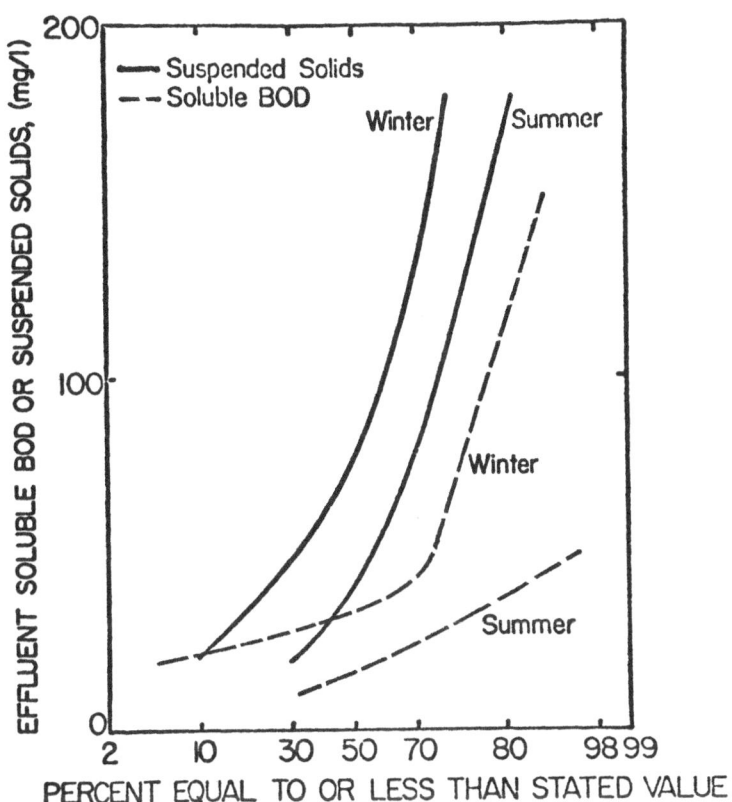

Fig. 24: Variation in BOD and Suspended Solids During Winter and Summer Operation

$$K_T = K_{20°C} \, \theta^{(T-20)} \qquad \qquad ....(21)$$

Equation (21) generally applies over the range of 4° to 30°C. This range is defined as the mesophilic range. Biological activity can also occur in the thermophilic range which has an optimum temperature of 55°C. At low temperatures (4°C) psychrophilic organisms predominate. The temperature effect on aerobic biological activity is shown in Table 5.

One explanation for the wide variation in θ can be rationalized by considering the nature of the process. In the activated sludge process at low loadings, BOD removal and oxidation depend on diffusion of oxygen, consequently a large portion of the floc is aerobic. At high temperatures, the high respiration rate depletes the oxygen rapidly, and only a small portion of the floc is aerobic.

TABLE 5: Temperature Dependence of Rate Constant For
Activated Sludge

Process	Temperature Range	Θ
Removal Rate of Acetate	0 - 10°C 20 - 25°C	1.042 1.214
Removal Rate of Phenol	0 - 10°C 10 - 20°C	1.131 1.056
Mixed Organic Chemical Effluent		1.055

It can be assumed that a large mass of organisms at a low respiration rate (summer), and hence the coefficient Θ is low.

By contrast, at high organic loadings, the floc tends to become dispersed (bulking sludge), and each organisms is more directly affected by changes in temperature. The coefficient, Θ, therefore increases. In the case of domestic sewage, where a major portion of the BOD is in suspended or colloidal form, removal in the presence of flocculated sludge is largely physical and hence relatively independent of temperature. In aerated lagoons, at a low solids level, the organisms are more dispersed, and the temperature coefficient is higher. This is also true for the BOD bottle and the stream. Trickling filters are analogous to activated sludge except that oxygen diffusion is uniplaner into the film. Similar calculations as for activated sludge lead to a coefficient Θ of 1.035. Marked improvement in aerated lagoon operation during the winter months can be achieved by adding a recycle and increasing the solids level in the basin.

In many cases the effluent suspended solids will increase with decreasing wastewater temperatures. The maximum temperatures for effective aerobic biological activity is about 38°C (100°F). At higher temperatures, the reaction rate decreases and the effluent suspended solids will increase. The effluent suspended solids during winter and summer operation for an activated sludge plant treating an organic chemicals wastewater is shown in Figure 24.

Since a major portion of BOD removal from domestic sewage is by biocoagulation which is insensitive to temperature, Θ is very low as compared to soluble organic wastewaters. This comparison is shown in Figure 25.

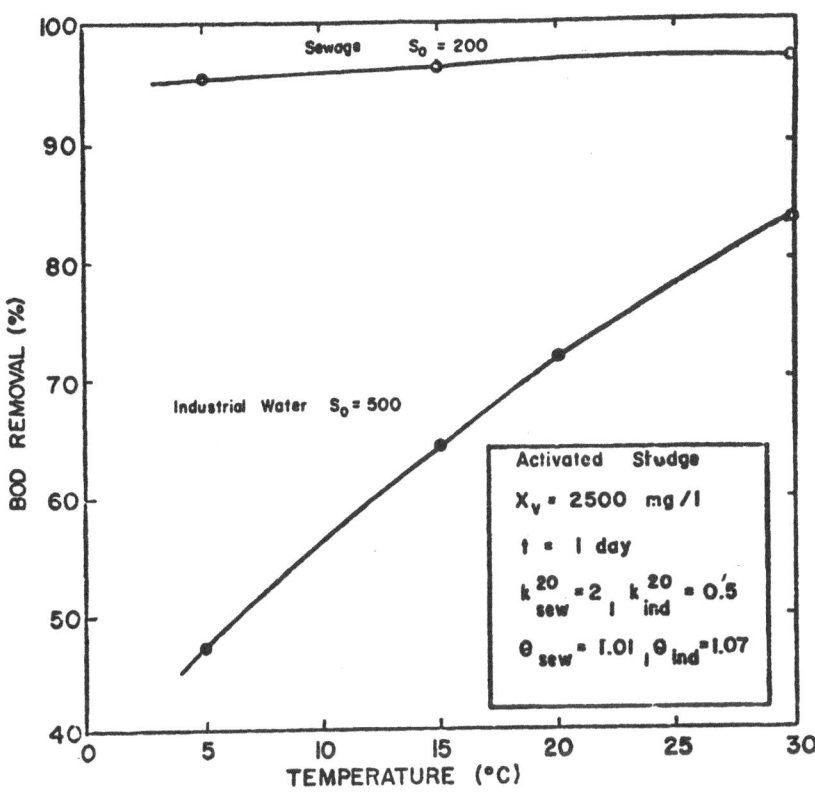

Fig. 25: Temperature Effect on BOD Removal in the Activated Sludge Process for Sewage and Industrial Wastewaters

EFFECT OF pH

Most bio-oxidation systems treating organic wastes have an optimum pH range of 6.5 to 8.5. The efficiency of the process falls off both above and below this range. It is important to note, however, that the pH of significance is that in the aeration basin rather than that of the influent wastewater since bacterial action will modify the pH in the basin. Several cases can be considered: caustic alkalinity present in the wastewater will react with the carbon dioxide produced by microbial respiration to produce bicarbonate which will buffer the process near pH 8.0. Approximately 0.5 kg OH^- as $CaCO_3$/kg BOD_5 (0.5 lb OH^- as $CaCO_3$/lb BOD_5) removed will be neutralized in the process.

Nitrification requires sufficient alkalinity to react with the hydrogen ions produced in the reaction.

In the case of organic acids, bio-oxidation converts the acids to CO_2 and H_2O, thereby increasing the pH. No preneutralization is required if the effluent BOD from a completely mixed reactor is less than about 25 mg/ℓ.

High concentrations of salts of organic acids, such as sodium acetate reacts to produce sodium carbonate. Organics containing sulfonates can yield sulfuric acid as a reaction produce requiring neutralization. Free mineral acidity requires external neutralization.

TOXICITY

Inhibition or toxicity to the biological process can be considered in several categories.

a) Organics which are toxic in high concentration but which are biodegradable in low concentration such as phenol. The concentration of significance is that in the basin in contact with the biomass. Therefore, the degradation rate must be sufficient to reduce the concentration of the substance in the basin to a value less than the inhibiting or toxic threshold.

b) Most heavy metals will exhibit a toxicity to aerobic and anaerobic biological systems. After acclimation, higher concentrations of metal ion can be tolerated, a portion of which will be removed by the biomass in the process. Metal removal by the activated sludge process treating a refinery wastewater are shown in Table 5.

c) Salts in high concentrations will inhibit biological activity. The limiting concentration of ammonia has been reported as 1600 mg/ℓ at pH 7.0. A limiting concentration of chloride of 16,000 mg/ℓ has been reported. Organics in saline water (Cl^- > 3 percent) have been successfully treated using marine organisms.

High salt concentration will result in poor flocculation and dispersion. This result is a high concentration of effluent suspended solids as shown in Figure 26. It should be noted that the increase in suspended solids are dispersed and non-settleable.

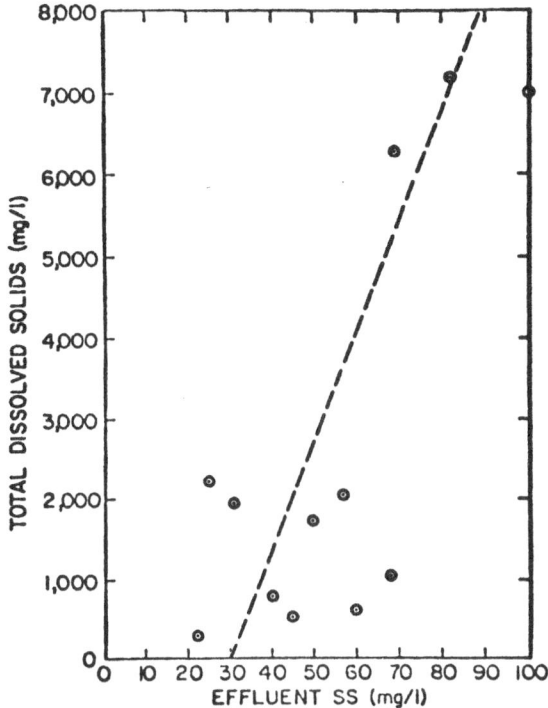

Fig. 26: Effect of Influent Total Dissolved Solids on
Effluent Suspended Solids

NITRIFICATION

The most work on nitrification in recent years has been reported by Downing [23] and Wuhrmann [24].   Nitrification results
from the oxidation of ammonia by Nitrosomonas to nitrite and the
subsequent oxidation of the nitrite to nitrate by Nitrobacter.
Since a build-up of nitrite is rarely observed, it can be concluded that the rate of conversion to nitrite controls the rate of
overall reaction as shown in Figure 27.

The reactions which occur are as follows. Ammonia is oxidized
to nitrite by Nitrosomonas:

$$2NH_4^+ + 3O_2 \rightarrow 2NO_2^- + 2H_2O + 4H^+ + \text{new cells}$$

Other organisms capable of oxidizing ammonia to nitrite are
nitrosococcus, nitrosospima, nitrosocystis and nitrosogloea,
although these are probably of secondary importance in wastewater

oxidation systems. The nitrite is then oxidized to nitrate by Nitrobacter:

$$2NO_2^- + O_2 \rightarrow 2NO_3^- + \text{new cells}$$

The nitrifiers are autotropic organisms and use CO_2 or HCO_3^- as a carbon source. The organisms can survive with an initial lag period under anaerobic conditions for at least 4 hr. The oxygen requirements for nitrification have been formulated by McCarty [25].

Nitrosomonas

$$55NH_4^+ + 5CO_2 + 76O_2 \rightarrow C_5H_7NO_2 + 52H_2O + 109H^+$$

Nitrobacter

$$400NO_2^- + 5CO_2 + NH_4^+ + 195O_2 + 2H_2O \rightarrow C_5H_7NO_2 + 400NO_3^- + H^+$$

On this basis, 3.22 mg/ℓ O_2 is required for oxidation of 1 mg/ℓ NH_4^+ to NO_2^- and 1.11 mg/ℓ O_2 for oxidation at nitrite to nitrate. The total O_2 required would be 4.33 mg/ℓ O_2/mg/ℓ NH_4^+ oxidized. The cell yield for Nitrosomonas has been reported as 0.05 to 0.29 and for Nitrobacter 0.02 to 0.08.

For effective nitrification to occur, the sludge retention period or sludge age, G, must be greater than the growth rate of the nitrifying organisms. Shorter sludge ages will result in a washout of these organisms. The results of Downing and Wuhrmann show that these resultant retention periods are usually sufficient to effect substantially complete nitrification. The data of Downing have been converted in terms of sludge age to effect nitrification. Several investigators have reported nitrification to occur over the temperature range of 5 to 45° with the optimum range being 25 to 32°C. Downing [23] reported a rate relationship.

$$K_n = 0.18 \cdot 1.128^{(T - 15)}$$

A rate K of 0.18 at 15°C was reported for Nitrosomonas. The temperature effect on the required suldge age using the data of Downing is shown in Figure 28. Nitrification as related to sludge age is shown in Figure 29.

Percent nitrification as defined in these figures is the percentage of total nitrate formed after complete oxidation. pH has significant effect on the growth of both Nitrosomonas and Nitrobacter. The optimum pH range for *Nitrosomonas* is 7.6 to 8.0 and for *Nitrobacter* 7.8.

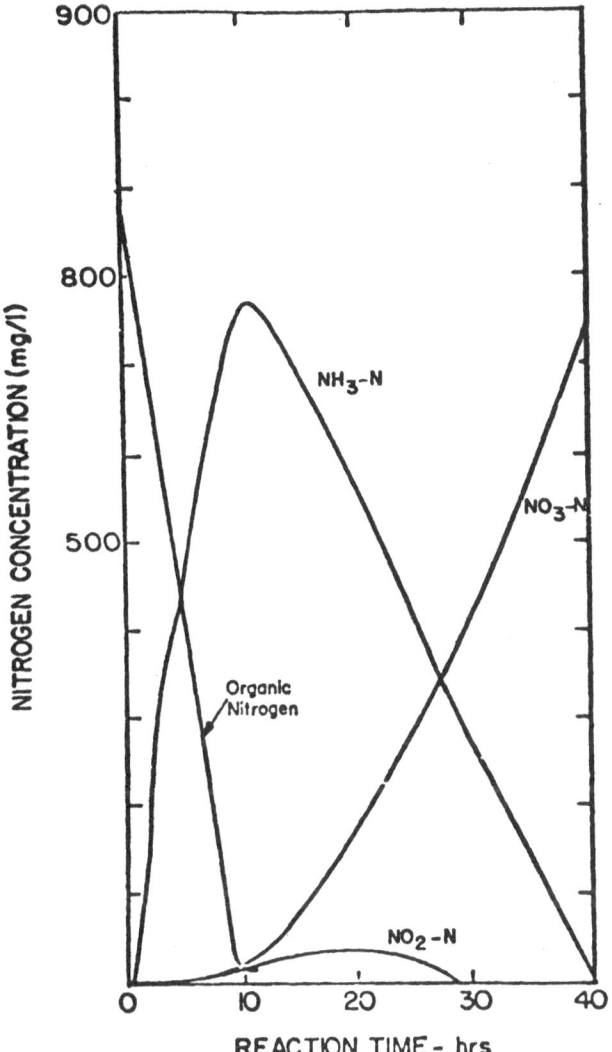

Fig. 27: Nitrification Kinetics

Since nitric   acid   is produced in the oxidation it is necessary
to provide 9.13 mg/ℓ alkalinity per mg/ℓ NH_3-N oxidized. The rate
of nitrification has been shown to be dependent on the dissolved
oxygen level at concentrations less than 2.0 mg/ℓ. The rate is
independent of ammonia concentration at levels in excess of approximately 0.5 mg/ℓ.

Heavy metals such as Cr, Ni, and Zn are toxic at concentrations
of 0.25 mg/ℓ. Some organic carbon compounds may be toxic at a
high concentration level.

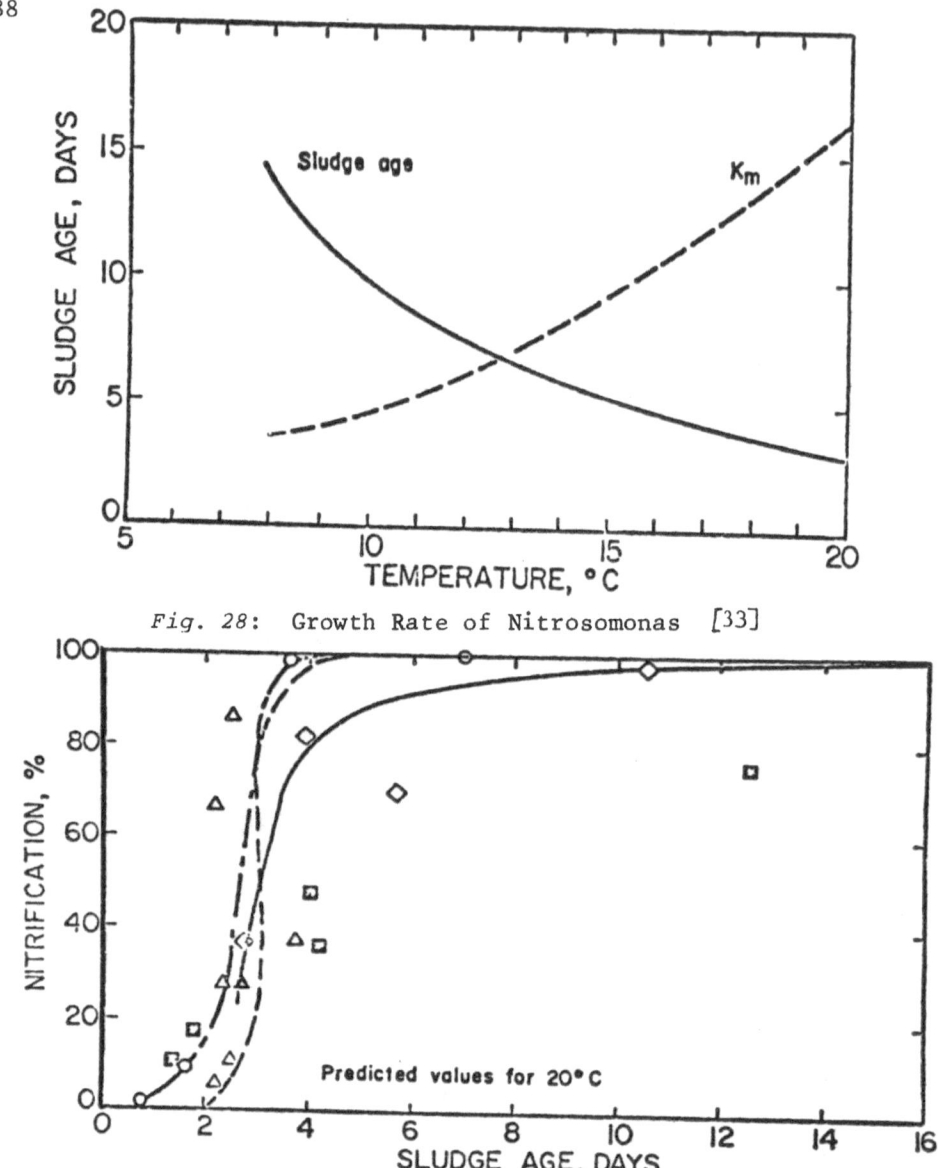

Fig. 28: Growth Rate of Nitrosomonas [33]

Fig. 29: Relationship Between Nitrification and Sludge Age
in the Activated Sludge Process

DENITRIFICATION

The most exhaustic study on denitrification has been reported
by Wuhrmann [21]. Many of the heterotrophic bacteria present in
activated sludge are facultative and can reduce nitrate. Nitrate
under these conditions will reduce to N_2 and small quantities of
N_2O.

The pH of the mixture profoundly affects the rates of denitrification relative to the presence of dissolved oxygen. pH values in the acid range will permit active denitrification in the presence of dissolved oxygen, whereas strict anaerobic conditions should be maintained to promote denitrification under alkaline conditions.

The denitrifying organisms are heterotrophic and require an organic carbon source for growth. It is possible, however, to use the endogenous by-products as a food supply for the denitrifiers. Denitrification rates would increase in the presence of available carbon such as supplied by untreated sewage. The rate of denitrification using an endogenous food source is shown in Figure 30. The data of Culp and Slechta [26] obtained from the Tahoe studies are also shown in this figure. The rate of denitrification with increasing concentrations of available carbon is shown in Figure 31.

The most common external carbon source is methanol. The reactions using methanol have been postulated:

First Stage

$$NO_3^- + \frac{1}{3} CH_3OH \rightarrow NO_2^- + \frac{1}{3} CO_2 + \frac{2}{3} H_2O$$

Second Stage

$$NO_2^- + \frac{1}{2} CH_3OH \rightarrow \frac{1}{2} N_2 + \frac{1}{2} CO_2 + \frac{1}{2} H_2O + OH^-$$

The overall reaction is:

$$NO_3^- + \frac{5}{6} CH_3OH \rightarrow \frac{1}{2} N_2 + \frac{5}{6} CO_2 + \frac{2}{6} H_2O + OH^-$$

74 mg/ℓ methanol is required to reduce 30 mg/ℓ NO_3^-
Methanol requirements have been reported by McCarty:

$$CH_4 = 2.47N_o + 1.53N_1 + 0.87D_o$$

Biomass produced

$$C_b = 0.53N_o + 0.32N_1 + 0.19D_o$$

25-30 percent excess is required for cell synthesis.

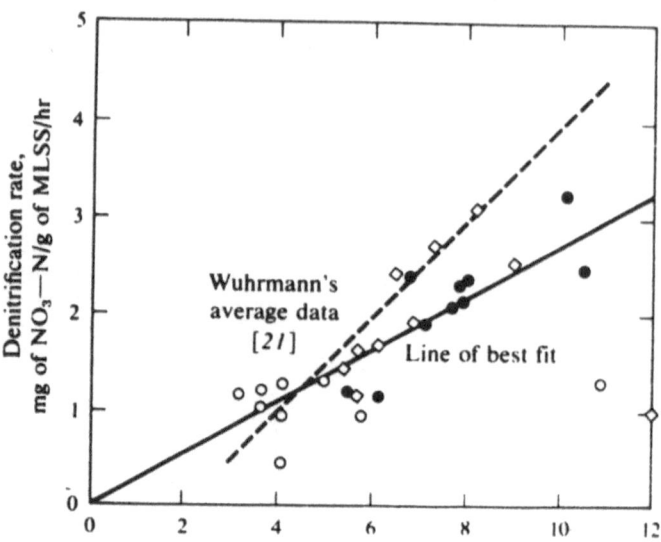

Fig. 30: Rate of Denitrification Using Endogenous Food Source

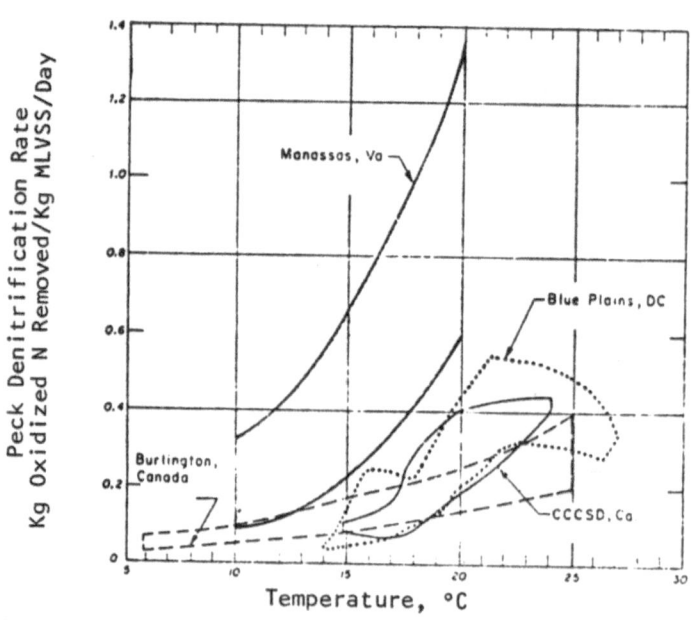

Fig. 31: Observed Denitrification Rates for Suspended Growth
System Using Methanol

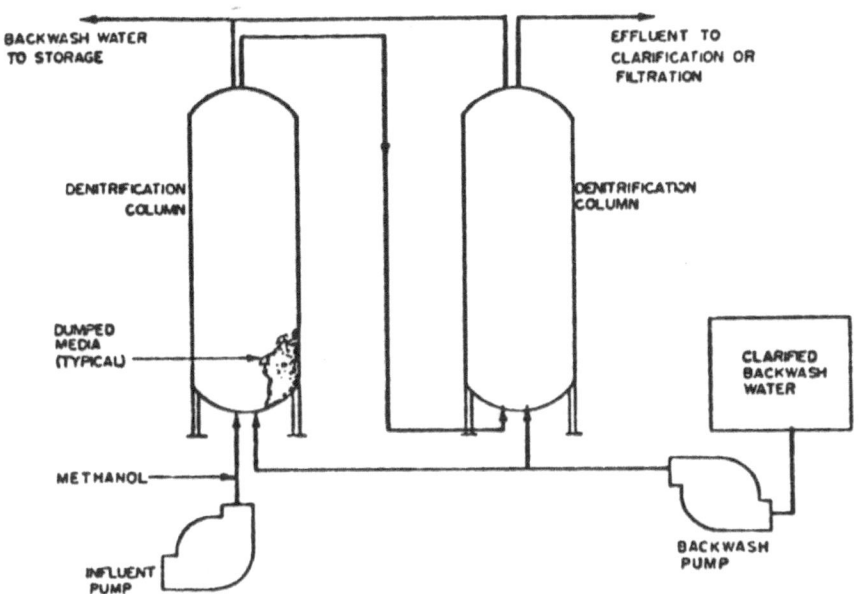

Fig. 32: Typical Process Schematic for Submerged High
Porosity Media Columns

There are several available process options for nitrification
and denitrifications as shown in Figure 33. Nitrification has
been and can be achieved in conjunction with carbonaceous BOD removal. In this case the sludge age must be controlled to permit
growth of the nitrifying organisms. A disadvantage to this process
is that the nitrifying population is controlled by the ammonia fed
to the system and variations in ammonia in the influent will result
in variations in the effluent. This problem can be minimized by
utilizing a two-stage process in which the first stage is optimized
for BOD removal and the second stage for nitrification. Since the
second stage reactor will contain substantially only nitrifying
organisms, they will always be in excess relative to the ammonia
fed and variations in effluent quality will be minimized.

If denitrification is required a carbon source such as methanol
must be fed in the single or two-stage system in order to employ
reasonable retention periods.

An alternative approach developed by Barnard [32] is shown in
Figure 33. In this process, a first-stage anaerobic basin receives
sewage and recirculated nitrified effluent from the second stage.
The nitrate is reduced under anaerobic conditions providing oxygen
for BOD removal. Nitrate passing through stage two is denitrified
in the stage three anaerobic basin. A final aeration stage preceeds

92

final clarification and recycle of the sludge. It was observed that the anaerobic-aerobic system preceeding the final clarifier resulted in a high biological removal of phosphate.

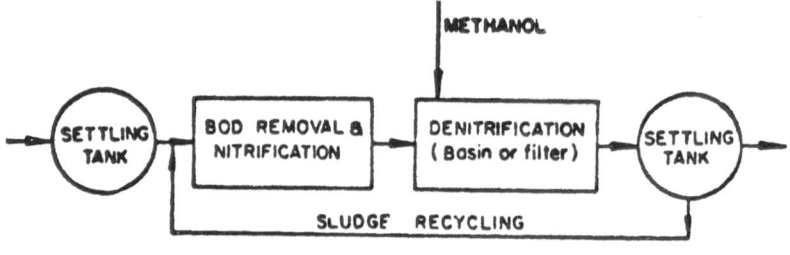

SINGLE STAGE

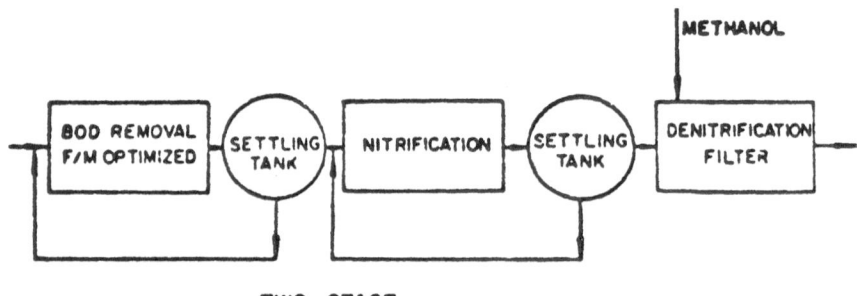

TWO STAGE

Fig. 33: Nitrogen Removal Processes

A similar process has been operating in Vienna. In this process, the sewage is being treated in an oxidation ditch. The placing of the cage aerators results in anoxic conditions, namely, dissolved oxygen across the aerator dropping to zero between the aerators. Under these conditions, nitrification and denitrification coexist in the aeration basin. Performance data for the Vienna plant is shown in Table 6.

KINETICS OF ANAEROBIC TREATMENT

Anaerobic treatment is employed for the degradation and breakdown of organic solids or for the breakdown of soluble organics to gaseous end products as shown in Figure 34. The volatile acids formed in the fermentation are acetic, propionic, and butyric. For most of the longer-chain volatile acids, a given species of methanol organisms converts the acid to methane, carbon dioxide, and a second volatile acid having a shorter carbon chain. The second volatile acid is then fermented in a similar fashion. Acetic acid is directly converted to CO_2 and CH_4. Thus, the overall conversion is the result of two or more reactions (Figure 35).

TABLE 6:   Nitrogen Removal at Blumental, Vienna

	Raw Sewage (mg/ℓ)	Effluent (mg/ℓ)	Removal (percent)
BOD	257.0	13.0	95
COD	475.0	50.0	90
TOC	153.0	14.0	91
TKN	13.8	0.4	–
NH_3-N	21.4	3.8	–
NO_2-N	0.2	0.0	–
NO_3-N	35.7	4.2	88

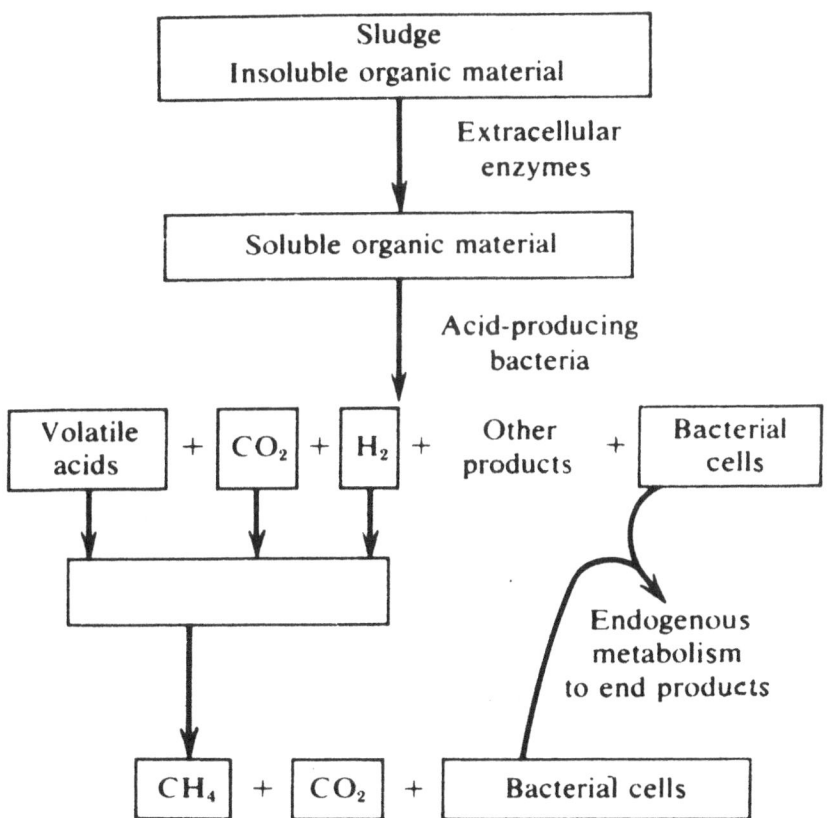

Fig. 34: Mechanism of Anaerobic Sludge Digestion

94

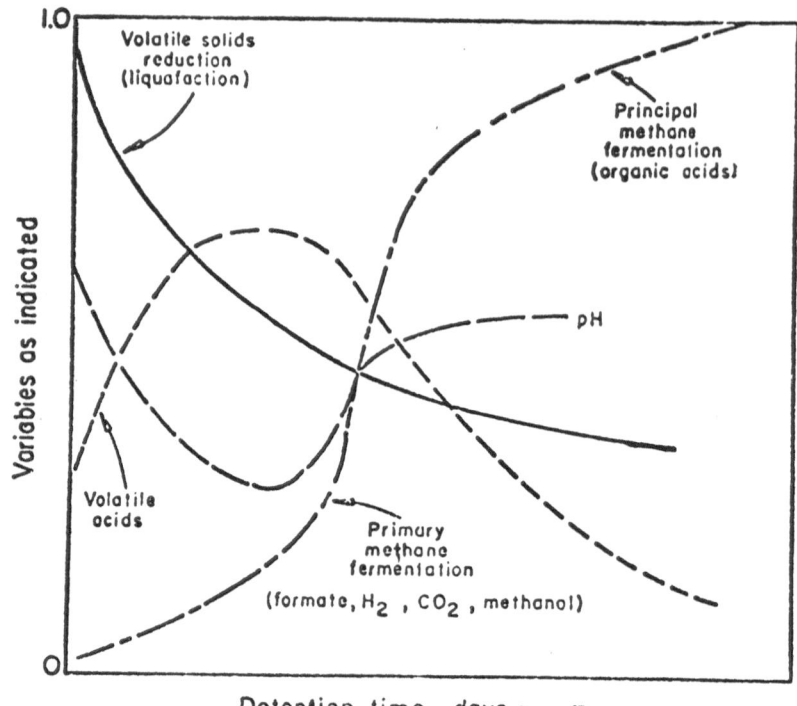

Fig. 35: Mechanism of Continuous Mixed Anaerobic Digestion

Successful anaerobic digestion depends upon maintaining a balance belween the various rates of reaction occurring in the digester. Since the rate of methane fermentation must control the overall rate to avoid process failure, further consideration of the rate of this fermentation is important.

To effect methane fermentation, sufficient time must be available in the reactor to permit growth of the organisms or they will be washed out of the system. In a completely mixed flowthrough digestion tank, this means that the detention time in the unit must be greater than the growth rate of the methane organisms. It is significant to note that there are several species of methane organisms active in a digestion system, all having different growth rates. Andres et. al. [4] have shown that some organisms with a high growth rate (< 2 days) can produce methane, probably from the fermentation of formate, methanol, CO_2 and H_2, and possibly some volatile acid fermentation. Other organisms require residence times of up to 20 days. Although data are limited, some results have been reported relativ, to the growth rate of methane organisms. These data are summarized in Table 7.

As shown in Figure 35, at low residence times there will be volatile solids reduction as a result of the liquefaction of the

solids and the subsequent conversion to volatile acids by acidification. During this period a small amount of methane fermentation
may occur (depending on environmental conditions such as pH) primarily due to reduction of formate, methanol, CO_2, and H_2. There
will be a decrease in pH and a corresponding increase in the volatile acid concentration. There will be very little COD reduction,
however, because the organics have merely been converted from a
solid form to a soluble form in the supernatant liquor. When the
detention time in the digester exceeds the growth rate of the
principal methane organisms there will be a rapid increase in
methane production with a corresponding decrease in volatile acid
concentration and an increase in pH. There are probably several
methane organisms responsible for the volatile acid conversion,
each of which will have a different generation time or growth
rate; the methane production curve is relatively flat, as shown
in Figure 35.

TABLE 7:   Growth Rate of Methane Organisms

Substrate	Temperature (°C)	Residence Time (days)	Reference
Methanol	35	2.0	[27]
Formate	35	3.0	[27]
Acetate	35	5.0	[27]
Propionate	35	7.5	[27]
Primary and Activated Sludge	37	3.2	[28]
Acetate	35	2- 4.2	[29]
Acetate	25	4.2	[29]
Propionate	25	2.8	[29]
Butyrate	35	2.7	[29]

The acetic acid at higher residence times are obtained from
two sources; direct fermentation, and the breakdown of higher
carbon acids to acetic. The major part of the methane production
comes from acetic acid fermentation, although some is generated
from the breakdown of the higher acids. At long residence periods,
substantially all of the volatile acids are converted to methane
and carbon dioxide. It should be noted that since all of the
volatile solids present are not degradable in the digestion unit,
a portion will remain, even after long periods of retention. For
sewage sludge, this fraction is approximately 40 percent.

GAS PRODUCTION

The major part of the gas produced in a sludge digester comes from the breakdown of volatile acids. Some gas is produced by the early stages of methane fermentation of CO_2 and H_2, methanol, etc., but this contribution is probably very small. The gas will be composed of CH_4, CO_2, and small quantities of H_2S and H_2. The percentage of CH_4 in the gas will depend in large measure on the residence time, the percentage of CO_2 being higher at the lower residence times with corresponding lesser numbers of methane bacteria. Lawrence and McCarty [27] have shown from theoretical considerations supported by experimental evidence that 1.11 cu m of methane gas will be produced per kg of COD reduced (5.62 cu ft/lb). The reported gas production for volatile solids reduction in a well-operating anaerobic digestion tank is 3.4 to 4 cu m/kg (17 to 20 cu ft/lb) of VS destroyed with a methane content of about 65 percent. This is about equivalent to 1 to 1.4 cu m CH_4/kg (5 to 7 cu ft of CH_4/lb) of COD destroyed, which is close to the value reported by Lawrence and McCarty [29]. It is significant to note at this point that these values are a maximum assuming complete conversion of the solids to methane. Volatile solids reduction can, of course, occur by liquefaction and conversion to volatile acids without any COD reduction. Under these conditions, the methane yield per unit of volatile solids reduction may be very low.

PROCESS OPTIMIZATION

Optimal performance from the activated sludge process related itself to maintaining a favorable sludge age (or F/M) and imposing constraints on the influent wastewater variability to avoid upsets to the process. It has been found that effluent quality from the activated sludge process can be related to sludge age [1][2] or F/M (Figure 36). When the sludge age is too low, filamentous and/or dispersed growth results yielding poor settling properties and a high suspended solids carryover from the final settling tank. This in turn increases the total BOD discharged from the plant. For soluble and total BOD removal was 94 and 88 percent removal, respectively, at an F/M of 1.0 (based on activated mass). The BOD contributed by the carryover suspended solids was 0.2 mg BOD/mg SS. Increasing the loading (F/M) to 2.0 reduced the soluble and total BOD removals to 90.5 and 76 percent, respectively, and increased the BOD contribution of the suspended solids to 0.5 mg BOD/mgSS. While there was only a small decrease in soluble BOD removal, the effluent deteriorated markedly due to increased carryover of suspended solids with a high active fraction.

When the sludge age becomes too high (or the F/M too low) the biological floc is oxidized and dispersed.

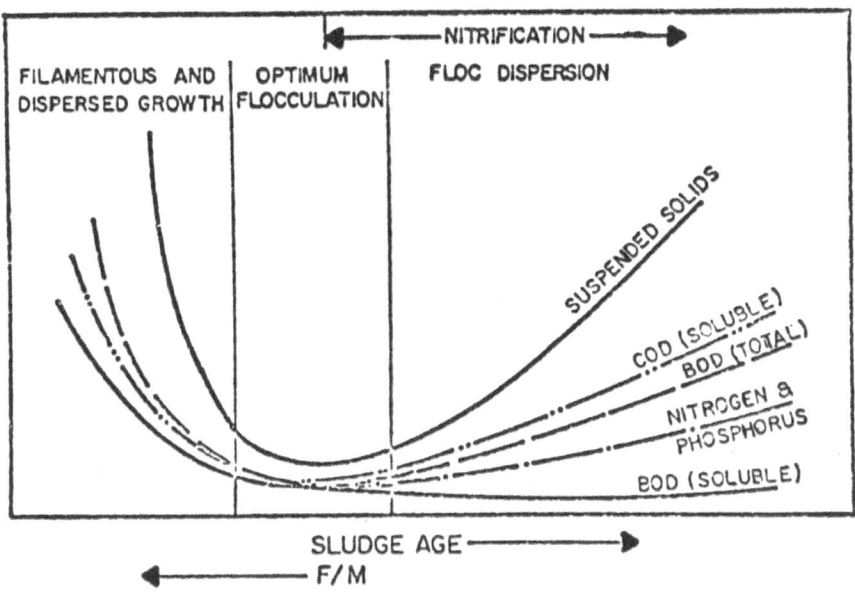

Fig. 36: Effluent Characteristics as Related to Sludge Age and F/M

Chudoba [30] has shown that the biological oxidation of degradable organic compounds yield refractory organics as a by-product and equal to 0.5 to 1.2 percent of the original COD. By contrast, other investigators have reported residual COD values of 5 to 15 percent of the original COD. It would appear that the residual BOD after treatment bears a relationship to the F/M employed in the process. As a result, while the effluent BOD will remain relatively constant with increasing initial concentration of biodegradable substrate, the effluent COD will increase due to the increased bio-resistant products of oxidation. This phenomena results in a changing BOD_5/COD ratio of 0.5 to 0.7. Activated process effluents will decrease to 0.03 to 0.2. Chudoba showed that after very long periods of aeration, the ratio further reduced to 0.007 to 0.04.

The COD of the effluent will therefore be composed of bioresistant materials present in the wastewater, refractory metabolic by-products and residual compounds resulting from cell lysis and auto-oxidation. Chudoba [31] showed that the residual COD of an activated sludge effluent increased with sludge age as a result of the release of refractory organic compounds to solution. While most of the residual COD and BOD is in suspension, the F/M of 0.2

to 40 mg/ℓ at a sludge age of 50 days and an F/M of 0.082. The soluble BOD remained constant over this period. Filtration of the biological effluent will effect a substantial reduction in COD and BOD due to removal of the finely suspended organics. Coagulation preceding filtration is necessary to remove the finely dispersed organics as indicated by the difference,between the coarse filtered and soluble COD. At high sludge ages, the auto-oxidation of the cell mass will release nitrogen and phosphorus back to solution. The nitrogen and phosphorus discharged in the effluent will therefore depend on the BOD/N and the BOD/P ratio in the wastewater and the sludge age in the process (See Table 8).

Table 8:   Effluent Quality Attainable From the Activated
Sludge Process

Parameter	Concentration, mg/ℓ	
Soluble BOD	< 10	(1)
Suspended Solids	< 30	(2)
Total BOD	< 20	(3)
COD	–	(4)
Nitrogen	–	(5)
Phosphorus	–	(6)
TDS	No significant change	

AEROBIC BIOLOGICAL TREATMENT PROCESSES

The various biological treatment processes are summarized below:

1.   Activated sludge should provide an effluent with a soluble BOD_5 of less than 10 to 15 mg/ℓ and a total BOD_5, including carry-over suspended solids, of less than 20 mg/ℓ. The process requires treatment and disposal for excess sludge and would generally be considered where high effluent quality is required, available land area is limited, and waste flows exceed 380 cu m/day (0.1 mgd).

2.   Extended aeration or total oxidation will provide an eff-luent with a soluble BOD_5 of less than 10 to 15 mg/ℓ and a total BOD_5 of less than 40 mg/ℓ. The suspended solids carryover may run as high as 50 mg/ℓ (high clarity, low solids, effluent will usually require post-treatment by filtration, coagulation, etc.). This process is usually considered for waste flows less than 7560 cu m/day (2 mgd).

3. Contact stabilization is applicable where a major portion of the BOD is present in colloidal or suspended form. As a general rule, the process should be considered when 85 percent of the BOD_5 is removed after 15 min of contact with aerated activated sludge. The effluent suspended solids are of the same order as those obtained from activated sludge.

4. An aerobic lagoon is only applicable where partial treatment (approximately 50 to 60 percent BOD_5 reduction) and a high effluent suspended solids are permissible. This process should be considered as a stage development, which can be converted into an extended aeration plant at some future date by the addition of a clarifier, return sludge pump, and additional aeration equipment.

5. An aerated lagoon will provide effluent soluble BOD_5 of less than 25 mg/ℓ with a total BOD_5 if less than 50 mg/ℓ, depending on the operating temperature. The effluent suspended solids may exceed 100 mg/ℓ. The system is temperature sensitive and treatment efficiencies will decrease during winter operation. Post-treatment is necessary if a highly clarifier effluent is desired. Large land areas are required for the process.

6. High-rate trickling filters will provide 85 percent reduction of BOD for domestic sewage. Roughing filters at high loadings provide 50 to 60 percent BOD reduction from soluble organic industrial wastes.

7. Anaerobic and facultative ponds for industrial waste treatment should only be considered if odors will not cause a nuisance. If high,degree treatment is required, these ponds must be followed by an aerobic treatment (aerated lagoons, activated sludge, etc.).

The various biological treatment processes and their constraints are summarized in Table 9.

TABLE 9: Design Criteria for Biological Processes

Process	Detention (days)	Depth (m)	X_V (mg/ℓ)	BOD Reduction	Conditions
Anaerobic	5-50	2.4-4.5	< 25	50-80	Loading 280-4500 Kg BOD/10000 m²-day $SO_4 < 100$ mg/ℓ
Facultative Lagoon	7-50	0.9-2.4	< 25	70-95	Loading 22-56 Kg BOD/ 1000 m²-day 10-50 mg/ℓ algae
Aerobic Algal Lagoon	2- 6	0.18-0.3	< 50	80-95	Loading 112-225 Kg BOD/ 10000 m²-day 100 mg/ℓ Algae; periodically mixed; velocity 0.3-0.45 m/sec
Aerated Lagoon (aerobic)	0.5-3.0	2.4-4.9	$2\frac{1}{2}$ BOD removal	50-70	0.35/4 Kw/1000 m³; fully mixed conditions
Aerated Lagoons (facultative)	3-10	2.4-4.9	50-100	80-90	$0.8 < Kw < 1000$ m³ < 2.75
Activated Sludge	0.16-0.33	3.7-4.9	2000-3000	85-95	F/M for flocculation
Extended Aeration	0.5 -1.0	3.7-4.9	3000-4000	85-95	F/M < 0.2

REFERENCES

1. McCarty, P.L., Thermodynamics of Biological Synthesis and Growth, *Advances in Water Pollution Research*, Vol. 2, Pergamon Press, 1964.

2. Servisi, J.A., and Bogan, R.H., Free Energy as a Parameter in Biological Treatment, *Proc. ASCE*, 89, No. SA3, 17, 1963.

3. Eckenfelder, W.W., and O'Connor, D.J., *Biological Waste Treatment*, Pergamon Press, Oxford, 1961.

4. Andrews, J.F., et al., Kinetics and Characteristics of Multi-Stage Methane Fermentations, *SERL*, Rep. 64-11, University of California, Berkeley, 1962.

5. Porges, N., and Hoover, S., *Biological Treatment of Sewage and Industrial Wastes*, Vol. 1, Reinhold Publishing Co., New York, 1956.

6. Wuhrmann, K., *Hauptwirkungne and Wechsel Wirk Kugen einiger Betriebsparameter im Belebschlammsystem Ergebnissemehrijahirger*, Grossversuche, Verlag, Zurich, 1964.

7. Grau, P., Dohanyos, M., and Chudoba, J., Kinetics of Multi-Component Substrate Removal by Activated Sludge, *Water Research*, in press.

8. Wuhrmann, K., Factors Affecting Efficiency and Solids Production in the Activated Sludge Process, *Biological Treatment of Sewage and Industrial Wastes*, p. 49, Reinhold, New York, 1955.

9. Tischler, L.F., and Eckenfelder, W.W., Linear Substrate Removal in the Activated Sludge Process, *Advances in Water Pollution Research*, 361, Pergamon Press, Oxford, England, 1969.

10. Gaudy, A.F., et al., Sequential Substrate Removal in Deterogeneous Population, *WPCF Journal*, 35, No. 7, 903, 1963.

11. Tischler, L.F., A Mathematical Study of the Kinetics of Biological Oxidation, M.S. Thesis, University of Texas, Austin, Texas, 1969.

12. Adams, C.E., Eckenfelder, W.W., and Hovious, J., A Kinetic Model for Design of Completely Mixed Activated Sludge Treating Variable Strength Industrial Wastewaters, *Water Research*, 9, 1, 37, 1975.

13. McCauley, D.C., Union Carbide Research Report, October 1974.

14. Rickard, J., and Gaudy, A., Mixing Affects in Biological Axidation, *WPCF Journal*, 40, 1968.

15. Busch, A.W., *Aerobic Biological Treatment*, Oligodynamics Publishing Co., Houston, Texas, 1971.

16. Pasveer, A., Distribution of Oxygen in Activated Sludge Floc, *Sewage Industrial Wastes*, 26, No. 1, 28, 1954.

17. Mueller, J., Nominal Diameter of Floc Related to Oxygen Transfer, *J. San. Engr. Div.*, SA 2 4756, April 1966.

18. Zahardka, V., Role of Aeration in the Activated Sludge Process, *Advances in Water Pollution Res.*, Vol. 2, pp. 53 WPCF, Washington, D.C., 1967.

19. Okun, D., *Advances in Biological Waste Treatment*, Pergamon Press, Oxford, England, p.38, 1963.

20. Adams, P., The Use of Hydrogen Peroxide in the Control of Activated Sludge Bulking, *28th Industrial Waste Conference*, Purdue University, 1973.

21. Ford, D.L., Ph.D. Thesis, University of Texas, Austin, Texas, 1966.

22. Von Der Emde, W., *Advances in Biological Waste Treatment*, Pergamon Press, Oxford, England, 1963.

23. Downing, A., *Advances in Water Quality Improvement*, Vol. 1, University of Texas Press, Austin, Texas, 1966.

24. Wuhrmann, K., Nitrogen Removal in Sewage Treatment Processes, *Verhandl. Intern. Verein. Limnol.*, XV, 580-596.

25. McCarty, P.L., Biological Denitrification of Wastewaters by the Addition of Organic Materials, *24th Industrial Waste Conference*, Purdue University, 1969.

26. Culp, G., and Sletcha, A., Nitrogen Removal from Sewage Final Progress Report, U.S.P.H.S., Demonstration Grant 86-01.

27. Speeces, R.E., and McCarty, P.L., Nutrient Requirements and Biological Solids Accumulation in Anaerobic Digestion, *Advances in Water Pollution Research*, Vol. 2, Pergamon Press, Oxford, England, 1964.

28. Torpey, W.N., Loading to Failure of a Pilot High Rate Disaster, *Sewage Ind. Wastes*, 27, 121, 1955.

29. Lawrence, A., and McCarty, P.L., Kinetics of Methane Fermentation in Anaerobic Waste Treatment, Tech. Rep. 75, Department of Civil Engineering, Stanford Univ., Stanford, Calif., 1967.

30. Chudoba, J., Scientific Papers of the Institute of Chemical Engineering, *Technology of Water*, Prague, F-12, 1967.

31. Chudoba, J., Scientific Papers of the Institute of Chemical Engineering, *Technology of Water*, Prague, F-16, 1971.

32. Barnard, J.L., Biological Denitrification, Presented at South African Branch, Institute of Water Pollution Control, August 1972.

33. Downing, A., in *Advances in Water Quality Improvement,* Vol., University of Texas Press, Austin, Texas 1966.

ACTIVATED SLUDGE

W. Wesley Eckenfelder, Jr.

Distinguished Professor,
Vanderbilt University, Nashville, Tennessee

The activated sludge process is a continuous system in which aerobic biological growths are mixed and aerated with wastewater and separated in a gravity clarifier. A portion of the concentrated sludge is recycled and mixed with additional wastewater. The process should provide an effluent with a soluble BOD of 10-30 mg/ℓ, although the organic concentration of the effluent in terms of COD may be as high as 500 mg/ℓ, depending on the concentration of bioresistant compounds originally in the wastewater. There are many impurities in industrial wastewaters such as oil and grease, which must be removed or altered by preliminary treatment before subsequent activated sludge treatment can be considered.

Many modifications of this process have been developed over the decades, the principal ones being shown in *Fig. 1*. The conventional process employes long rectangular aeration tanks, which approximate plug-flow with some longitudinal mixing. This process is primarily employed for the treatment of domestic wastewater. Returned sludge is mixed with the wastewater in a mixing box or chamber at the head end of the aeration tank. The mixed liquor then flows through the aeration tanks, during which progressive removal of organics occurs. The oxygen utilization rate is high at the beginning of the aeration tanks and decreases with aeration time. Where complete treatment is achieved, the oxygen utilization rate will approach the endogenous level toward the end of the aeration tanks. The principal disadvantage of this system for the treatment of industrial wastewaters are:

1- the oxygen utilization rate varies with tank length and requires irregular spacing of the aeration equipment or a modulated air supply;

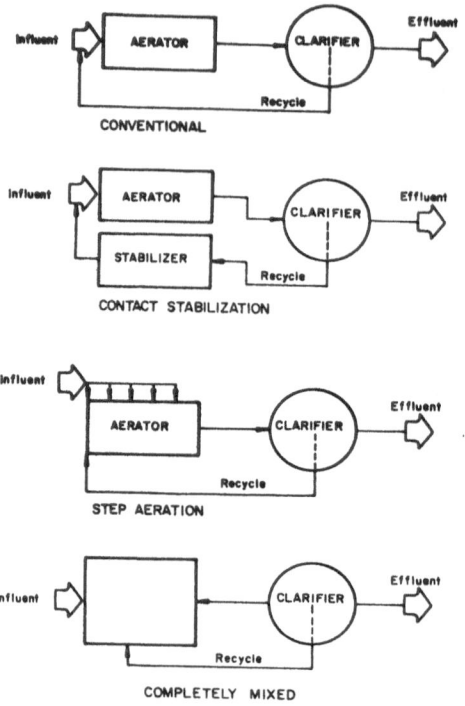

Fig. 1 : Activated Sludge Systems

2- load variation may have a deleterious effect on the acti-
    vated sludge when it is mixed at the head end of the aera-
    tion tanks;

3- the sludge is susceptable to slugs or spills of acidity,
    causticity or toxic materials.

In the completely mixed system the aeration tank serves as an
equilization basin to smooth-out load variations and as a diluent
for slugs and toxic materials. Since all portions of the tank are
mixed, the oxygen utilization rate will not vary with time and the
aeration equipment can be equally spaced.

Step aeration is a variant between the conventional process and
the completely mixed process and has been successfully employed
for the treatment of domestic wastewaters.

In the extended aeration process, sufficient aeration time is
provided to oxidize virtually all of the sludge synthesized from
the BOD present in the wastewater.

The contact stabilization process is applicable to wastewaters containing a high proportion of the BOD in suspended or colloidal form. Since bio-adsorption and flocculation of colloids and agglomeration of suspended solids occur very rapidly, only short retention periods (15 to 30 minutes) are required to affect clarification. After the contact period the activated sludge is separated in a clarifier by sedimentation. A sludge re-aeration or stabilization period is required to stabilize the organics removed in the contact tank. The retention period in the stabilization tank is dependent on the time required to assimilate the soluble and colloidal material removed from the wastewater in the contact tank.

Effective removal in the contact period requires sufficient activated sludge to remove the colloidal and suspended matter and a portion of the soluble organics. The retention time in the stabilization tank must be sufficient to stabilize these organics. If it is insufficient, unoxidized organics will be carried back to the contact tank and the removal efficiency will be decreased. If the stabilization period is too long, the sludge will undergo excessive auto-oxidation and will loose its initial high removal capacity. Increasing retention period in the contract tanks will increase the amount of soluble organics removed and decrease required stabilization time. A large increase in contact time will negate the requirement for sludge stabilization. In this case, the process becomes the same as the conventional activated sludge process as shown in *Fig. 2*.

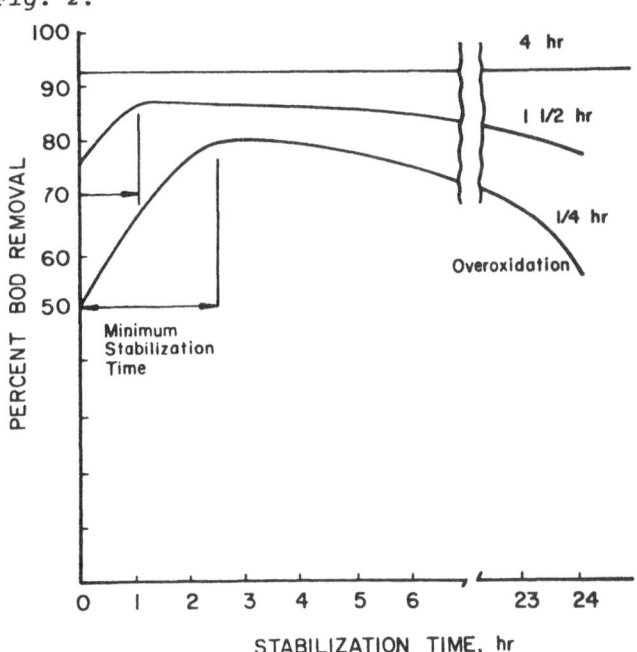

Fig. 2 : Schematic Representation of the Contact-Stabilization Process

Increasing the biological solids level also decreases the stabilization time requirements since the organic loading for unit solids becomes less. The total oxygen requirements in the process are those required for synthesis of the organics removed and for endogenous respiration. The split of this oxygen between the contact tank and the stabilization tank depends on the solids level carried in both units and on the retention period in each tank. Increasing the contact tank solids level or retention period will increase the percentage of total oxygen to that unit.

A distinction must be made when considering the activated sludge process for the treatment of domestic sewage as compared to soluble industrial wastewaters, based upon the physical composition of the wastewaters and the resulting mechanisms of removal. The organic content of domestic sewage consists of three components, suspended organics resulting from ground garbage, paper, rubber, etc., colloidal matter, and soluble organics consisting mainly of carbohydrates and some nitrogenous material. Most of the organics are in the form of particulates, Hunter and Heukelekian have shown 35%, 40% and 25% of the total COD was present as suspended, colloidal, and soluble material, respectively.

When sewage is mixed with activated sludge, several things immediately occur. Suspended solids are enmeshed in the biological floc, colloidal solids are adsorbed on the flow interface, and some soluble organics are adsorbed by enzymatic action. The net effect of these reactions is to attain near 90% BOD removal in less than 15 minutes of contact time of the wastewater with the sludge. This is illustrated in *Fig. 3.* It is significant to note that while these organics have been removed from the wastewater by physical and biological mechanisms, achieving ultimate stabilization requires a considerably longer period of aeration.

The soluble organics which have been removed from solution are most readily assimilable by the microbial cell and are probably synthesized in less than one hour of aeration. The colloids must first be broken down by extracellular enzymes in order to be available to the cell so that complete stabilization of these organics requires a longer aeration period. The suspended solids must also undergo a slow hydrolytic breakdown so that it is probable that a portion of the organic suspended matter entering the aeration tanks is disposed of as excess sludge in the detention periods normally encountered in the activated sludge process. The one exception is the extended aeration process in which the biodegradable portion of the suspended solids are oxidized.

The various phases of the activated sludge process treating domestic sewage is shown in *Fig. 4.* When the return sludge is first contacted with the sewage, the suspended and colloidal, and a por-

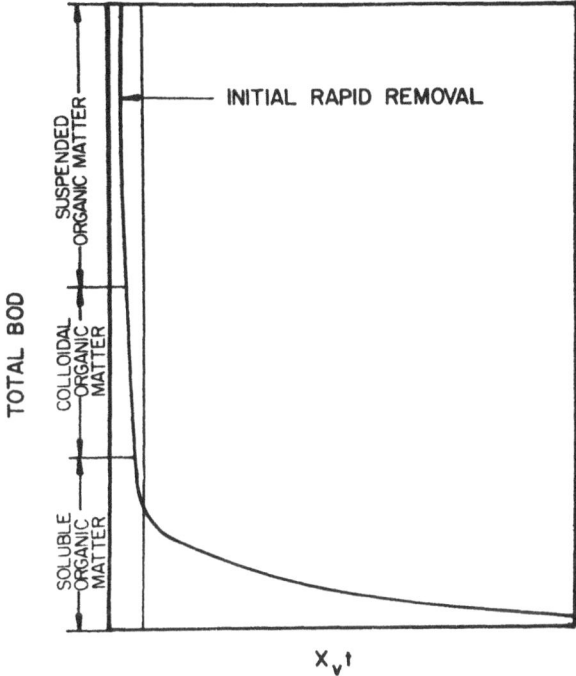

Fig. 3 : Removal of BOD from Domestic
Sewage by the Activated Sludge
Process

110

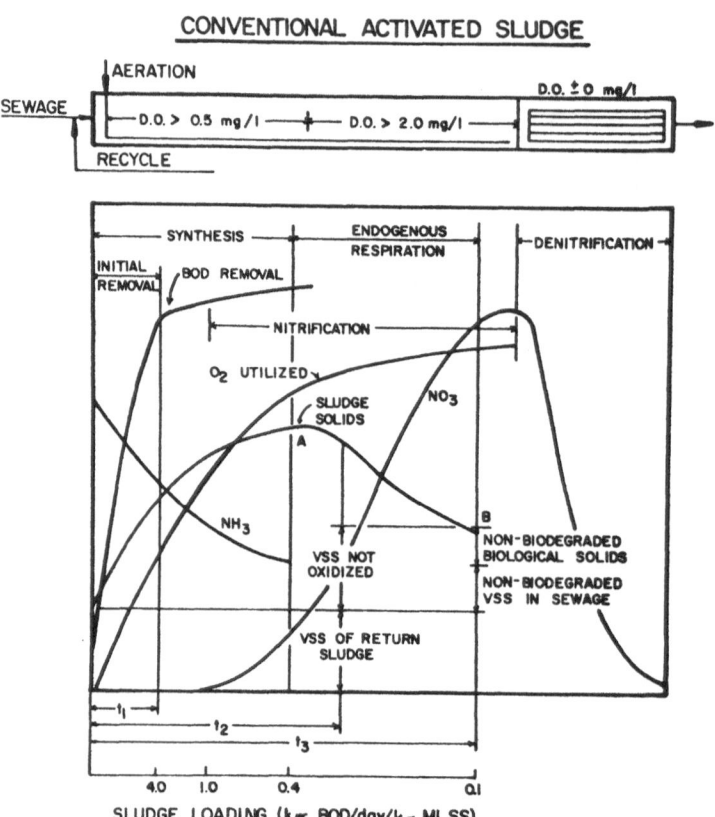

Fig. 4 :   Schematic Representation of the
Activated Sludge Process

tion of the dissolved organics are removed as denoted by time T_1. As aeration continues additional soluble organics are removed. Maximum synthesis, and, hence, the highest sludge yield, will occur after exhaustion of the substrate (point A, Figure 4). Nitrogen in the form of ammonia will be utilized by the organisms during this period for synthesis of new cells. Extending the aeration time beyond A results in the auto-oxidation of synthesized sludge and the continuing oxidation of suspended solids present in the wastewater. A portion of the synthesized biological sludge is not biodegradable in the time periods under consideration. In addition, some of the volatile solids initially present in the sewage will not be biodegraded and will accumulate as a residue. The total non-biodegraded solids are shown as point B in Figure 4. The rate of oxygen utilization is rapid during assimilation and decreases during endogenous respiration.

Nitrification will occur during the activated sludge process when ammonia is oxidized to nitrite which is then oxidized to nitrate. A sludge age greater than the growth rate of the nitrifying organisms is necessary for nitrification.

Due to the relatively high temperature coefficient, nitrification is more affected by winter operation than is carbonaceous BOD removal. Denitrification is the reduction of nitrate to gaseous end products (N_2, N_2O) under anaerobic conditions. Many of the facultative organisms present in activated sludge will use the nitrate as an oxygen source in the absence of dissolved oxygen. Denitrification may occur in final settling tanks resulting in rising sludge. The increasing emphasis on nutrient removal makes nitrification-denitrification one of the economical processes for nitrogen removal from domestic wastewater.

The high purity oxygen system is a series of well-mixed reactors employing concurrent gas-liquid contact in a covered aeration tank. Feed wastewater, recycle sludge, and oxygen gas are introduced into the first stage. The oxygen gas is fed at a pressure of only 1.4 inches of water above ambient, as shown in Fig. 5. Two gas-liquid contacting systems can be employed; submerged turbine aeration and surface aeration. With turbine aeration, recirculating gas blowers pump the gas through a hollow shaft to a rotating sparger. The pumping action of the impeller on the same shaft as the sparger promotes adequate liquid mixing and yields relatively long residence times for the dispersed oxygen bubbles. Gas is recirculated within a stage at a rate that is usually higher than the rate of gas flow from one stage to another. A slight pressure drop occurs from stage-to-stage with no gas backmixing. Since the relative liquid mixing and oxygen transfer requirements vary from stage-to-stage, each stage is equipped with an indepedent mixer-compressor combination designed to provide the required level of

112

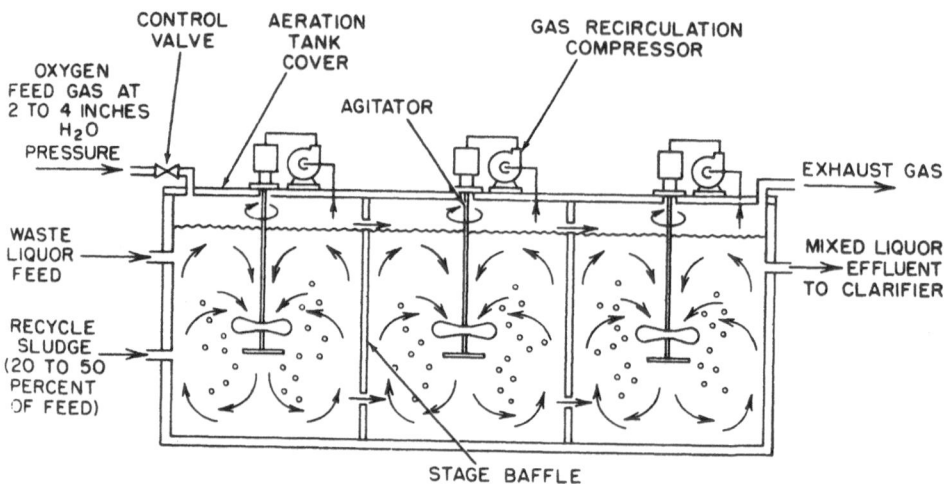

(a) Three-stage oxygenation system utilizing rotating spargers and recirculation compressors

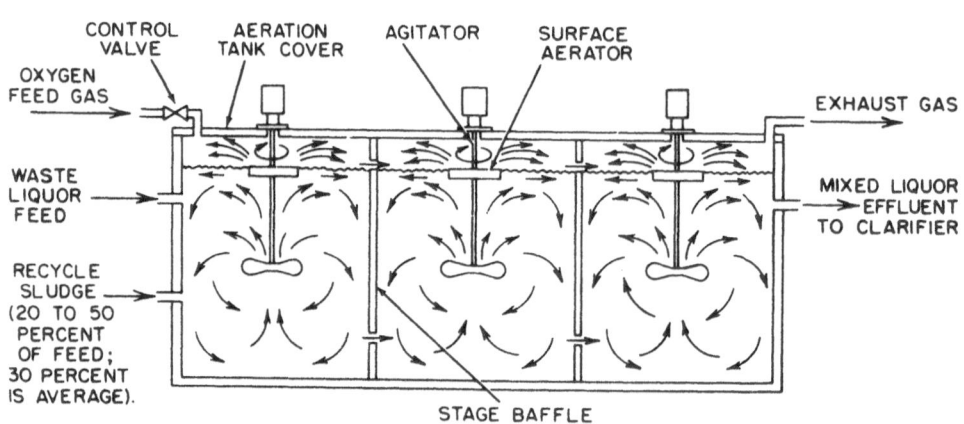

(b) Three-stage oxygenation system utilizing mechanical surface aerators

Fig. 5 : Schematic Diagram of
"Unox" System

mixing and oxygenation.

Utilizing surface aerators, the gas-liquid contact eliminates the need for gas recirculating compressors with associated piping. The required level of bulk fluid mixing to maintain the sludge in suspension and ensure a uniform liquid composition is provided in an efficient pumping, slow-speed low shear impeller. Oxygen gas is automatically fed to either system on a pressure demand basis with the entire unit operating, in effect, as a respirometer. As the organic loading increases the pressure decreases resulting in an automatic increase in feed-oxygen flow. Due to the high mixed liquor solids maintained in the oxygen system, the major portion of soluble BOD removal, and thus the highest oxygen demand, occur in the first stage requiring the highest mixer and compressor horse-power. The additional stages are then utilized to stabilize a sludge, with an oxygen demand decreasing in the later stages due to increasing degrees of sludge stabilization. Effluent mixed liquor from the system is settled conventionally and the activated sludge is returned to the first stage for blending with the feed. A restricted exhaust line from the final stage vents the essentially odorless gas to the atmosphere. Normally the system will operate most economically with a vent-gas composition of about 50% oxygen. Due to the net transfer of gas to the liquid, the vent-gas flow rate will be a fraction (10% to 20% of the gas feed rate). Based upon economic considerations about 90% of oxygen utilization with on-site oxygen generation is desired. Two basic oxygen generator designs are employed. A traditional cryogenic air separation process for large installations (283, 875 to 378,500 m^3/day) and a pressure swing adsorption (PSA) system for smaller installations. With the larger installations deep tank construction with submerged turbine aeration will normally be preferable, while with the smaller plants a surface aerator (PSA) combination will be the most cost effective. The power requirements for the surface and turbine aeration equipment vary from 15.8 to 27.6 kilowatts per thousand cubic meters, depending on the waste strength, mixing requirements, feed oxygen purity and the rate of capacity of the aeration equipment. At peak load conditions the oxygen systems are designed to maintain 6 mg/ℓ DO in the mixed liquor. During unusually severe peak loads, additional oxygen can be transferred to the liquid if the DO level decreases to 1 mg/ℓ. Liquid oxygen storage is designed for back-up purposes with the same supply capacity as the installed plant. It is possible to double the feed oxygen flow to the aeration tank upon need. This results in an increased gas phase oxygen partial pressure and increased transfer, but reduced oxygen utilization. Although this is not an economic mode for operation over extended time periods, it is quite effective for short-term operation.

For small installations of one or two shift operations such as a dairy, a batch activated sludge process may be employed. Operations at a milk processing plant is shown in *Fig. 6*. The process operates as an extended aeration plant with a mean design F/M of 0.1-0.2. Accumulated sludge is withdrawn on a bi-weekly or monthly basis for disposal. A major advantage of this process is its simplicity of operation, since each stage of the process operation is regulated by time controls. Because of the variable tank volume, the most efficient aeration equipment is an eductor-induced air system.

ACTIVATED SLUDGE DESIGN PROCEDURE

PARAMETERS

1. Flow Rate, Q, m^3/day
2. Temperature of Waste, T_i, °C
3. Ambient Air Temperature, T_a, °C
4. Influent BOD, S_0, mg/ℓ
5. Influent SS, S_i, mg/ℓ
6. Non-biodegradable Fraction, f, of influent SS
7. Influent NH_3-N, mg/ℓ
8. Influent Phosphorus, mg/ℓ
9. Average and maximum effluent, BOD, mg/ℓ
10. Volatile Fraction of Aeration Basin MLSS, f_v
11. Correlation of F/M vs Zone Settling Velocity (ZSV)
12. Correlation of F/M vs Effluent SS Concentration
13. Correlation of mg BOD/mg VSS vs Sludge Age
14. Correlation of ZSV vs Concentration of SS
15. Specific BOD Reaction Rate Coefficient, K, (1/day)
16. Oxygen Requirement Coefficients, a' & b' (1/day)
17. Sludge Production Coefficients, a & b (1/day)
18. Temperature Coefficients, θ, for K and b

PROCEDURE

1. Select an underflow solids concentration and compute the solids flux, G_L, as shown in Example. Data should be for winter operating conditions.
2. Select a recycle rate, R. This is usually 0.25-0.35 of the wastewater flow Q.

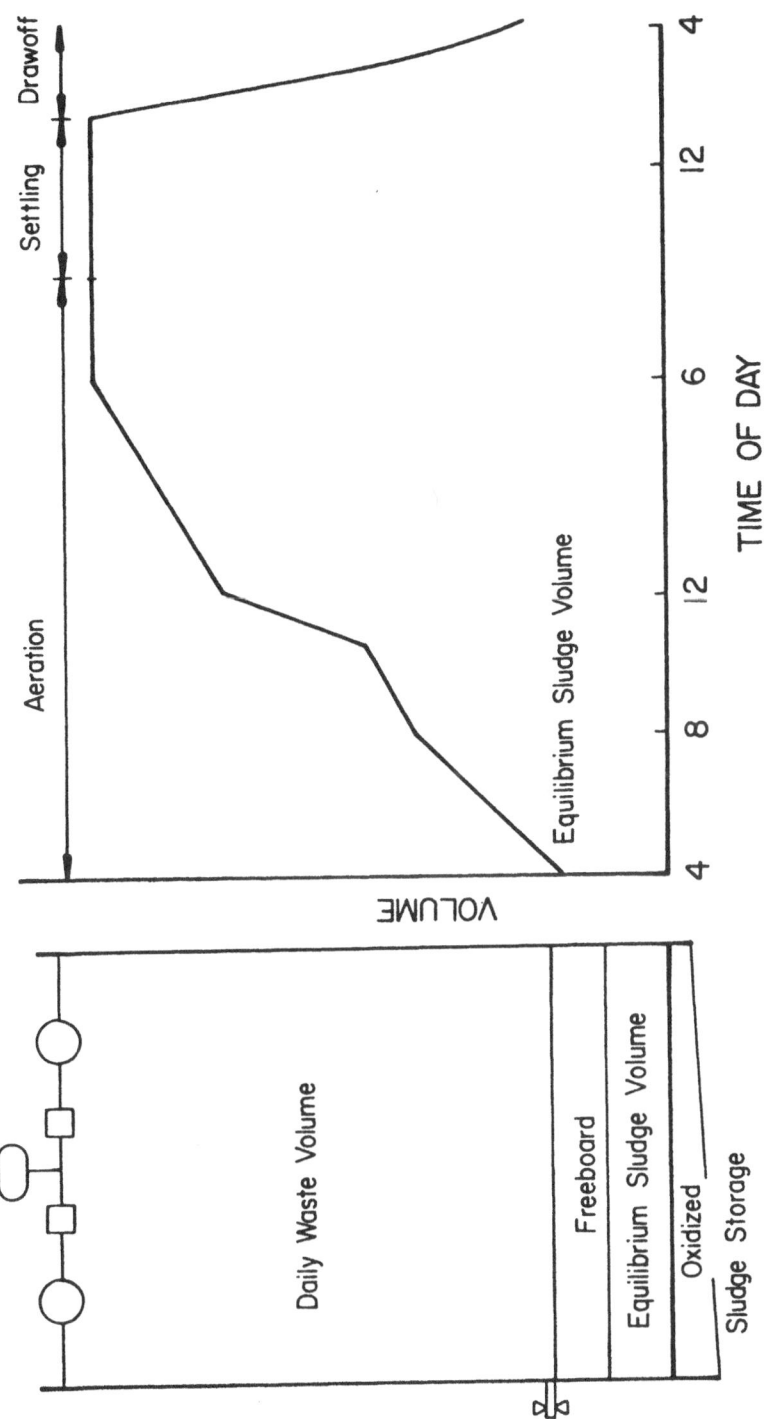

Figure 6: Batch Activated Sludge Process

3. Calculate X_a, the MLSS in the aeration basin:

$$X_a, \text{ mg/}\ell = \frac{RX_R \text{ (mg/}\ell\text{)}}{(Q+R)}$$

4. The clarifier area required based on thickening:

$$A(m^2) = \frac{(Q+R) m^3/\text{day} (X_a) (\text{mg/}\ell) 4.22 \cdot 10^{-5}}{G_L (Kg/m^2/\text{day})}$$

5. Area required for clarification under winter operating conditions:

$$A(m^2) = Q(m^3/\text{day})/O.R. \ (m^3/\text{day/}m^2)$$

in which the overflow rate is:

$$O.R. \ m^3/\text{day/}m^2 = ZSV \ m/hr \ (0.7)(24)(\text{maximum MLSS})$$

the larger of the two areas should be used. If the area is controlled by clarification the MLSS at the selected average recycle rate [Step 2] should be recomputed for this area [Step 4].

6. The Kgs BOD removed/day under maximum conditions:

$$S_r \text{ Kgs/day} = (S_0 - S_e)(m^3/\text{day}) \cdot 10^{-3}$$

7. Estimate the H.P.

$$KW = S_r/27.4 \text{ Kg. BOD removed/day/KW}$$

8. The winter temperature in the aeration basin (based on low speed surface aerators):

$$(T_i - T_w)(0.555 \text{ Kg-cal/Kg-°C}) Q = (T_w - T_a) 844.8 \text{ kcal/KW-HR-(KW)}$$

9. Correct the BOD reaction rate K:

$$K_{Tw} = K_{20 °C} \; \theta^{(Tw-20)}$$

10. The basin detention time for BOD removal under winter conditions, the soluble BOD is:

$$S_{e\ total} - f \frac{mg\ BOD}{mg\ SS} = S_{e\ soluble}$$

Only the soluble BOD is used in the kinetic relationship.

$$t,\ days = S_0\ (S_0 - S_e)/KX_v S_e$$

11. The basin detention time for the maximum F/M:

$$t,\ days = \frac{S_0}{X_v F/M}$$

12. If F/M controls, the effluent S_e should be recalculated [Step 10]. If the reaction rate controls, the operating F/M should be recalculated [Step 11].

13. The biodegradable fraction, x, for the operating F/M:

$$x = \frac{aS_r - bX_v - \sqrt{(aS_r - bX_v)^2 - 4bX_v(0.77aS_r)}}{2bX_v}$$

14. The maximum allowable influent BOD to meet a maximum effluent limitation:

$$S_0 = \frac{S_e + \sqrt{S_e^2 + 4KS_e X_v t}}{2}$$

(this provides the basis for equalization design)

15. The average summer operating temperature:

$$T_w = \frac{T_i Q + 1864 \ KW \ T_a}{1864 \ KW + Q}$$

16. The summer reaction rate:

$$K_{Tw} = K_{20} \circ_c \theta^{(Tw-20)}$$

17. The summer effluent BOD based on the maximum influent BOD

$$S_e = \frac{S_0^2}{S_0 + KX_v t}$$

18. The biodegradable fraction at the summer temperature, correcting the endogenous coefficient b to summer conditions:

$$b_{Tw} = b_{20} \circ_c \theta^{(Tw-20)}$$

19. Oxygen requirements:

$$O_2/day = a' \ S_r + b'_{Tw} \ xX_v$$

20. Oxygen transfer efficiency:

$$N = N_0 \ \frac{(\beta C_{sw} - C_L)}{9.1} \ \alpha \ \theta^{(T_w-20)}$$

21.
$$Kw = \frac{Kgs \ O_2/hr \ [Step \ 19]}{N \ Kgs \ O_2/Kw-HR}$$

22. Check against assumed K_w in [Step 7] and return steps 7-22 if significantly different.

23. Check power level for mixing:

$$K_w/10^6 \ell > 508$$

24. Excess biological sludge production - summer conditions:

$$\Delta X_v = aS_r - b_s x X_v$$

25. Nutrient requirements - summer conditions:

$$lb \ \text{N/day} = 0.123 \ \frac{x}{0.77} \ \Delta X_v + 0.07 \ \frac{(0.77 - x)}{0.77} \ \Delta X_v$$

$$lb \ \text{P/day} = 0.026 \ \frac{x}{0.77} \ \Delta X_v + 0.01 \ \frac{(0.77 - x)}{0.77} \ \Delta X_v$$

26. Excess biological sludge - winter conditions, Step 24 with endogenous coefficient b and x adjusted for temperature.

27. Nutrient requirements - winter conditions, Step 25 adjusted for winter conditions.

TRICKLING FILTRATION

Klaus R. Imhoff

Director of Ruhrverband and Ruhrtalsperrenverein

The trickling filter process has been invented around 1900 in Birmingham, England. By industrialization the inhabitants had mushroomed to a number of 400,000 and wastewater receiving rivers were very small. Still nowadays one can imagine the emergency.

The process has been copied from nature. It is known that organically polluted water undergoes rapid selfpurification in small brooks. Also the packings of trickling filters cover with bacteria and higher organisms which fed themselves on organic pollution and consume oxygen. Conventional rock filled trickling filters are endangered by clogging. Therefore a good primary clarification is necessary. The sludge content of the trickling filter influent should not be higher than 0.3 mℓ/ℓ. Surplus bacterial sludge is flushed through the filter and is retained in a secondary clarifier. With normal domestic wastewater conditions it is useful to recirculate dilution water from the secondary clarifier into the trickling filter influeht. The recirculation rate should be selected to an order that the influent BOD is approximately 150 mg/ℓ.

One can assume that the purification capacity of trickling filter is dependent on its bacterial mass contained per unit volume. This amount is more or less constant for a given filter material. Therefore it is dependent on the wastewater load if full treatment with nitrification and oxidation of organic carbon or only partial treatment is achieved.

According to Heukelekian [1] the high loaded trickling filter has a sludge content between 3,200 and 6,300 g of dry solids per m^3. With a BOD_5- loading of 600 g/m^3·d a sludge loading results in the order of 0.10 to 0.19 g BOD/g DS·d. In low loaded trickling filters which are flushed at a rate of 0.1 to 0.2 m/h Heukelekian measured between 4,700 and 7,200 g DS/m³. Assuming a load of 200 g of BOD/m³·d a sludge load results between 0.03 and 0.04 g BOD/g DS·d. Therefore the high loaded trickling filter can be compared to the low loaded activated sludge process while the low loaded trickling filter corresponds to the activated sludge process with aerobic sludge decomposition.

OPERATION RESULTS

Many operational results have been evaluated by Rincke [2]. The very important dependency between BOD volume loading and degradation efficiency is shown in *Fig. 1*.

The measurements have been made with rock filled trickling filters. If no 24 h samples were available the daily BOD₅ load was calculated to be 15 times the average of the analyzed day hours. With contact times of less than one hour between biological slimes and wastewater it would be better to describe the process by the hourly BOD volume loading. This can be demonstrated by the results of shock load tests. Therefore 600 g BOD/m³·d can only be assumed for a good oxidation of the carbon compounds if the usual load pattern is given and only 200 g BOD/m³·d can be applied if nitrification is to be achieved.

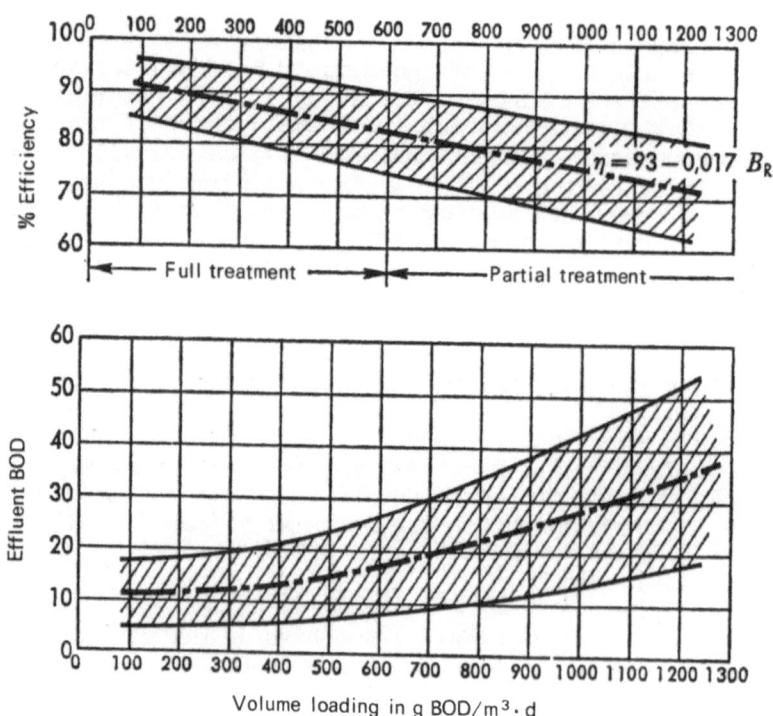

Fig. 1 : BOD Volume Loading, Degradation Efficiency and Effluent BOD of Trickling Filters

The result of a shock load has been obtained by tests in a plant for 4,500 capita. The primary stage consists of a screen, a grit channel and an Imhoff tank. For secondary treatment wastewater is mixed with dilution water from a pond and pumped on a trickling filter. The secondary clarifier is a Dortmund tank with vertical flow. Its effluent is discharged into the fish pond. To stimulate a shock load, 3.5 m³ of septic tank sludge were added to the plant influent between 11 and 12 o'clock. By this procedure 1 ℓ/s septic sludge mixed with the ordinary flow of 12.2 ℓ/s. The effect has been plotted in *Fig. 2*.

By sedimentation in the Imhoff tank a considerable part of the influent load could be retained but residual pollution had an order that the plant effluent deteriorated. The effluent BOD increased from 15 mg/ℓ to 45 mg/ℓ and it lasted several hours until the trickling filter recovered.

Assuming normal loading conditions a trickling filter can be sized with 600 g BOD/m³·d for 90% degradation of carbon compounds and with 200 g BOD/m³·d for nitrification. The height of the trickling filter is selected between three and four meters. In

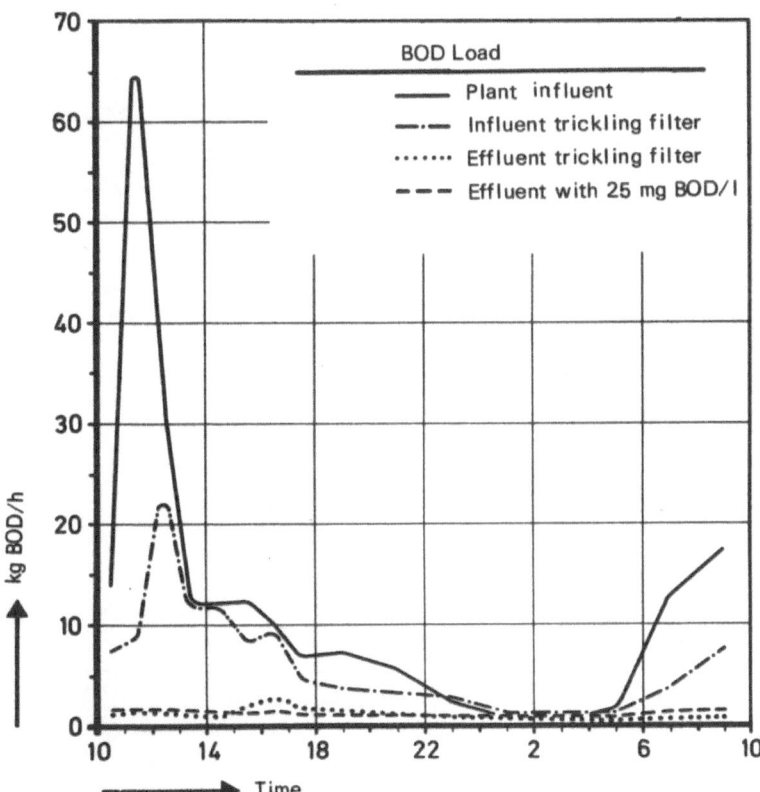

Fig. 2 : Shock Load by Septic Tank Sludge Addition
to the Influent of a Trickling Filter Plant

124

higher trickling filters a longer contact time between biological slimes and wastewater is given. Therefore higher trickling filters have a better operational performance.

NITRIFICATION

Biological oxidation of nitrogen compounds is performed by nit-rosomonas and nitrobacter. These bacteria have a slower growth rate than carbon oxidizing bacteria. Therefore carbon compounds have to be decomposed to a considerable degree before nitrification can start.

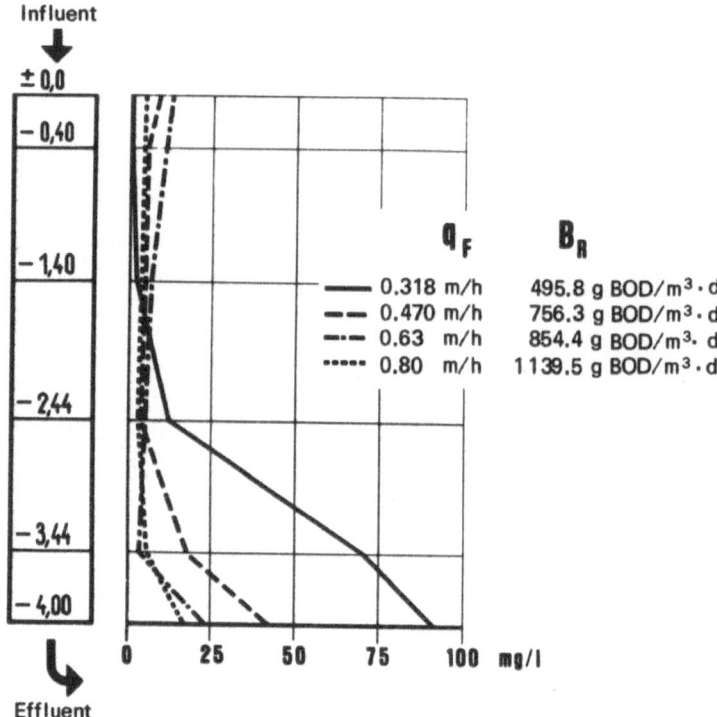

Fig. 3 : Increase of Nitrate Content in a Trickling Filter at Different Loadings

In *Fig. 3* the nitrate content of wastewater has been plotted for different levels of the trickling filter dependent on the BOD volume loading. One recognizes that nitrification starts with a BOD volume loading of 500 g/m^3·d and that nitrification only oc-curs in the lower part of the trickling filter.

Because of anaerobic zones which are existing in each trickling
filter, nitrification and denitrification are given in the same
process. As shown by *Fig. 4* the overall nitrogen removal can be
increased from 25% to about 40% by lowering the BOD volume loading
from 600 to 200 g BOD/m^3·d.

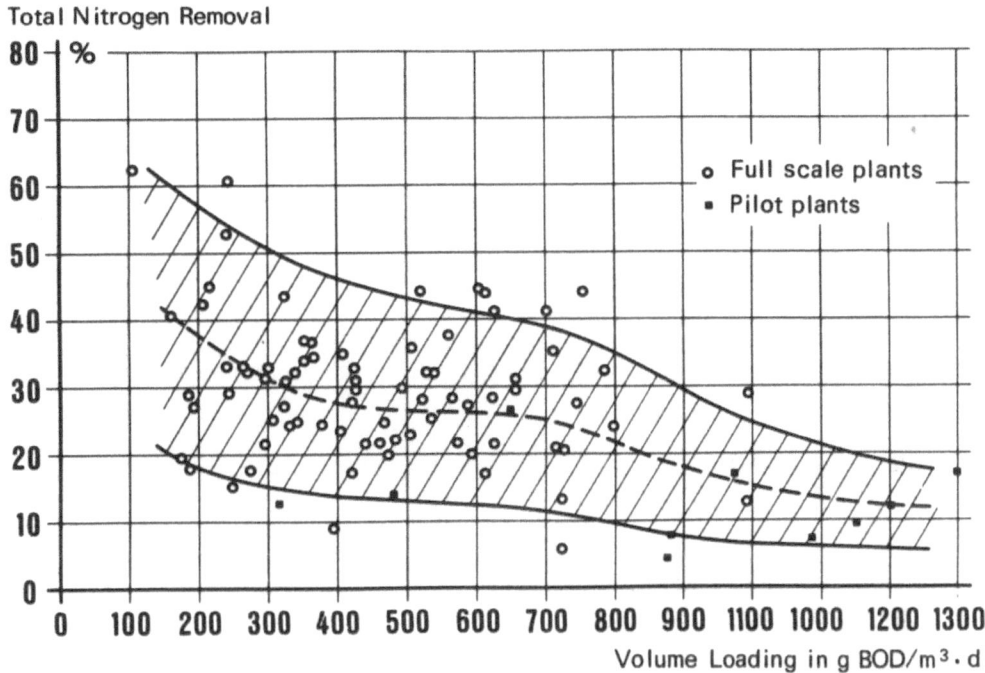

Fig. 4 :   Percentage of Nitrogen Removal in
           Trickling Filters

HYDRAULIC SURFACE LOADING

A trickling filter has to be desludged continuously by a proper hydraulic loading. Halvorson concluded from tests that about
0.8 m^3 of water should be loaded per square meter of area per hour
to desludge a trickling filter from surplus substance. The cited
figure gives only an overall hint for the flushing force because
neither the number of rotating arms (a) nor the rotation speed (r)
are considered. The more distributing arms are given and the faster the sprinkler turns the smaller the hydraulic flushing force
becomes. Usually two, four or six distributing arms are applied
and rotation speeds between 50 and 400 revolution per hour are selected. The effective flushing force can be adjusted according to

following equation

$$F = \frac{q_A}{a \cdot r}$$

F = flushing force (mm/dose)
q_A = hydraulic load (mm/h)
a = number of arms
r = rotation per hour

The flushing force should have an order between 2 and 6 mm/dose. To maintain the flushing force in case of shortage of water, frequently two flushing arms are applied which start to sprinkle the filter. In case of increasing flow the additional arms get their water by weir overflow.

Another possibility to maintain a sufficient flushing force is intermittent dosage. This is performed by an inverted syphon Fig.5.

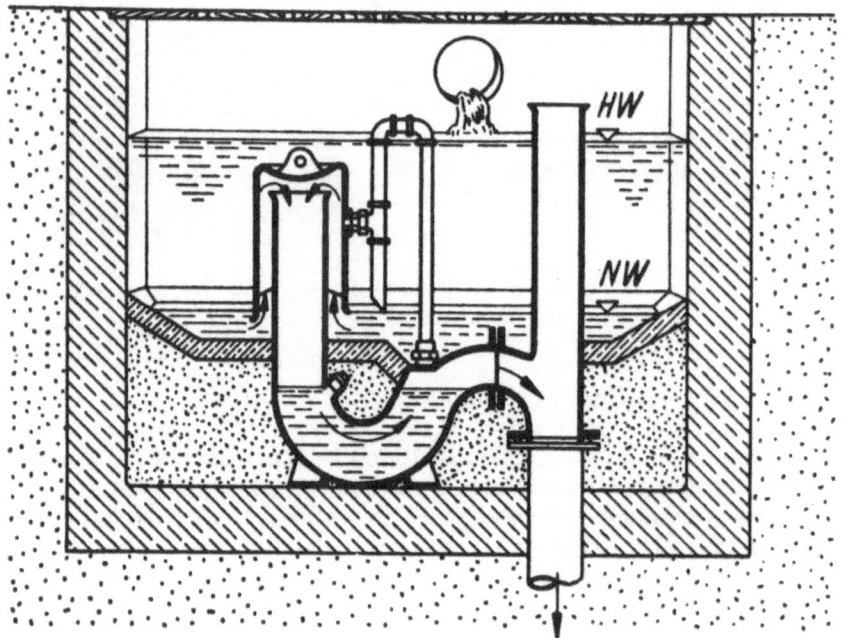

Fig. 5 : Inverted Syphon

At high water level air is replaced and the syphon is starting to flow. At low water level air is allowed to enter the pipe and the water column is disrupted. By this procedure a sufficient flow can be achieved at least for a certain time.

In this chapter also the dependency between contact time and hydraulic loading shall be mentioned. According to several researchers the contact time can be formulated in the following way:

$$t = k \cdot H \cdot q_A^{\frac{2}{3}}$$

$t =$ contact time (min)

$k = 25.8$

$H =$ height (m)

$q_A =$ hydraulic load (m/h)

From this formulation the following conclusions can be drawn:

1. If you double the water flow the contact time is only diminished by 37%.

2. If one wants to maintain the same contact time with 200% increase of flow the trickling filter height has to be increased by 60% or the surface of the trickling filter has to be increased by 100%.

3. With the same trickling filter volume and the same amount of flow the contact time is increased by 25% if you double the trickling filter height and if you consequently also double the hydraulic load. From this it can be concluded that higher trickling filters have the better performance compared to lower trickling filters.

PUMPING and RECIRCULATION

Since the development of submerged pumps, pumping stations have become simple structures. Consequently pumping has not to be avoided under all circumstances and a sufficient trickling filter height can be selected. By recirculating wastewater from the effluent of the secondary clarifier many advantages can be obtained.

1. Concentrated wastewater can be diluted and clogging can be avoided.

128

2. Shock loads can be diminished.

3. The hydraulic load can be maintained also during times
    when the plant influent is considerably diminished.

4. With recirculation the overall contact time is increased
    and degradation is improved.

5. The development of the psychoda fly can be controlled.

6. Odours can be controlled and too long retention times
    can be diminished in the primary clarifier. At the same
    time denitrification takes place and nitrogen compounds
    are eliminated to a higher degree.

With recirculation the following disadvantages are associated.

1. Higher energy consumption.

2. Bigger size of primary or secondary clarifier.

For each specific case it has to be decided whether recirculation is advantageous or not.

NUTRIENTS and TOXIC SUBSTANCES

In the bacterial mass organic carbon (C): nitrogen (N): phosphorus (P) are incorporated in the ratio of 50:8:1.   General domestic wastewater has a ratio of C:N:P in the order of 25:10:1.
For a complete reaction there is a lack of carbon or a surplus of
nitrogen and phosphorus. Consequently nitrogen and phosphorus are
only removed by some 50%.

The combined treatment of domestic sewage and of wastewater
from industry can be advantageous if nutrients supplement and if
no toxicity is involved. This must be carefully investigated by
pilot tests. If there is a lack of phosphorus and nitrogen the
bacterial population will change and clogging of stone filled trickling filters may result. Also the degradation of BOD or total organic carbon is impared as shown by *Fig. 6*.

Toxic substances have to be removed by pretreatment of wastewater to certain levels as indicated in *Table 1*.

The biological process is less impared by toxic substances if
the organic loading of the unit is increased and the sludge age is
diminished. If two trickling filters are available it is advantageous to run them in series rather than in parallel. The sequence
can be changed after some time.

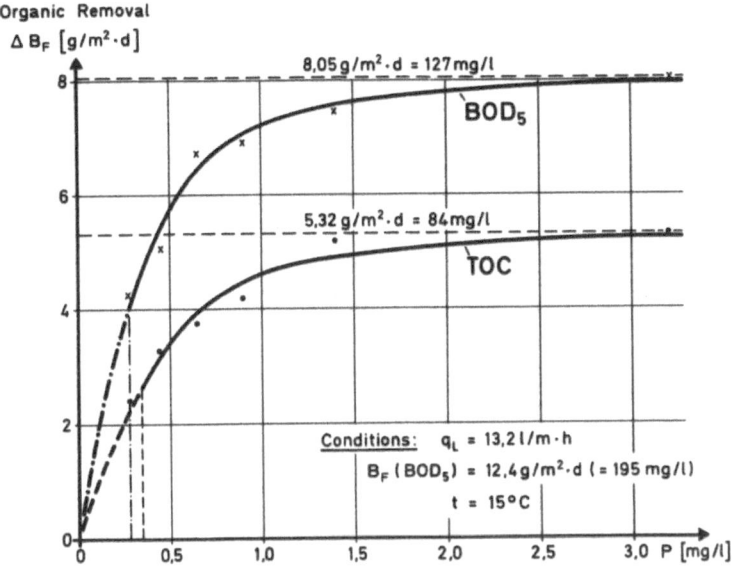

Fig. 6 : Phosphorus Supply and Degradation in a Pilot Trickling Filter According [4].

TABLE 1 : Threshold Concentrations for Toxic Substances

Toxic Substance	Trickling Filter or Activated Sludge Process (mg/ℓ)	Percent Removal in the Plant	Digester (mg/ℓ)
Copper	1	20	100
Cyanide	1-2	–	2
Chromium (VI)	3	40	100
Nickel	3	10	100
Zinc	5	50	1000

TRICKLING FILTER MEDIA

Trickling filter media have to be weather and wastewater resistant and should be of uniform size. For crushed stone a size between 3 and 8 cm can be chosen. The material should be selected according to [5] and should be washed before filling the filter to remove the sand. During packing the material has to be handed carefully. Artificial compression has to be avoided.

Since 10 years different plastic packings have been developed and successfully applied.

TABLE 2 :   Comparison of Different Trickling Filter Media

Name	Material	Specific Weight (kg/m^3)	Specific Surface (m^2/m^3)	Void (%)
Surfpac (DOW)	Prlystyren	64	82	94
Flocor (ICI)	PVC	37	85	98
Mini-Flocor	PVC	45	180	98
Cloisonyl	PVC	80	220	94
Bioprofile (Babcock) } 32 mm	PVC	40	160	98
Bioprofile } 42 mm	PVC	32	120	99
Hydropack } (Uhde)	PVC	–	200	96
Crushed Lava Stone	∅ 5 cm	1350	105	50

As can be seen from *Table 2* plastic packings have a very low specific weight, approximately 200% of specific surface and 200% of voids compared to the stone packing. Since plastic medium trickling filters are much more expensive per volume unit than stone filters the main application of plastic packings is given for the pretreatment of highly concentrated wastes. In *Fig. 7* different results of pilot plant tests have been plotted.

The diagram is valid for one passage of wastewater without recirculation. If recirculation is applied the degradation efficiency can be increased.

In case of a carcass utilization plant a plastic medium trickling filter has been constructed for the first biological treatment stage. It had a volume of 250 m^3, a height of 10 m and a diameter of 6 m. For the first packing Cloisonyl material had been selected. The hydraulic loading was 1.35 m^3/m^2·h, the BOD volume loading

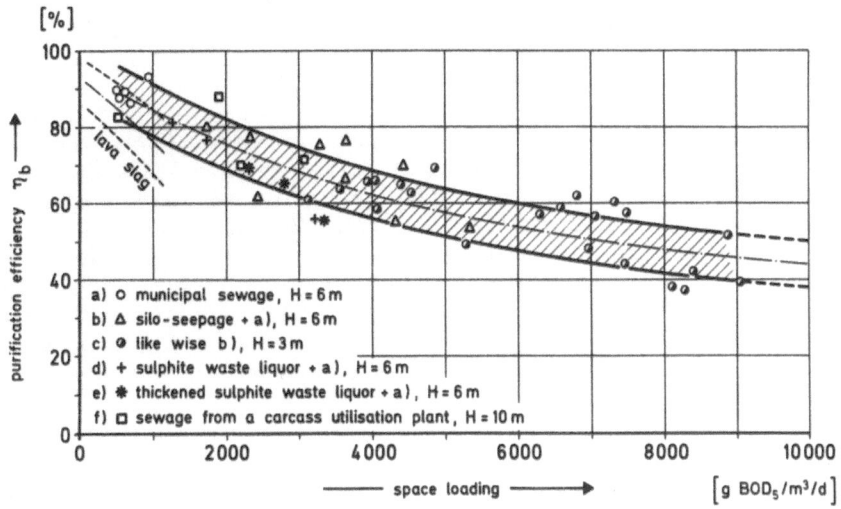

Fig. 7 : BOD Removal in Plastic Medium Trickling
Filters for Different Types of Wastewater

(without recirculation) 1.65 kg BOD/m^3·d. With 300 to 400% of recirculation the influent BOD was diminished from 4,200 to 650 mg/ℓ.
As a second biological stage an oxidation ditch was applied with a
volume of 300 m^3. Its volume loading was 0.60 kg of BOD/m^3·d. The
sludge loading was 0.19 kg BOD/kg DS·d. In this unit the BOD was
diminished from 650 to 50 mg of BOD/ℓ.

The higher a trickling filter is loaded the more likely an o-
dour nuisance may occur. Therefore plastic medium trickling filters have to be covered. By the cover the ventilation can be reduced to that degree which is necessary for sufficient oxygen supply. If still an odour nuisance is left the spent air has to be
treated by a soil filter. It also can be washed by activated
sludge.

In general it is not necessary to cover trickling filters and
to ventilate the process artificially. Only in cases of cold climate it is useful to cover the filter and to avoid heat losses.

CONSTRUCTION ASPECTS

Due to the most successful form of wastewater distribution by
rotating sprinklers trickling filters have in general a cylindrical shape. They get a bottom of special hollow stones which allow

air to enter and treated wastewater to flow out. The wall can be constructed in prestressed concrete or by plastic or metal sheets. In the latter case the static force has to be carried by metal rings comparable to a wooden barrel.

By the frequent change of temperature a considerable tension can be introduced into the trickling filter wall. With increasing temperature also the diameter of the trickling filter is increased and stone material can consolidate. When the wall becomes cold it must be strong enough to counter-balance the passive earth pressure to reach its original position. Otherwise after many temperature cycles, cracks will result. In case of a lava stone, we calculate in Germany a specific weight of 1.2 tons/m^3 and an angle of action in the order of 42.5 degrees which equals three times the active stone pressure.

OPERATION

Compared to the activated sludge process, trickling filters are much easier to operate. Only the dosage rate can be adjusted by recirculation. By pulling a sort of cake plate across the filter and by measuring its filling time one should control if the hydraulic load is equally distributed. If there is a local concentration of outflow some holes have to be closed and additional ones have to be drilled at another location. To avoid ponding the flushing force has to be recalculated according to the given formula.

In case of ponding also sodium nitrate can be applied. The sprinkler is put out of operation and a rate of 400 g of sodium nitrate is distributed per square meter surface and slightly irrigated with wastewater. Dissolved nitrates become an oxygen source for the anaerobic sludge accumulation. The resulting nitrogen gas loosens sludge particles which can be flushed out later on. Several hours after the application of sodium nitrate the trickling filter can be put back to operation.

Every day the distribution arms have to be cleaned. For this purpose there has to be a platform at one side of the trickling filter and an opening in the wall where a brush can be introduced into the sprinkler arms.

In cold climate difficulties may result from the formation of ice. During such times the ventilation holes will be closed and the recirculation rate will be diminished. One also opens outlets at the end of the distribution arms that warm wastewater may splash against the trickling filter wall and may dissolve ice at a location where it starts to grow.

Biological sludge has to be removed continuously from the secondary tank. Generally the sludge water is discharged to the influent of the primary sedimentation tank. The amount of sludge is in the order between 15 — 20 g DS/capita and day. This sludge has much better sedimentation properties than activated sludge.

SPECIAL FILTER STRUCTURES

In this paragraph a combined sedimentation trickling filter structure shall be described which has been applied successfully in the Federal Republic of Germany for more than 1,000 times.

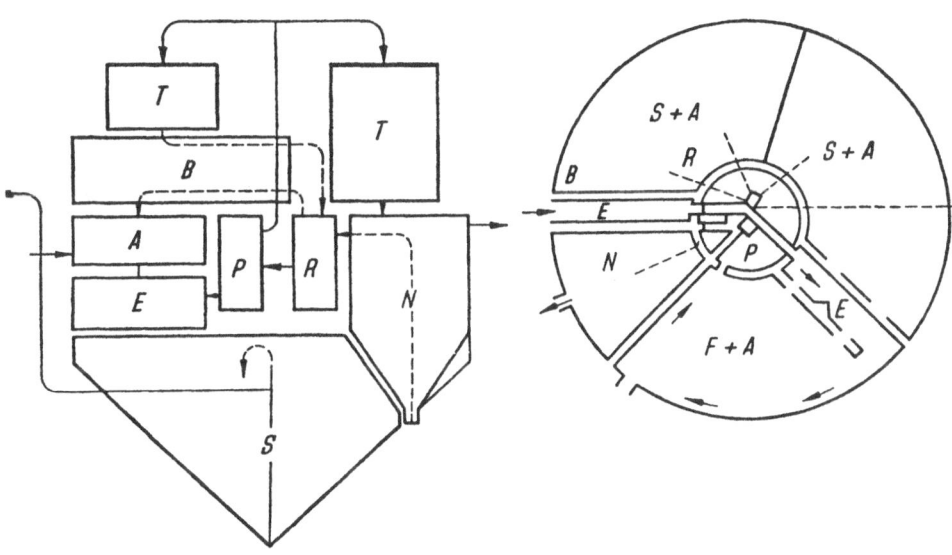

Fig. 8 : Schreiber Package Trickling Filter Plant

Primary treatment is achieved by an Imhoff channel (E) and a long time sedimentation compartment which also is used to equalize the flow (F + A). By the pumps (P) the trickling filter (T) gets an intermittent dosage. The intervals can be as long as ten minutes. By the recirculation pump (R) sludge and recirculation water is withdrawn from the final sedimentation tank (N) and discharged to the Imhoff channel (E). Here sludge will settle and fall down into the sludge compartment (S). Letter B signifies the compartment where the operator can enter.

The effect of load equalization has been measured. Results are given in *Table 3*.

TABLE 3 :   Effect of Load Equalization on the Treatment
            Efficiency of a Trickling Filter According
            to Schreiber

		With Flow Equalization Trickling Filter		Without Flow Equalization Trickling Filter	
		Influent	*Effluent*	*Influent*	*Effluent*
BOD concent.	min	65	10	100	15
(mg/ℓ)	max	113	15	160	41
BOD load	min	6.0	0.54	7.0	0.51
(kg/h)	max	10.5	0.87	16.0	2.74

The combined trickling filter can be applied for sizes between 250 and 15,000 inhabitants. The structure is very compact which helps to minimize heat losses in winter time. Possible foundation problems are restricted to a small area.

Related to the trickling filter is the disc filter. It consists of a horizontal axle to which circular disks are vertically attached.

The axle is 25 cm above the water level. The disks are turning through the wastewater in the same direction as the wastewater is flowing to increase contact time. The diameter of the disks ordinarily is two or three meters. The space between two disks is 20 mm. Taking plastic material, per linear meter of axle 30 disks of 3 m diameter or 34 disks of 2 m diameter can be applied. The energy consumption is 75 watt per meter axle with three meter disk diameter and 50 watt per meter axle with 2 m disks.

To avoid clogging the BOD surface loading should be less than 100 g BOD/$m^2 \cdot$d. Full treatment can be obtained at a loading of 10 g BOD/$m^2 \cdot$d. To gain sufficient contact time two or three axles should be in a sequence. More details are given in (2) and (6).

Fig. 9 : Biodisc Filter

METHODS for INCREASING
the TREATMENT EFFICIENCY

By the low loaded activated sludge process 95% BOD removal can
be obtained and effluent BODs between 10 and 15 mg/ℓ can be achieved. These results are superior to those of trickling filters
which reach 90% BOD removal and effluent BODs between 15 and 25
mg/ℓ. For upgrading a trickling filter effluent two possibilities
shall be mentioned. Sludge recirculation within the final clarifier and the application of a short time activated sludge process.

As pointed out in *Fig. 10* sludge can be recirculated from the
bottom of the Dortmund tank to the wastewater inlet. To increase
the oxygen content also compressed air can be applied to the influent cylinder. Its sludge content shall be in an order between 60
and 80 mℓ/ℓ. after 30 minutes sedimentation time. With such a system the BOD concentration can be diminished by 10 to 15 mg/ℓ [7].

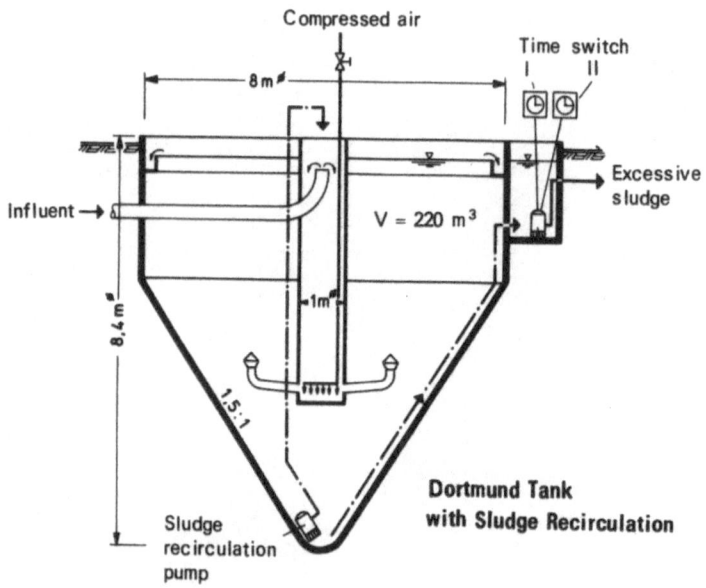

Fig. 10 : Dortmund Tank with Sludge
Recirculation

Without intermediate sedimentation the trickling filter effluent can also be discharged into an activated sludge tank. With 30
min retention time and BOD loadings below 0.8 kg BOD/m^3·d respectively sludge loadings below 0.5 g BOD/g DS·d effluent BODs below
15 mg/ℓ can be obtained.

EXAMPLE

Dimensioning of a biological treatment plant for domestic
wastewater.

Population inclusive equivalents:

At present time   4,600 P
In 30 years   8,400 P

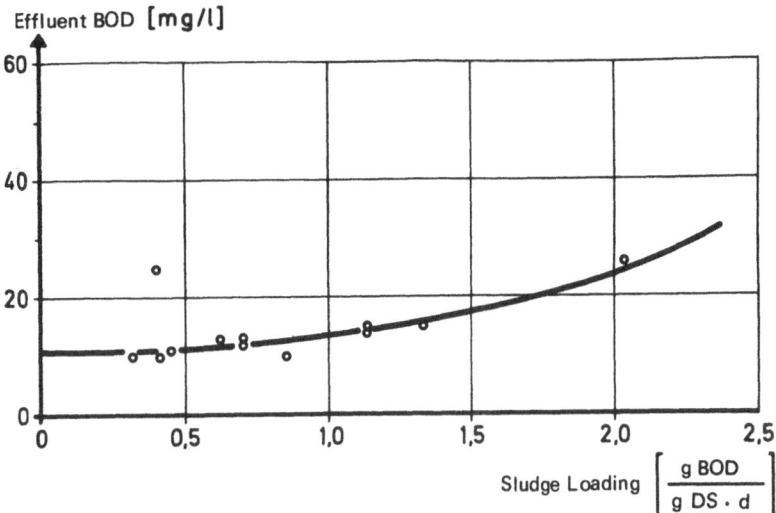

Efficiency of the Activated Sludge Process after a Trickling Filter

Fig. 11 :   Efficiency of the Activated
              Sludge Process After a Trickling
              Filter

Wastewater discharge:

4 to 4.5 ℓ/s per 1,000 pop. equiv.

$$\frac{4.2 \; \ell/s \cdot 8,400 \; P}{1,000 \; P} = 35 \; \ell/s \cong 126 \; m^3/h \cong 2,016 \; m^3/d$$

(for an average operation of 16 h/d)

$\underline{BOD_5, \; sed.:}$ 40 g/P $\cdot$ d

$$\cong 336,000 \; g/d \cong \frac{336,000}{15} = 22,400 \; g/h$$

(the BOD_5- load is calculated according to the 15-h-average)

BOD_5- concentration:

$$\frac{336,000}{2,016} = 167 \; mg/\ell \quad (24 \; h\text{-average})$$

$$\frac{22,400}{126} = 178 \text{ mg/}\ell \quad \text{(average of the day hours)}$$

Since the treated wastewater is discharged into a small creek, about 95% of BOD_5- removal is imparative. This can be achieved with certainty only by a two stage biological treatment. The available head of inflowing wastewater allows the application of a gravity flow trickling filter as first biological stage. The second activated sludge stage shall only be operated at critical low flow conditions. During higher flows of the receiving water the trickling filter effluent will be discharged directly into the final sedimentation tank.

Site of the plant:

- Quite remote from human settlings.

TOPOGRAPHICAL STRUCTURE:

- Hilly region, ample head.

Screen:

- Automatically controlled counterflow screen.

Grit removal:

- One aerated grit channel with mechanical sand removal device.
  Dimensions:width 1.2 m, depth 2.4 m, length 15.0 m.

Primary sedimentation:

Volume 280 m³
Detention time 2.2 h at dry weather flow (later) or 0.4 h at maximal storm flow (later).

Trickling filter:

Recirculated flow 100%

Space loading 600 g $BOD_5/m^3 \cdot d$

Required volume $\frac{336,000}{600} = 560 \text{ m}^3 \cong 600 \text{ m}^3$

Selected: One trickling filter of 2.75 m height and $\emptyset =$

$$\sqrt{\frac{4 \cdot 600}{\pi \cdot 2.75} + 2.28^2} = 16.8 \text{ m} \approx 17 \text{ m diameter}$$

Sprinkler with four distributing arms since hydraulic surface loading is equal to 1.14 m/h.

Efficiency $\sim$ 82%   (acc. to *Fig. 1*)

Effluent BOD_5 = 0.18 · 178 = 32 mg/ℓ.

ACTIVATED SLUDGE TANK

Selected:

Detention time 1.98 h, volume 250 m^3

Space loading: $\dfrac{32 \text{ g } BOD_5 \cdot 2,016 \text{ m}^3}{m^3 \cdot d \cdot 250 \text{ m}^3} = 258 \dfrac{\text{g } BOD_5}{m^3 \cdot d}$

Sludge content: 2,400 g of dry solids/m^3

Recirculation flow: 32%

Sludge Loading: $\dfrac{258}{2,400} = 0.11 \dfrac{\text{g } BOD_5}{\text{g } DS \cdot d}$

Sludge age: 16 days

A safety factor of 50% is chosen for dimensioning of aeration devices, because of varying BOD - concentrations of trickling filter effluent.

Required oxygen intake:

$$\frac{32 \text{ g } BOD_5 \cdot 126 \text{ m}^3 \cdot 1.5 \text{ g } O_2}{m^3 \cdot h \cdot g \text{ } BOD_5} \cdot 1.5 = 9,070 \text{ g } O_2/h$$

An assumed oxygen yield of 1.5 kg O_2/kWh under operational conditions leads to

$$\frac{9.07 \text{ kg } O_2 \cdot \text{kWh}}{h \cdot 1.5 \text{ kg } O_2} = 6 \text{ kW aerator capacity,}$$

$$\cong \frac{6,000 \text{ W}}{250 \text{ m}^3} = 24 \text{ Watt/m}^3$$

specific energy input for aeration and revolving.

A mechanical aerator with a two speed gear and adjustable submergence is chosen.

FINAL SEDIMENTATION

The maximum discharge to the biological stage under storm conditions is restricted to twice the dry weather flow:

$$2 \cdot 126 \ m^3/h \ = \ 252 \ m^3/h$$

The required surface for permissible hydraulic loading of $3.23 \ m^3/m^2 \cdot h$ is $78 \ m^2$.

Selected: one Dortmund tank (with upstream flow) of

$$\emptyset \ = \ \sqrt{\frac{78 \cdot 4}{\pi}} \ = \ 10 \ m \ \text{diameter}$$

Depth 8.70 m, volume 250 m^3

Detention time at dry weather flow: $\frac{250}{126}$ = 1.98 h

Dry solid load of the surface:

$$\frac{1.62 \ m^3 \cdot 2.4 \ kg \ DS}{m^2 \cdot h \cdot m^3} \ = \ 3.9 \ \frac{kg \ DS}{m^2 \cdot h}$$

SLUDGE TREATMENT

Digester (Imhoff tank):
From 54 g DS/P·d and a safety addition of 20% for sludge intake from storm water and local septic tank cleaning, some 13.6 m^3/d of sludge result for the final stage in 30 years. At this time the selected volume of the unheated digester of 1.070 m^3 provides a detention time of about 80 days.

SLUDGE DRYING BEDS

Sludge drying beds of 960 m^2 surface can be constructed, which corresponds to 8.75 pop. equiv./m^2.

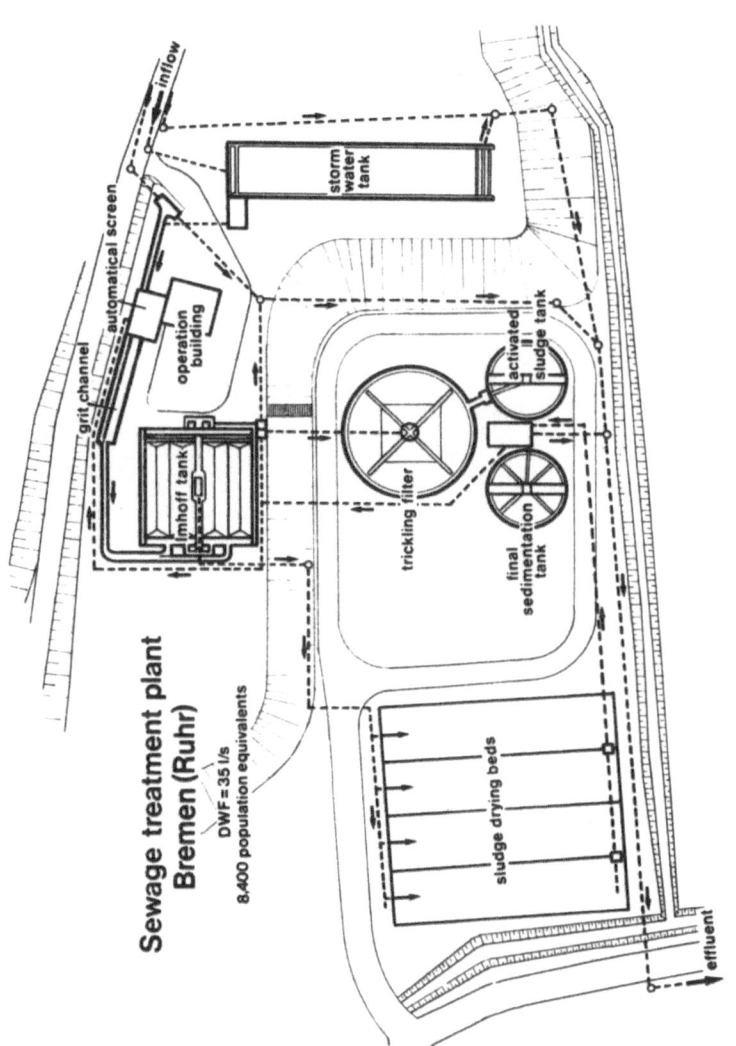

Fig. 12 : General Site

142

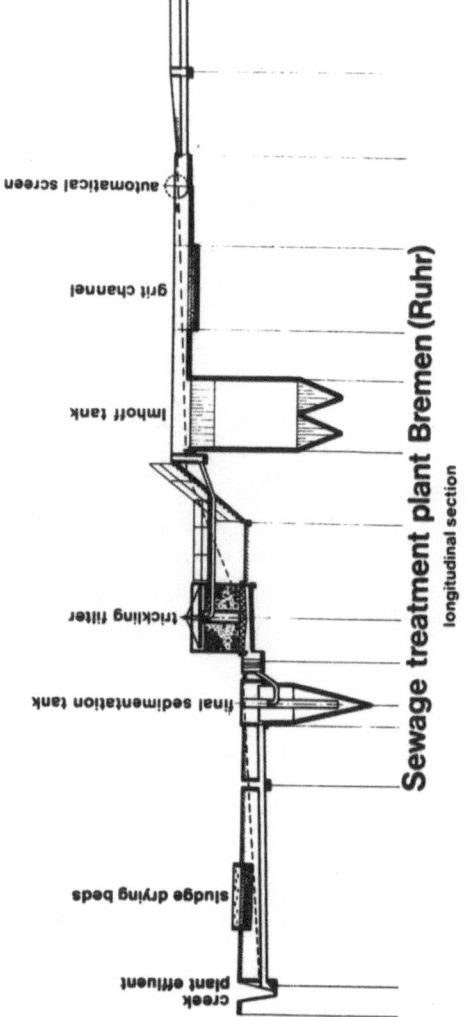

Sewage treatment plant Bremen (Ruhr)
longitudinal section

automatical screen

grit channel

Imhoff tank

trickling filter

final sedimentation tank

sludge drying beds

creek
plant effluent

Fig. 13: Trickling Filtration

REFERENCES

1. Heukelekian, H., The Relationship between Accumulation, Bioche-
   mical and Biological Characteristics of Film and Purification
   Capacity of a Biofilter and a Standard Filter, *Sewage Works
   Journal,* 23, 269, 516, 743, 1945.

2. Rincke, G., Trickling Filtration, *Lehr- und Handbuch der Abwassertechnik,* Band II, 2. Auflage, Verlag W]lhelm Ernst u. Sohn,
   Berlin, 1975.

3. Imhoff, K. R., Disposal of Septic Tank Sludge, *Gewässerschutz-
   Wasser - Abwasser,* Bd. 4, 141, 1971.

4. Rinche, G., Wolters, N., Technology of Plastic Medium Trickling
   Filters, 5th International Water Pollution Research Conference,
   San Francisco, II, 15, 1970.

5. DIN 1957: Füllstoffe für Tropfkörper; Anforderung, Prüfung,
   Einbringen. Beuth-Vertrieb, Berlin, Juli 1965.

6. Krauth, K.h., The Efficiency of Disk Filters, *Gas- und Wasserfach,* 34, 1973.

7. Imhoff, K. R., Increasing the Degree of Treatment by Operational
   Adjustments, *Gewässerschutz - Wasser - Abwasser,* Bd. 19, 475,
   1975.

CONVENTIONAL TRICKLING FILTERS

E. de Fraja Frangipane

Full Professor and Director
of the Sanitary Engineering Institute
Polytechnic of Milan, Italy

1. FOREWORD

As known, in a water body, self-purification is caused by the development of some species of micro-organisms, especially bacteria (aerobic bacteria requiring free oxygen), which use - and decompose - the organic matter present in water as a source of energy and of substances for the synthesis of a new cellular material (section 3.1); in this way, a series of biochemical reactions occur through which complex compounds are gradully separated into other more simple ones up to stable final products.

Biological filtration is one of the treatment processes that artificially cause and enhance the above mentioned natural phenomenon.

Biological filtration is obtained with *trickling filters*. These must be structured and operated in such a way as to allow the occurrence of the particular conditions enabling a rapid biological transformation of the complex organic matter present in the sewage; precisely:

- a support for both sprouting and development of aerobic micro-organisms;
- an aerobic environment (presence of free oxygen);
- the supply of organic matter and water (sewage)

2. THE CONFIGURATION of TRICKLING FILTERS

The trickling filter *(Fig. 1)* consists of a crushed rock or field stone layer (bed), a few meters high, placed within a generally cylindrical perimetric wall. The sewage (previously clarified in the sedimentation tank) is sprayed over the filter surface by various fixed or mobile distribution systems (more frequently mobile). The sewage percolates through the crushed stone layer and flows along the surface of each element constituting the bed; the room among the various elements is never submerged to allow the presence of air and hence of free oxygen. After passing through the stone layer, the sewage is collected at its base; then it flows outside the percolating filter through a system of drainage channels. In this way, micro-organisms sprout and develop on the crushed stone layer: they are fed with the sewage (water + organic substance) that flows through the bed into an environment where, due to the continuous atmospheric exchange, the presence of free oxygen is guaranteed.

3. MECHANISM of BOD REMOVAL

At the initial operating phase of a trickling filter, the sewage passes through the filter practically without undergoing any transformation. The surface of each crushed stone element of the filter is gradually covered with an active jelly-like biological film (zooglea), which is formed owing to the natural growth of micro-organisms (bacteria and other biota). The biological process that develops in the trickling filter is usually considered of an aerobic nature; the biological membrane (zooglea), in the presence of a sufficient amount of oxygen of air, breaks and stabilizes the decomposable organic compounds in the sewage with formation of carbon dioxide, water, nitrates, sulfates, phosphates. In the practice, the phenomenon is more complex *(Fig. 2)*. At the beginning, the predominant action is aerobic, but, with the gradual development of the microorganisms that form the biological membrane, this last becomes thicker, in this way, an anaerobic layer is formed which comes into contact with the surface of the medium where an anaerobic activity takes place, which is as much important as the aerobic one simultaneously occurring along the outer surface of the membrane itself.

3.1 THE ACTION of MICROBIC FLORA

The microorganisms present in the trickling filter mirror the environmental conditions that occur in the filter itself. The predominant micro-organisms are *bacteria*: aerobic, facultative, anaerobic. They are for the greatest part facultative [1] since they live under aerobic conditions as long as free oxygen is avail-

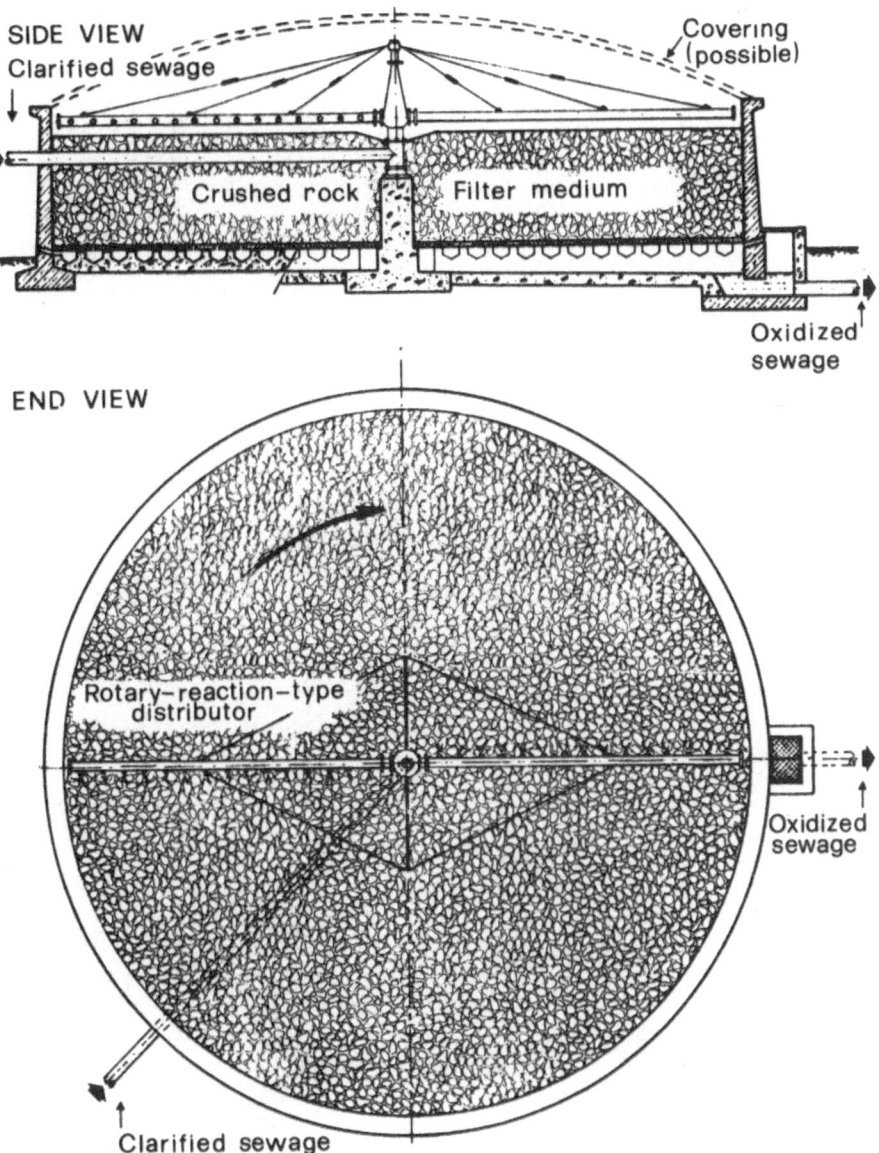

SIDE VIEW
Clarified sewage
Covering (possible)
Crushed rock
Filter medium
Oxidized sewage

END VIEW
Rotary-reaction-type distributor
Oxidized sewage
Clarified sewage

Fig. 1 : Example of a Trickling Filter with Rotary-
Reaction-Type Distributor.

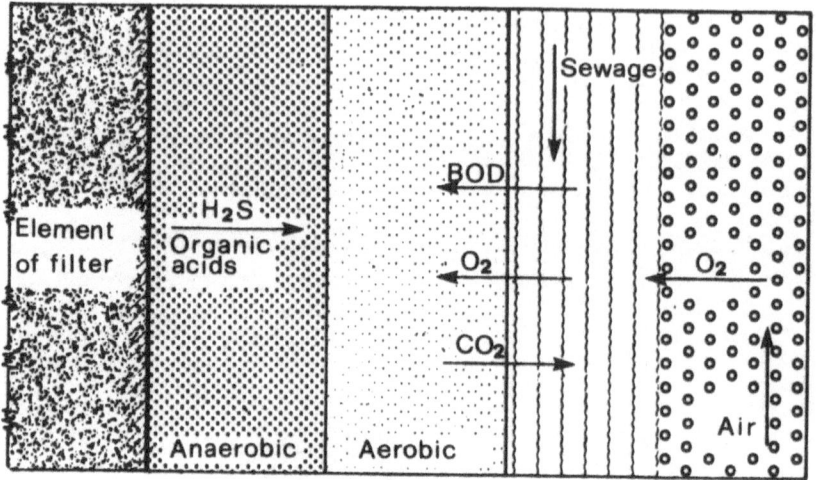

Fig. 2 : Mechanism of the Biological Process
that Occurs in the Membrane of the
Trickling Filter.

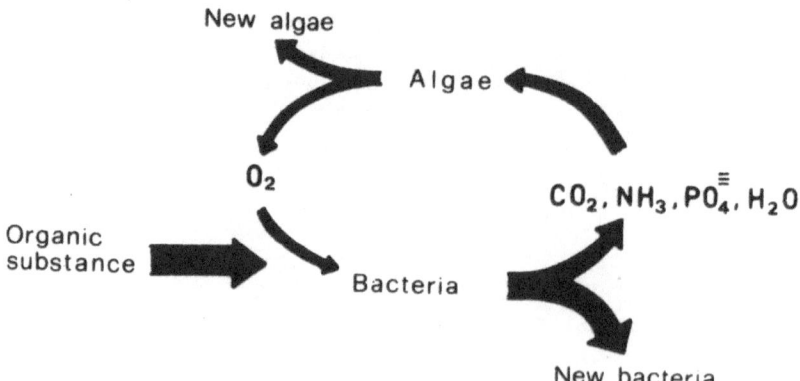

Fig. 3 : Example of the Connection between the
Activity of Bacteria and that of Algae.

able and under anaerobic conditions when oxygen is absent. Such heterotrophic bacteria develop by acting on the complex organic compounds present in the sewage and by producing - through a series of reactions - the new cellular material (protoplasm) for the formation of new cells. Such an activity is developed in two phases: bacteria utilize the organic compounds both to draw the elements required for the formation of the new molecules, and to obtain the energy requires for such transformations.

Energy is provided by *oxidation reactions*. In this case, oxidation of the organic material does not occur by direct oxygen addition, but rather by an indirect action of hydrogen removal. By applying the concept of hydrogen transfer, every biological oxidation may be considered as a particular case of the following general reaction:

$$A \cdot H_2 + B \nrightleftharpoons A + B \cdot H_2 \qquad \qquad \dots \quad (1)$$

where $A \cdot H_2$ is the oxidized substance yielding hydrogen, whereas B is the hydrogen-acceptor reduced substance[1].

If the hydrogen-acceptor substance is oxygen, the biological transformation is aerobic[2] and equation (1) may be written as:

$$A \cdot H_2 + \tfrac{1}{2}O_2 \rightarrow A + H_2O \qquad \qquad \dots \quad (2)$$

Redox reactions do not occur spontaneously but need a particular catalyst. Such catalyst are known as *enzymes* and are reduced by the bacteria themselves. Enzymes also enhance the hydrolysis of complex organic compounds, making them assimiable from the bacterial cell.

In any case, the redox process does not occur in one phase only, as equation (1) might suggest. In the practice, the process occurs through a series of several, separate and successive redox reactions each being ruled by a specific enzyme catalyst; on the whole they are expressed by equation (1).

---

[1] Implicitly, when a substance is oxidized, another substance is simultaneously reduced.

[2] If the hydrogen-acceptor substance is not oxygen, the reaction will be anaerobic.

Redox reactions (respiration process) are those supplying the energy required from the subsequent phase of synthesis for the formation of new cellular material.

Respiration (redox reactions) and synthesis are not sharply separated. In a number of cases, in fact, the final products of respiration processes are used for the synthesis of a new cellular material; hence the two-fold activity occurs by a single process.

The interpretation given so far of energy conservation and transfer from redox reactions (producing it) to synthesis reactions (requiring it) is quite complex and do not completely defined. However, it allows a logical explanation of this phase of the process.

The energy supplied by redox is not heat, this last being not usable by micro-organisms for the synthesis of the new cellular material. Other biochemical phenomena occur, which cause some coupled chemical reactions, which develop as an intermediate phase between the initial redox reaction and the subsequent reaction of synthesis. In this intermediate phase, the ADP co-enzyme accumulates the available energy, by absorbing inorganic phosphorus and by transforming itself into the ATP co-enzyme, which is reduced again to the ADP by supplying the accumulated energy when required from the reactions of synthesis. The reactions of synthesis lead to the formation of new cellular material thus allowing bacteria to develop and reproduce.

Protoplasm (cellular material) consists of carbon (C), hydrogen (H), oxygen (O) and nitrogen (N)[1]; hence the reactions of synthesis may occur only if the substrate supplies these four elements in the desired amount[2].

As known, the sewage forming the substrate of the bacterial flora of a trickling filter, contains considerable amounts of complex organic substances from which bacteria may draw the four elements necessary for protoplasm formation. Therefore, organic

---

[1] Bacteria protoplasm corresponds to the general formula, $C_5 H_7 O_2 N$.

[2] The reactions of synthesis require, in addition to C, H, O, N, other elements too, such as phosphorous (P), sulfur (S), sodium (Na), potassium (K), calcium (Ca), magnesium (Mg), iron (Fe), molybdenum (Mo), cobalt (Co), Mangenese (Mn), Zinc (Zn) and copper (Cu). These elements are required in very low amount, and they are present in the sewage too.

substances are simultaneously removed from the sewage, and the removal extent is the higher the more active is the development of bacterial flora. However, the growth of bacterial flora is strictly determined by the environmental conditions, which therefore are of the utmost importance for the purification degree achievable by sewage treatment in trickling filters.

Bacteria generally reproduce by binary fission, i.e. they separate into two new organisms, each behaving like the original cell. Fission time may vary from a few days to a few hours down to twenty minutes or even less; it depends on the species and on the environmental conditions. By supposing a fission time of 20 minutes, in a 10-hr period each bacterial cell will yield more than 1,000,000 new bacteria [2].

Bacteria reproduction is actually limited by several factors, such as amount and nature of nutrients, temperature, natural death of bacteria , accumulation of toxic secondary products produced by biochemical reactions and elimination due to life competition of other species of micro-organisms. However, bacterial cells always show an extraordinary reproduction power; thus it is easy to understand why, by developing in a trickling filter, they may exert such a surprising action as it is the transformation of the organic matter when sewage percolates through the bed.

Fungi[1], are also present in percolating filters. Being aerobic organisms, they may develop only where the presence of free oxygen is guaranteed. Fungi use the organic matter as a source of carbon and energy. Hence they are in competition with bacteria for the substances they need as feed; this, under the environmental conditions generally occurring in a trickling filter, extensively limits their development: as a matter of fact, they are overcome by the latter that are predominant.

Unlike bacteria, fungi can develop even in slightly damp environment and at a very low pH[2]. Such a property explains why fungi predominate over bacteria in the aerobic biological processes involved in composting of refuse (slightly damp environment with less than 60% water content) and in the trickling filters treating industrial effluents with low pH values.

---

[1] The protoplasm of fungi would correspond to the general formula $C_{10}H_{17}O_6N$

[2] It is known that, in order to obtain fungi cultures in the laboratory, substrates with a pH brought to 4.5 are adopted. At normal pH, in fact, bacteria predominantly develop.

In the upper part of the filter, on the light-exposed surface, algae develop too[1]; as known, they differ from bacteria and from fungi in the photosynthetic power they are endowed with. Such a property allows them to use the energy of light; in the presence of sunlight, algae transform the inorganic substances present in water into organic matter for the formation of protoplasm.Hence algae are autotrophic organisms that do not directly participate in the transformation of organic matter that takes place in the trickling filter. However, a connection exists between the activity of bacteria and of algae. These two types of micro-organisms are not in competition as to the feed sources (organic matter for the former, inorganic ions for the latter), but in many circumstances their activities are interdependent. Such a symbiosis may be described as follows (Fig. 3): bacteria metabolize complex organic compounds in the presence of oxygen to produce new cells, carbon dioxide, water, ammonia and other inorganic compounds. Algae use these products(carbon dioxide, ammonia and other inorganic compounds)to produce new cells, oxygen, which on its turn is used by bacteria in the aerobic processes of decomposition of the organic substance. In a trickling filter such a symbiosis occurs only in the filter surface, where the environmental conditions are fit for algae growth.

Pluricellular organisms (metazoa) such as *nematodes, insect larvae* and others may be easily found in trickling filters. These species feed with micro-organisms present in the filter and live in the filter upper layer. Their contribution to sewage purification is low. However, they are useful because they decrease all hazards of filter clogging, often caused by an excessive growth of the biological membrane.

4. CHARACTERISTICS and CLASSIFICATION of TRICKLING FILTER PLANTS

Trickling filters constitute the simplest and easiest oxidation treatment: its performance and control do not require a numerous and specialized staff.

Trickling filters are not overcome, as it is sometimes said,by the activated sludge system. Each system finds preferential applications - each offers its own advantages and disadvantages. The operating costs of trickling filters is certainly lower; furthermore no particular control is necessary, while only a load loss of a few meters is required. On the other hand, the activated sludge system requires less ground area and the purification degree obtained is higher. The importance of such factors depends on circumstances. The advantages of the activated sludge systems are ever

---

[1] An analysis for the algae Chlorella has given the general formula $C_5H_7O_2N$ for the protoplasm.

more predominant with the increase in the number of inhabitants served, whereas those of of trickling filters become even more determinant with the decrease in the plant size.

4.1 CLASSIFICATION of TRICKLING FILTER PLANTS

Trickling filters may be either low loaded or high-rated depending upon the hydraulic load and on the organic load applied. They may be arranged either in parallel or in series (many-stage-system) if the plant comprises from more than one unit.

The *organic loading* is the daily amount of organic matter expressed in kg of BOD_5 brought on the trickling filter (clarified sewage) per 1 m^3 filter media (Kg/m^3 x day).

The *hydraulic loading* is the daily volume of sewage (the recirculation flow possibly inclusive) expressed in m^3, which daily passes through 1 m^2 of filter surface (m^3/m^2 x day).

The trickling filter receives clarified sewage, i.e. free from the solid matter that may be separated by sedimentation. Before being sent to the trickling filter, the sewage must undergo a clarification treatment in a sedimentation tank, after which the sewage still contains that fraction of solid organic matter that cannot be separated by sedimentation (especially colloidal substances) and the whole organic matter in the dissolved state. Under such conditions the sewage is sent to the trickling filter, where the described biochemical process of transformation of the organic substances occurs.

In its run in the trickling filter, the liquid stream will convey the relieving film of the biological membrane (living, dead or disintegrated substance) which will be found in the retain effluent. It follows that it is necessary to retain the settleable solids sloughed from the filter by a subsequent final sedimentation.

This phenomenon is differently influenced depending on whether trickling filters are low-rated or high-rated. As a matter of fact, while in the former case the relieving films are easily held in the filter by the biological membrane of new formation, which will disintegrate them[1]; in the high-rate filters, the fairly considerable liquid stream will remove all such films, by continuously conveying them outside the trickling filter. Consequently, the amount of material that is found as a sludge in the final sedimentation tank is higher than that of low-loaded filters.

---

[1] In some periods of the year (spring and autumn), the filter is abundantly relieved. In such periods, even very high sludge volumes are reached.

This phenomenon explains why high-rate filters may purify higher amounts of sewage. The sewage stream that continuously takes off the jelly-like films alleviates the disintegration action of the biological living membrane of the filter, which otherwise would also work for the disintegration of the film themselves. In this case, unlike in low-rate filters, the biological membrane must exclusively act on the organic substances present in the sewage, whereas the disintegration process of the relieving films will take place in another plant area, and precisely in the sludge digestion phase, reached by such films after final sedimentation.

To conclude, the low-rate trickling filter decomposes a greater amount of organic matter (of the sewage and of the biological film itself) and yields an effluent with a lower amount of sludge to be treated in digestors. In high rate filters, instead, a part of this task is appointed to digestors.

4.1.1 Low-Rate Trickling Filters

A plant for the oxidation treatment with low-rate trickling filters comprises of (Fig.4) the primary and final sedimentations, in addition to the pre-treatments (screen, grit chamber, floatation) and sludge treatment (digestion, drying).

The filter will be dimensioned depending upon the organic and hydraulic loadings. The specific loading values indicated by the various authors (Table 1) or applied by various designers vary within fairly wide limits. The organic loading ranges between a minimum of 0.08 and a maximum of 0.40 Kg/m^3 x days (kg BOD_5 of clarified liquid/day x 1 m^3 filter). The hydraulic loading ranges from 1 to 5.6 m^3/m x day (m^3 of sewage/day x 1 $m^3 < 2$ filter). In rainy periods and in the case of combined sewers, the hydraulic loading on trickling filters cannot be increased by more than one time and a half, if almost the same degree of purification is desired.

Obviously, the higher the loadings, the lower is the yield of the treatment process, i.e. BOD removal. For not too high loadings, BOD reduction after a complete treatment ranges between 80% and 95% over the BOD of raw sewage. Final effluents may be obtained with BOD values of 30 and even of 20 mg/ℓ.

Imhoff [3] indicates the value of 0.175 kg/m^3 x day as an admissable organic loading. In the design phase, however, the sewage BOD to be treated is not always known and reference must be made to the BOD value g per capita per day. By taking into account the values in Table 2, a BOD_5 value of 35 g per capita and per day may be evaluated for clarified sewage (after primary sedimentation). With the admissable loading indicated above, each cubic meter of filter may serve 5 persons (175:35 = 5 inhabitants).

The calculation done on the basis of the organic loading allows the determination of the filter total volume; the dimensioning of the trickling filter (diameter and height) can be subsequently performed by assigning a given value to the height of the filtering layer (media), which is chosen by taking into account either the material size and the values recommended by different authors *(Table 1)* or else the designer's personal experience. The filter height may be also drawn from the hydraulic loading value, which allows the definition of the filter surface and hence its

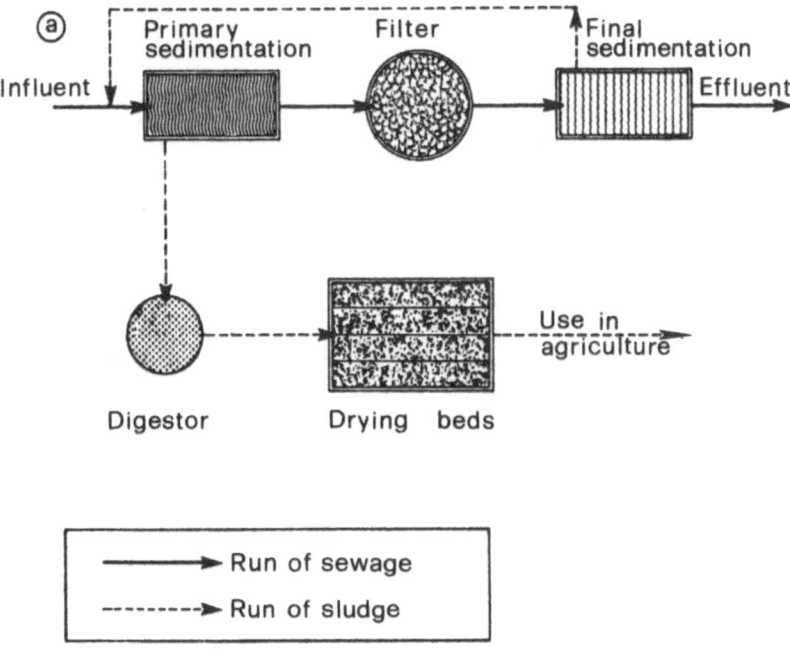

Fig. 4 :   Schematic Drawing of the Treatment with a Low-Rate One-Stage Trickling Filter.

TABLE 1 : Values of the Organic and Hydraulic Loadings for the Dimensioning of Low-Rate Trickling Filters

Loading	Imhoff [3]	Ippolito [4]	Eckenfelder O'Connor [5]	Fair, Geyer [6]	McKinney [2]	FSIWA [1] [7]	TWSWA [2] [8]
Organic kg/m³ x day	0.175	0.15+0.20			0.16-0.32	0.80-0.40	0.16
Hydraulic m³/m² x day	-	1.0-2.0	1.9-5.6	1.9-5.6	1.9-3.7	1.0-4.0	-
per inhabitant no. inhabitants/m³	5	4.3-5.7	-	-	4.6-9.2	2.3-11.5	4.6
Bed height (m)	-	2.00-3.00	-	1.8-3.00	-	1.8-2.50	1.35-2.15

(1) Federation of Sewage and Industrial Wastes Association (USA).

(2) Texas Water and Sewage Works Association (USA).

TABLE 2: Average Amounts of Mineral and Organic Substances per Inhabitant, under European Conditions and for Separate Sewers [3]

Values expressed in g/inhabitant *x* day.

Physical State	Substances			
	Mineral	*Organic*	*Total*	*BOD$_5$*
Suspended Materials				
- separable by sedimentation	10	30	40	20
- not separable by sedimentation	5	10	15	10
Dissolved Materials	75	50	125	30 }40
Total Materials	90	90	180	60

height, the total volume deduced from the organic load being known.

In order to clearly explain what stated above, a numerical example is given:

An urban area with a population of 10,000 inhabitants must be served with a water supply of 200 ℓ/inh. x day. The sewer is of the separate type and the percentage of water volume reaching the sewer is 75%. Hence the daily input into the sewerage per inhabitant will be

$$\frac{75 \times 200}{100} = 150 \ \ell/\text{inh. x day.}$$

The daily flow entering the treatment plant expressed in m³ will be:

$$Q = \frac{10,000 \times 150}{1,000} = 1,500 \ m^3/\text{day}$$

The organic matter flowing to the trickling filter every day (Table 2) expressed in kg BOD, will be (after primary sedimentation):

$$BOD = \frac{10,000 \times 40}{1,000} = 400 \ kg/\text{day}$$

Wishing to apply an organic loading of 0.175 kg/m³ x day (value recommended by Imhoff [3]) to the trickling filter, the total filter volume will be:

$$V = \frac{400}{0.175} \ \frac{kg/day}{kg/m^3 \ x \ day} = 2,285 \ m^3$$

In order to determine the filter height, a hydraulic loading of 1,5 m³/m² per day (average of the values recommended by Ippolito [4] is applied; the total surface of the filter will be:

$$S = \frac{1,500}{1,5} \ \frac{m^3/day}{m^3/m^2 \ x \ day} = 1,000 \ m^2$$

The filter height will be:

$$h = \frac{V}{S} = \frac{2,285}{1000} \ \frac{m^3}{m^2} = 2,28 \ m$$

Two trickling filters are adopted with a surface of 500 m² each (inner diameter ∅ 25 m)
with filter height of 2,28 m or else 4 percolators of 250 m² each (inner diameter ∅ 18 m)
always with a filter height of 2,28 m.

The former solution is the cheapest: as a matter
of fact, the filter volume being the same (total
surface and height), the perimetral wall will
be 157 m long in the case of two filters and 226
m in the case of the 4-filter system. Even the
hydraulic equipments (distributors and dosing
syphons) are more expensive in the 4-filter system.

The oxidation treatment in the low-rate filter is
preceded by primary sedimentation and followed by final sedimentation (for removing the filter relieving film from the effluent).

The final sedimentation sludge is usually sent to
the primary setting tank so that it may settle there with fresh
sludge *(Fig. 5)*. Jointly with this last, it is finally sent to digestion and then to drying.These different treatment phases are dimensioned on the basis of criteria that are not
the same as those adopted for high-rate filters or actived
sludge units, owing to the different requirements.

4.1.2 High-Rate Trickling Filters

High-rate filters, in comparison with the low-rate
ones, operate with much higher loading *(Table 3)*. Whereas the calculation based on the hydraulic loading is meaningless for low-rate
filters, hydraulic loading per surface unit acquires some importance
for the high-rate ones. As a matter of fact, their operation requires a load not below a certain limit (0.8 m³/m² x hr)[3]. In fact
(section 4.1) the liquid flow must be such as to continuously remove the disintegrated material of the biological film, which will
be abundantly found as jelly-like flakes in the high-rate effluent
in a far higher amount than occurring with low-loaded filters.This
explains why high-rate filters can decompose higher amounts of the

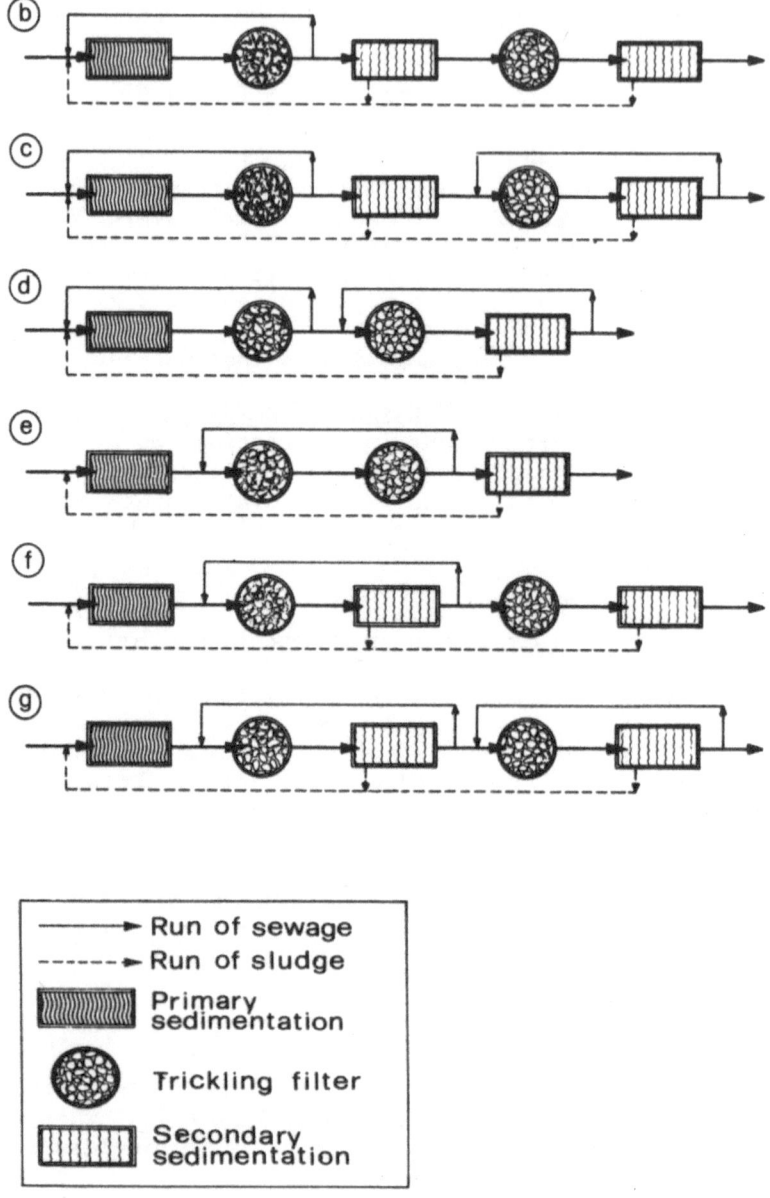

Fig. 5 : Schematic Drawings of High-Rate One-Stage
Trickling Filters (the digestor and drying
beds are not represented).

organic substance present in the sewage; actually it must not decompose the organic substance of the disintegrated film of the biological membrane, as it occurs, instead, with low-rate filters.

The sludge volume that will be found in the final sedimentation tank will be larger and consequently the digestor too must be larger sized.

The higher load brought on the high-rate filter (even more than 10 times as that of a low-rate one) allows one – due to the phenomenon described above – to remove a higher amount of BOD per filter unit volume. However, the effluent still exhibits a high organic matter content. In order to increase the efficiency of BOD removal, an effluent *recirculation* is carried out, by causing the sewage to pass through the filter several times. By the action of recirculation, the organic matter present in the sewage is contacted with the biological membrane more than once. This increases the contact time and allows a more extensive purification.

Furthermore, recirculation leads to an increase in the hydraulic loading (a positive factor in high-rate filters) which keeps high even at times of reduced influent rate and decreases the influent concentration, with the final result that the purification degree of the effluent is improved. The higher is the initial concentration (BOD) of the influent sewage, the higher will be the recirculation ratio.

It is defined as *recirculation ratio* the value:

$$R = \frac{Q' - Q}{Q} = \frac{Q_r}{Q}$$

where:

Q = influent sewage flow to the purification plant

Q_r = flow of recirculation sewage

Q' = $Q + Q_r$ = sewage flow that actually passes on the percolator

The (average) theoretical number of times that flow Q reaching the plant passes through the bed will be:

$$F = \frac{Q + Q_r}{Q} = 1 + \frac{Q_r}{Q} = 1 + R$$

TABLE 3: Values of the Organic and Hydraulic Loadings for the Dimensioning of High-Rate Filter

Loading	Imhoff [3]	Ippolito [4]	Eckenfelder O'Connor [5]	Fair, Geyer [6]	McKinney [1]	FSIWA (4) [7]	TWSWA (5) [8]	TSS (6) [9]
Organic kg/m^3 x day [1]	0.875	-	-	-	1.44	0.40-4.80	0.70	1.7
Hydraulic m^3/m^2 x day [2]	>0.8 m^3/m^2 x hr	4 - 10	9.35 - 28	14-28	9.3-37.4	8-40	-	-
per inhabitant no. inhab./m^{3} [3]	25	-	-	-	41	11.5-137	20	49
Bed height (m)	-	3.00	-	1.80-3.00	-	-	0.90-2.40	-

[1] kg of BOD$_5$ of clarified sewage per day per 1 m^3 bed.

[2] m^3 of sewage per day per 1 m^2 bed

[3] Number of inhabitants per 1 m^3 bed

[4] Federation of Sewage and Industrial Wastes Association (USA)

[5] Texas Water and Sewage Works Association (USA)

[6] "Ten State Standards", Upper Mississippi River Board of Public Engineers and Great Lakes Board of Public Health Engineers (1954).

where F is the *recirculation factor*.

If, for example, the recirculation flow Q_r equals the influent flow Q, the recirculation ratio will be R=1, whereas the recirculation factor will be F=2.

As a matter of fact, BOD removal decreases on increasing the number of passages. Such a condition may be considered as if the number of passages were lower than the real one; hence in the expression of the recirculation factor, F, a reduction coefficient f<1 must be introduced.

The actual recirculation factor F' may be then given by the expression:

$$F' = \frac{1+R}{(1+(1-f)R)^2}$$

where, with f=1, the previous formula is found:

$$F' = 1 + R = F$$

If F' and R are the dependent variables, f being constant, F' reaches a maximum value for

$$R = (2f-1)/(1-f)$$

Being f=0.9 for percolating filters the average number of *actual* passages will be:

$$F' = \frac{1 + R}{(1+0.1 \cdot R)^2}$$

The recirculation ratio required to obtain the highest value of actual passages will be:

$$R = (2 \cdot 0.9-1)/(1-0.9) = \frac{0.8}{0.1} = 8$$

hence

$$F' \; max = \frac{1+8}{(1 + 0.1 \cdot 8)^2} = 2.78$$

$$F = 1 + 8 = 9$$

164

That is, it will be necessary to let the sewage pass through the percolating filter 1+8 times in order to obtain a BOD removal corresponding to an average number of *actual* passages equal to 2.78 with efficiency:

$$F' \ max/F = \frac{2.78}{9} = 0.31$$

The above statements constitute a mathematical proof of how the recirculation ratio R= 8:1 represents the maximum convenience limit, above which the effluent BOD will practically be the same as that of the influent.

In practice, the recirculation ratio is kept at such values (0.5 + 3.0) that the BOD of sewage passing through the percolating filter (recirculation inclusive) does not exceed that required in the effluent by three times [9].

Sewage recirculation necessarily requires a pumpage conveying the effluent upstream of the filter; the plant performance thus becomes more expensive in comparison with low-rate filters, provided that the topographic soil configuration allows these last to operate by gravity.

Sewage recirculation may occur in different ways. *Fig. 5* shows a schematic drawing of some possible solutions. Some of them (d-e-g-) draw the recirculation capacity either downstream of the final sedimentation tank or directly from them (b); in the former case, the final sedimentation tank must be dimensioned according to the total sewage flow passing through the filter (the recirculation yield inclusive).

As to the solutions b), c), e) and g), the recirculation flow is conveyed upstream of the primary sedimentation tank; hence, they must be dimensioned in view of the whole sewage flow, recirculation inclusive.

Only solution f) foresees the withdrawal of the recirculation flow upstream of the final sedimentation and forwarding it downstream of the primary sedimentation; in this way, the inconvenience arises that the filter relieving films are brought back to the filter, but the primary and final sedimentation tanks are just dimensioned for the influent sewage flow. As to the construction cost, solution f) is the most economical one.

In solutions d), f) and g), sludge recirculation must be carried out by a system of pumps that are independent of those foreseen for sewage recirculation.

Hence, scheme g) is the most expensive. Tests carried out at plants operating according to schemes b), c) and d) have shown that the three recirculation systems give equally satisfactory results [2]. A comparative investigation of systems d) and f), simultaneously operated under the same conditions, has shown no significant differences in the final result [2].

The choice of the scheme and the recirculation system of the sewage and sludge is made on the basis of the designer's experience, who will choose, depending on the particular conditions, the cheapest solution (evaluated from both the construction and operation costs) guaranteeing the desired purification degree.

The primary and final sedimentation tanks must be dimensioned according to the sewage capacity that actually circulates in the single unit, depending upon the plant scheme adopted. Hence, the detention time will be referred to the actual sewage flow circulating in the tank.

4.1.3 Two- or More- Stage Trickling Filters

In particular cases, when it is desired to guarantee a better effluent quality, a treatment in series is done by successively passing the sewage through two or more filters.

The two-stage system (two filters in series) is preferred over the multiple-stage system (with more than two filters in series). The two-stage system, consisting of two filters one after the other, is equivalent to the treatment in one single filter with the same diameter, and height equalling the sum of the heights of the two single filters. The separation into steps offers advantages with regard to aeration, due to the lower height obtained in the filters.

The treatment schemes may be of two types: with just one final sedimentation tank (*Fig. 6, scheme h*) or with two sedimentation tanks, each after every single step (*Fig. 6, scheme i*). Obviously, the latter solution is more expensive as it foresees one additional sedimentation tank.

Schemes h) and i) of *Fig. 6* refer to low-rate filter. In these, the periodic run inversion is allowed, so the 2nd filter (less loaded) may be used as first and vice-versa.

The schemes shown in *Fig. 7* refer to plants with high-rate trickling filters (with recirculation), which may be dimensioned by the criterion that the sewage BOD on the second fil-

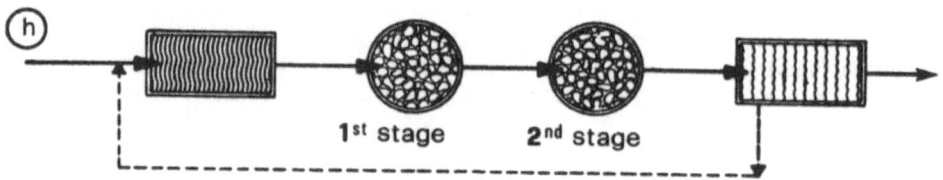

(h)

1st stage   2nd stage

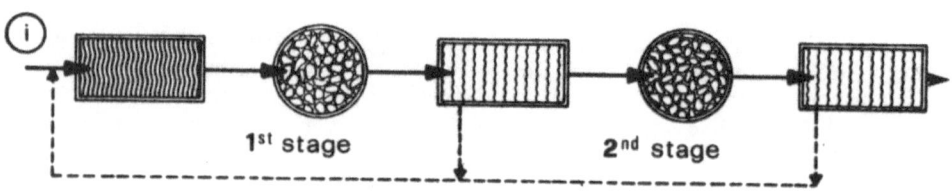

(i)

1st stage   2nd stage

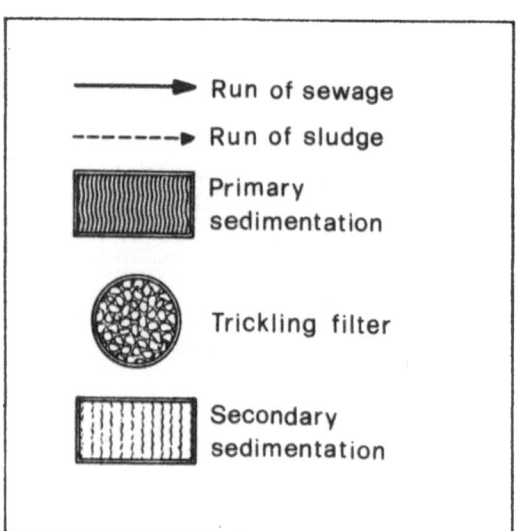

→ Run of sewage

----→ Run of sludge

Primary
sedimentation

Trickling filter

Secondary
sedimentation

Fig. 6 : Schematic Drawings of Low-Rate Two-Stage
Trickling Filters (the digestor and drying
beds are not represented)

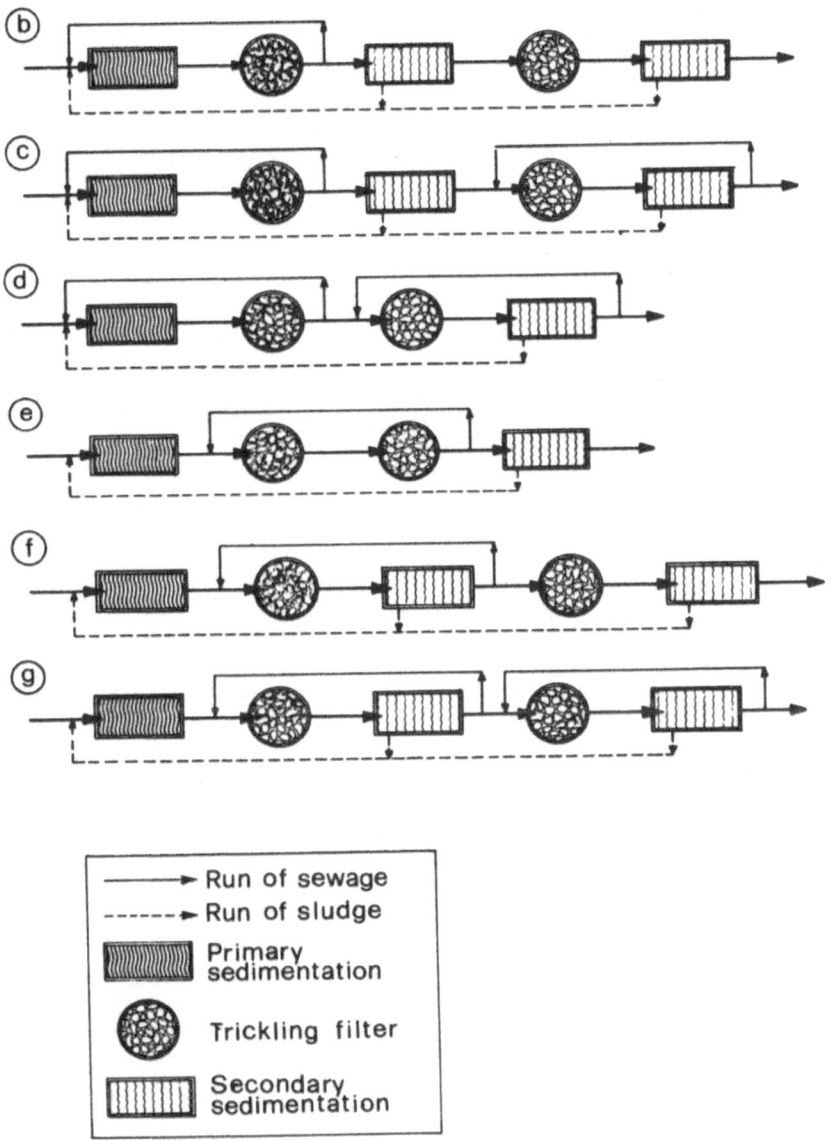

Fig. 7 : Schematic Drawings of High-Rate Two-Stage
Trickling Filters (the digestor and drying
beds are not represented).

ter, recirculation inclusive, must not exceed the double of the wanted BOD value in the effluent [9]. Such a criterion may be applied for hydraulic loadings not above 28 m^3/m^2 x day.

The problem of scheme choice is essentially economic; its evaluation is the designer's task, who, also on the basis of his own experience, may adopt the most convenient system to obtain the purification degree wanted.

4.2 EFFICIENCY of the TREATMENTS

The height of the low-rate trickling filter varies from a minimum of 1.35 m to a maximum of 3 m. However, the filter top layer exhibits an intense development of the biological membrane only for the upper 60-90 cm, whereas on the bottom it is quite reduced [2]. In this sense, the filter might be considered over-dimensioned. The filter effluent (after final sedimentation) is stable with a free oxygen content above 50% saturation[7]. The excess oxygen the sewage is loaded with from the filter upper layers and the conversion of the whole nitrogen excess into ammonia, allow the development of autotrophic nitrifying bacteria in the filter lower layers, by fixing ammonia and giving rise to the formation of nitrites and subsequently of nitrates. This is the reason why the effluent from a low-rate trickling filter shows excess nitrogen as nitrates; this is the condition giving a high stability to the effluent.

On this regard, the behaviour of high-rate filters is quite different. As a matter of fact, the effluent from high-rate filters (final sedimentation inclusive) is as clear as that from low-loaded filters, generally has a higher BOD and almost the same dissolved oxygen content, but it is not so stable as its nitrate content is very low or even nil.

In some particular cases, e.g., when a treated effluent is disposed into a lake, the continuous discharge of nitrates in large amounts, as it happens with low-rate filters, causes a gradual increase in the nitrate concentration in the water body and hence an abnormal and excessive development of algae, for which nitrates, as well as phosphates, constitute the nutrient salts. From this point of view, high-rate filters are more convenient than the low-rate ones. In fact, the nitrates that are abundantly present in low-rate effluents mainly derive from the relieving film decomposition, which, as known, mostly occurs inside the filter by the action of the biological membrane; in high-rate filters, instead, the films that are continuously removed by the liquid stream as flakes, are found in the sludge and correspond to as many substances removed from the final sedimentation.

Apart from such phenomena that constitute a quite particular feature of the water pollution problem, the main purpose of
domestic sewage purification is both the elimination and the transformation of the organic substances originally present in it.There-
fore, the purification plant capacity is especially measured as a
function of the percent BOD removed. A quite important factor is
also the decrease in suspended material and bacterial content; Imhoff (1955) gives the comparison data reported in *Table 4*.

5. CONSTRUCTION CHARACTERISTICS
of the TRICKLING FILTERS

The construction of the trickling filters does not arise any
particular difficulty both with regard to civil engineering works
and to mechanical equipments.

Trickling filters generally have a circular plan (more seldom
hexagonal or octagonal) with diameters ranging from several meters
to some tens meters[1]; they are equipped with *a rotating distributor* consisting of two *(Fig. 1)*, four or more horizontal pipes supported by a central column.

In a long series of large scale runs, the best results were obtained with rotation rates allowing winch revolution in 15 - 30 minutes [11]. In order to obtain a distribution of this type, the
normal spontaneous rotations system resulting from the reaction effect of sewage spouts should be replaced by a mechanical rotation
fed by a small engine that continuously and slowly moves the distributor. However, although such a distribution system offers some
advantages, it presents the inconvenient that the rotary-reaction
type distributor must be replaced by a more complex mechanical system, which requires a small engine and consequently energy consumption.

The rotating distributor that rotates spontaneously is kept in
motion by the reaction effect caused by the sewage flow leaving
the holes made on one side only of each arm constituting the winch.

In order that the spout thrust may be enough to cause winch rotation, the filter must be often fitted with a dosing-syphon *(Fig.
8)*, which intermittenly operates the winch. The syphon is often
necessary to guarantee the distributor rotation, especially when

----

[1]Dimensions of some of the largest circular trickling fil-
ters with one central rotating distributor: Las Vegas,
Nevada, USA: 3 percolators with diameter of 53 m; Green Bay
Plant, Winconsin, USA: two 42 m diameter percolators, Berne plant,
Switzerland: 4 percolators (+2 future ones) with diameter of 36 m.

the reduced sewage flow (night hours) is not enough to provoke the necessary thrust to overcome the resistance to winch rotation.

The dosing-syphon is dimensioned for output times of a few minutes at 5 - 10 min intervals [4].

The plant equipped with dosing-syphon requires a head-loss of about 1 - 2 m [1]. Furthermore, by taking into account the filter height (winch+bed+drainage canals), the overall head-loss is high (usually more than 5 m) and the topography of the siting plant area not always allows that the natural difference of level may be exploited. Therefore, it is quite often necessary to foresee a pumpage upstream of the percolator (either before or after the primary sedimentation vessel) making the plant more complex and its construction and performance more expensive.

In recirculation plants or when the amount flowing in is kept within values allowing the performance of the rotating distributors, the dosing-syphon may be eliminated (Fig. 9); in this way,the level difference required is reduced. However, in this case too, either an initial (Fig. 10) or an intermediate pumpage may be often necessary; the former is essential in level-ground areas, where the collector necessarily reaches a certain depth in respect of the soil plane.

Intermediate pumpage (predominantly from 3 to 6 m) is always required in *recirculation plants* with high-rate filters (Fig. 11).

Recirculation systems are very many: the most frequently adopted one consists of two or more pumps operated either automatically or manually. Recirculation may be carried out:

- just when the influent flow is low
- when the flow is constant over the 24 hours
- when the flow is proportional to the sewage influent amount
- at two or more constant capacities

Instead of rotating distributors, a series of *fixed distributors* may be applied too. However, they are used only when the filter surfaces are as large as to recommend the use of square or rectangular percolating filters instead of the circular ones. Fixed spraying nozzles are less frequently adopted; in fact, even in the case of rectangular-shaped filters, to and fro mobile distributors are frequently used (Great Britain).

---

[1] The head-loss may be considerably reduced, especially in small plants by means of particular syphons of low height and low load on the winch.

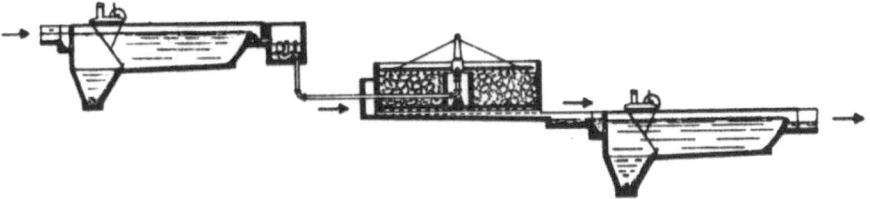

Fig. 8 : Longitudinal View of a Trickling Filter
Fed through a Dosing-Syphon.

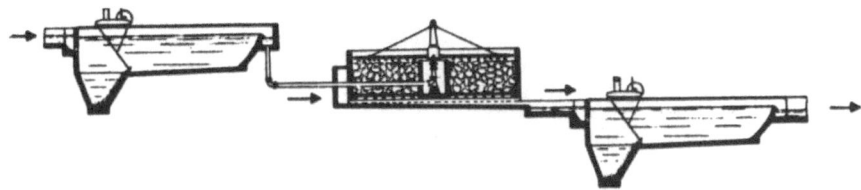

Fig. 9 : Longitudinal View of a Trickling Filter
continuosly Fed without Dosing-Syphon.

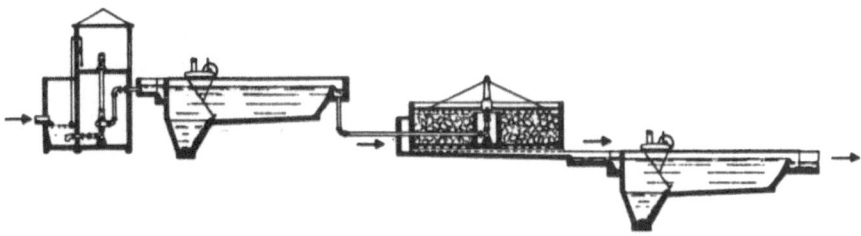

Fig. 10 : Longitudinal View of a Trickling Filter
with Initial Pumpage and continuous
Feeding without Dosing-Syphon.

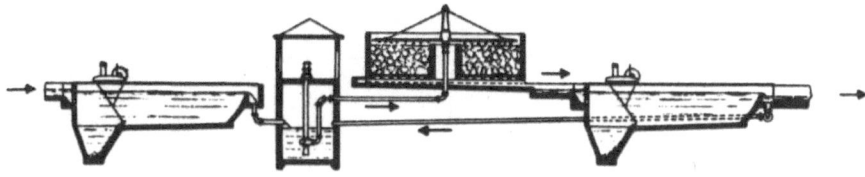

Fig. 11 : Longitudinal View of a Recirculation Plant
with High-Rate Trickling Filter.

TABLE 4: PERCENT REDUCTION of POLLUTING LOAD AFTER PURIFICATION TREATMENT

Purification Treatment	Reduction %		
	BOD	Suspended Materials	Bacterial Content
Chlorinated sewage (raw or clarified)	5 – 10	5 – 20	20
Clarified sewage (primary sedimentation treatment only)	25 – 40	40 – 70	25 – 75
High-rate trickling filters (with primary and final sedimentation)	65 – 85	65 – 92	40 – 80
Low-rate trickling filters (with primary and final sedimentation)	75 – 95	85 – 95	90 – 98
Active Sludge treatment (with primary and final sedimentation)	75 – 95	85 – 95	90 – 98
Chlorination of Purified Effluent (after secondary treatment)	90 – 95	85 – 95	98 – 99

The *perimetric wall* of trickling filters is built either in re-
inforced or simple concrete, or else in brick or stone walling. A-
long the perimeter, at the wall base, aeration canals are opened;
they connect the outer space with the drainage canals in the areas
below the bed; in this way air circulation is allowed in either direction depending upon the temperature differences between external and internal air (section 6.3) in order to guarantee the required air exchange inside the filter.

As to *filtering media,* rough materials must be preferred;their
size must be chosen on the basis of the criteria shown in section
6.2. As a general rule, the material must be hard, clean, long
lasting and resistant to the action of sewage; furthermore it must
not freeze. It must not be a friable material in order to avoid
the formation of smaller-sized materials, which would cause clogging phenomena, due to an excessive reduction in free spaces.Hence
all flaky or lamellar rocks must be discarded. Crushed stones obtained from the crushing of various rocks (preferably granites) are
frequently used; moreover use is also made of volcanic materials
(lava), blast furnace slags, coke, anthracite, bricks or pieces of
bricks. In some cases, ceramic materials, wood or plastics have
been adopted.

Provided that the general characteristics quoted above are fulfilled, the filtering media must be usually chosen among on-the-
spot materials, thus being easily available and more economical.

The filtering bed bottom consists of a series of *drainage channels,* which collect the sewage and let it out of the filter after
its passage through the bed. Such drainage channels must be capable of bearing the static load of the bed above them; obviously,
they must be fitted with openings enabling sewage collection and
simultaneously hindering the crushed stone fall inside them. They
must be arranged with a slope ranging from 0,5 to 5% depending u-
pon the bed size. The drainage channels usually flow into a central channel conveying the sewage out of the filter. All channels,
both the secondary and the main one, must be checked for rates of
at least 0.60-0.90 m/sec in order to avoid that the relieving film
of the bed may settle already in the canals inside the filter.

Drainage channels not only convey the sewage flow, but also allow the passage of air that must circulate in the bed. Therefore,
the sewage, at its maximum level, must not completely fill them.
Under the maximum filling conditions, at least half channel section
should be preferably left free for air passage in low-rate filters
and two thirds of it high-rate filters.

6. FACTORS INFLUENCING
the TRICKLING FILTER CAPACITY

The trickling filter capacity is strictly connected with the behaviour of the biological process and hence to the development and to the action of the biological membrane micro-organisms.Therefore, the factors conditioning the trickling filter capacity must be looked for among those influencing the bacterial activity.

As a number of such factors, the mechanism of their participation in the biological process has been investigated and defined; many others have not been thoroughly interpreted yet, especially because it is impossible to control all interdependent variables participating in the phenomenon.

Some of such factors must be considered more carefully since they affect the design criteria related to trickling filters:

- sewage composition and characteristics (BOD, organic or inorganic industrial effluents, oils, synthetic detergents, temperature);

- nature of the filtering bed;

- natural aeration and forced ventilation.

6.1 SEWAGE COMPOSITION
and CHARACTERISTICS

The purification degree attained in a trickling filter is affected by the sewage amount and quality. A usual domestic sewage does not exhibit particular characteristics that make it unfit for treatment in a trickling filter. The organic substances present in sewages mainly consist of proteins, greases and carbohydrates[1] that undergo the biochemical action of bacteria and of the other micro-organisms forming the biological membrane of the trickling filter. The biological process being aerobic, the biochemical oxygen demand (BOD) characterises the sewage better than any other parameter; the filter dimensioning may be done on the basis of the BOD value in that is allows - the capacity being known - the calculation of the organic load that may be applied to the filter. The BOD measure also allows one to check the purification efficiency.

---

[1] *Proteins* are high-molecular-weight nitrogenous organic substances; the general formula of a typical protein is as follows: C, 52%; H, 7%; O, 22%; N, 16%; S, 4-2%; P, 0-2%. *Greases,* predominantly water insoluble, are esters of glycerol with higher fat acids. *Carbohydrates,* are carbon, hydrogen and oxygen compounds.

Since the process is essentially aerobic, in view of the process behaviour and consequently of the treatment efficiency, the sewage must reach the filter under non-septic conditions. In fact, the sewage might be in a state of advanced decomposition and under *septic conditions* when it remains in the sewer for too long (too long runs and/or a too slow speed), or else when the primary sedimentation time should be excessively long. Such conditions may more easily occur by night when the considerable decrease in the line influents causes a speed decrease in the sewer and an increase in the residence time in the primary sedimentation tank.

The possibility that, under such conditions, the sewage may become septic and thus reach the trickling filter, may be obviated in different ways:

- by previous aeration (before primary sedimentation) in a convenient tank, which might also be used for the separation of oils and greases;

- by recirculating the sewage leaving the filter and by forwarding it upstream of the primary sedimentation tank. The recirculated sewage reduces the sewage residence time in the sedimentation tank and, being oxidized, simultaneously contributes to prevent the fresh sewage from septic state. If recirculation is continuous, there will be a true recirculation system. Otherwise, recirculation may be limited to the night hours, just to avoid the occurrence of septic conditions.

The metabolic activity of the different species of microorganisms is conditioned, among other factors, by the environmental *temperature* too. In the case of trickling filters, the environmental temperature is the sewage temperature rather than the atmospheric one; hence, the biochemical activity of the biological membrane (and hence the trickling filter capacity) may be considered a function of the sewage temperature too.

Out in rainy weather periods and in the case of combined sewers, such a temperature does not extensively vary during the day ($2-3°C$) and usually by no more than $10-15°C$ (limit values) in the course of the year. On the contrary, the changes of atmospheric temperature, are, as known, much more large *(Fig. 12)*. That is the reason why the influence of temperature on the biochemical process is less important than it might be supposed if reference were made to the temperature changes of the atmosphere and not of the sewage.

The sewage temperature increase causes an increase in the metabolic activity corresponding to an increase in the filter efficiency with a higher BOD removal. On the contrary, the metabolic

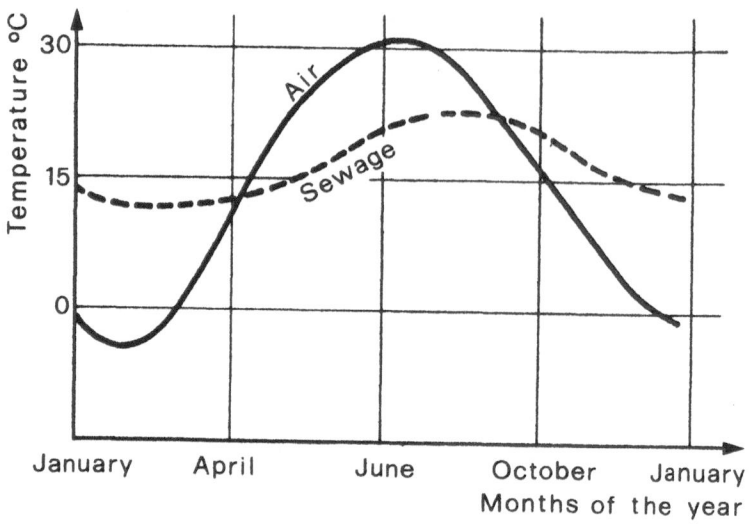

Fig. 12 : Example of Temperature Change of Air
and of Sewage in a Trickling Filter.

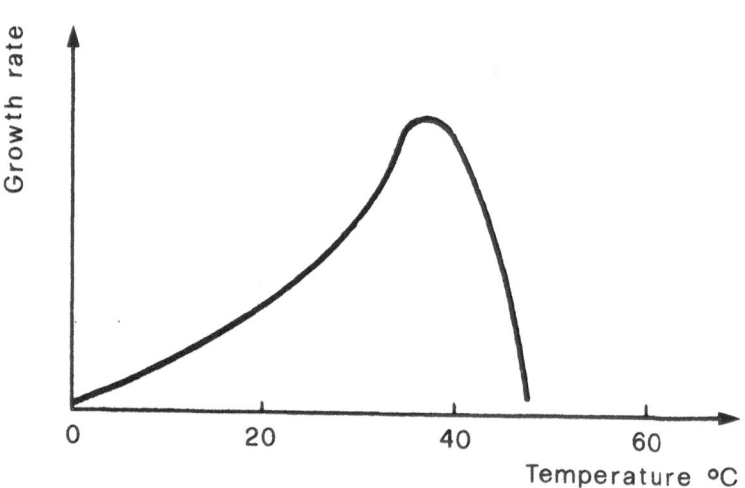

Fig. 13 : Rate of Growth of Mesophyllic Bacteria
with Temperature Increase.

activity decreases on decreasing temperature. Therefore, the treatment efficiency will be higher in summer than in winter. A variation in the BOD removal efficiency by about 0.32% has been observed
for every °C of temperature change. This value is in contrast with
what usually assumed, i.e. that the bacterial development doubles
every 10°C of temperature increase (Fig. 13), at least up to temperatures near that of species survival.

The mechanism determining the optimal temperature of micro-organisms is still unknown and in many cases depends on other
environmental factors. As a matter of fact, it could not be explained why the temperature change affects high-rate filters less
than the low-rate ones. In any case, on increasing temperature, at
least within the change limits usually occurring in the sewage, the
trickling filter capacity improves (BOD reduction).

Quite often, the sewage reaching the treatment plant does
not exclusively derive from domestic sources; it is even more frequent the case of sewers collecting industrial effluents too. Such
industrial effluents may even be of a quite different type and may
affect the sewage even quite extensively by varying its characteristics, often to a considerable extent.

Industrial effluents of an organic nature   cause no
inconvenience in the filter operation, as it happens for example
with the discharges derived from dairies, food industries, a few
pharmaceutical and textile industries, as well as from some fermentation process. The overall organic load (domestic sewage + industrial effluents) must be taken into account, dimensioning the
trickling filter in consequence of it.

However, industrial effluents, although organic, often contain particular substances that, beyond certain concentrations,
exert such a toxic effect as to remarkably affect the behaviour of
the biological process [2].

The presence of industrial inorganic effluents may affect
the sewage pH and cause the immission of toxic substances that,
beyond given concentrations, may either delay or hinder the biochemical action.

Not to upset the biological treatment efficiency, the influent pH should not be lower that 5.5 or higher than 9.0 (better
if it is $6.0 \leq pH \leq 8.0$). Most micro-organisms cannot survive at
pH values lower than 4.0 and higher than 11.0.

Both acids and strong bases, i.e. the strongly dissociated
ones, exert a high bactericidal action. Furthermore, both acids
and bases, even at concentrations below the toxic levels for bacteria, may exert indirect actions; for example the bactericidal ac-

tivity of a number of salts may be increased by the presence of acids or alkalis (probable synergic action).

However, independently of pH, a number of salts are toxic as regards the oxidative biological process. It may derive a partial or a total inhibitory effect, depending on the substance and on its concentration, on the dissociation degree of the salt, on the anion nature, on the metal ion valence and molecular weight. In general, bivalent cations are more toxic than the monovalent ones and heavy metal salts are more toxic than the light metal ones. The bactericidal activity of heavy metal salts results from the affinity of cations to the proteic material; when the proteic constituent of the bacterial cell is precipitated as an insoluble proteinate, the cell dies[1].[12].

The ion concentrations either delaying or inhibiting the biochemical process of a trickling filter are shown in Table 5 [13].

In some cases, the biological population becomes acclimatized up to certain concentration of the toxic substance. Such an acclimation may be caused either by a neutralization of the toxic substance by effect of the biological activity of the macroorganisms or by a selective growth of cultures just with some species of microorganisms capable of metabolism even in the presence of toxic substance [5].

TABLE 5 : Ion Concentrations with a Toxic Effect on the Biological Process of Trickling Filters

S u b s t a n c e	Toxic Concentration (mg/ℓ)
Boron	1
Chromium	3
Copper	1
Cyanide	1 − 2
Iron	5
Lead	0,1
Nickel	1 − 3

[1] The oligodynamic action applied in water disinfection treatments probably results from salt formation of the metal in solution.

It may frequently happen that the industries discharge cooling water, which - although free from toxic substances - cause a temperature increase in the sewage. If temperature increases within the limits allowing the biological process (35°C) the presence of such effluents is not noxious, but accelerates the activity and growth of microorganisms; hence the trickling filter efficiency is improved (*Fig. 13*).

6.2 FILTER BED

As known, the filter elements constitute the support for both sprouting and growth of the biological membrane. Therefore, the characteristics of the surface of the filtering media may somehow affect the biological process. To that purpose the materials with a rough surface must be preferred, since on them the initial accumulation of the organic matter solid particles present in the sewage occurs more easily. However, gravel or pebbles, which exhibit a smooth surface, have been sometimes used.

The material *size* is quite important too. As known, in fact, the filter through which the sewage percolates, must allow air to pass freely. Hence from this point of view, a large-sized material should be preferred; actually, due to the large volume of empty spaces it includes, it allows air to pass abundantly. On the contrary, the organic matter is transformed by the action of the biological membrane growing on the surface of all elements constituting the filtering media. Consequently, the larger is the biological membrane surface the higher might be the organic load of the percolator. Therefore, lower-sized materials should be more efficient because, the overall volume being the same, they exhibit a larger contact surface.

Imhoff reports the example of a bed consisting of materials with an average size of 6 cm (4 - 8 cm): the contact surface is 95 m^2 per m^3 of crushed stone. When dimensions range between 2.5 and 4 cm (average 3 cm), the contact surface is 190 m^2; hence purification, if considered as a function of the surface, should be double as that of the previous case. However, the free space of the elements with a 3-cm granulometry is eight times lower than that supplied by 6-cm elements.

Therefore, the two exigencies must be reconciled both by choosing the lowest possible size guaranteeing a free and sufficient passage of air, and preventing the hazard of cloggins, which may more frequently occur the smaller is the material size. As a matter of fact, since the two exigencies are in contrast, when the size is below a given value, the available space will not be enough to simultaneously allow the biological membrane growth, sewage percolation and passage of air.

A series of tests carried out at the Water Pollution Research Laboratory (Great Britain) with filters of different size
(2.5 - 6.5 cm) and of different nature have shown that the best
results are obtained with materials with smaller size and rough
surface. However, after the third operating year, the filter consisting of lower-sized material gave rise to an excessive accumulation of biological membrane; hence, under such conditions, some
trickling filters with larger-sized filtering media supply better
results [11].

In the United States, larger-sized materials, especially
from 6 to 10 cm, are preferably used. Specifications for material
supply [14] commonly require that 95% or even more of material may
pass through a 10 cm-mesh screen and not through a 6-cm one [15].
In Great Britain [16], instead, beds with a granulometry not higher than 4 cm are preferred.

In any case, a quite important factor is that the size of
the crushed stone is as much uniform as possible.

The bed *height* obviously depends on the organic load,but
is limited by ventilation requirements as a function of the size
of the filtering media. In the U.S. the heights commonly applied
vary from 1.50 m to 2.50 m with an average granulometry of 8 cm
[15], whereas in Great Britain a height of about 1.80 m is adopted
with granulometry of 4 cm. However, higher heights are frequently
adopted, but with them a sufficient ventilation must be guaranteed.
In such cases, larger-sized filtering materials and even a forced
ventilation are generally required. In the Berne plant (Switzerland), put in operation in 1967, trickling filters are 4.50 m high and
are equipped with forced ventilation.

6.3 NATURAL AERATION
and FORCED VENTILATION

Air circulation inside the trickling filter is important
so that the biological process along the active membrane surface
may be aerobic, that is it may occur in an environment where the
presence of free oxygen is guaranteed.

Air circulation may be either natural or forced. Natural
ventilation, which is the most commonly adopted takes place vertically along the filter, in either direction depending on the different density of air in the outer atmosphere in respect of that present in the trickling filter. Such a difference in the air density
is caused by the temperature difference in the two environments.
It may be assumed to a good approximation that air inside the trickling filter nearly has the same temperature as that of the sewage
circulating in the filter. Obviously, sewage temperature in winter

is higher and in summer is lower than that of the outer atmosphere. Hence, in winter the density of internal air will be lower than that of the outer atmosphere and circulation through the trickling filter will occur from bottom to top. On the contrary, circulation in summer will occur in the opposite direction. However, especially in summer, external temperatures may exhibit such night and day changes as to cause a phenomenon inversion even on the same day.

It has been found [17] that a temperature difference of 6°C between external air and sewage (hence internal air) causes a downcast air stream through the trickling filter of 0.3 m^3/m^2 x min (0.3 m^3 air/m^2 filter x min) equalling 18 m/h (0.3 m^3/m^2 x min = 0.3 m/min = 5 mm/sec = 18 m/h). When such a temperature difference falls to 2°C, the air stream stops and when the outer air is colder than the sewage, the air stream becomes upcast *(Fig. 14)*. The air stream stop, that takes place in spite of a 2°C - difference in temperature may be explained by the density variation that, independently of temperature, occurs owing to other phenomena connected with the biological activity and affecting the air state and composition inside the trickling filter. The phenomenon of upcast or downcast air stream caused by the temperature difference between the trickling filter inside and outside may be influenced by other factors, such as wind. In fact, if a strong horizontal wind blows at the filter top, there may occur a pressure variation drafting air from the bottom to the top or vice-versa; in this way the natural circulation caused by temperature difference is either increased or hindered.

However, the air circulation direction inside the filter is of no importance for practical purposes; as a matter of fact, nothing changes whether air circulates in one direction or in another. At a first sight, air circulation from the top to the bottom might seem more logical to guarantee a higher oxygen supply to the upper layers of the trickling filter, where a more active biochemical process occurs since the sewage there contains all its original organic load. In practice, the oxygen amount supplied by the air circulating in the trickling filter is in high excess in respect of that required by the aerobic biochemical processes.

In fact, a cubic meter of air weights 1294 g at 0°C and at a pressure of 760 mm, in the dry state (i.e. without steam). Hence, the 209.4 ℓ of oxygen that is present in 1 m^3 of air weight 300 g. Under average temperature and pressure conditions, it may be assumed that 1 m^3 of air weights 1250 g and that the oxygen content is of 280 g. Let us suppose that the sewage brought on the trickling filter (after primary sedimentation) has BOD_5 of 200 mg/ℓ O_2 (section 2.1), i.e. BOD_{20} = 1.46 x 200 = 292 mg/ℓ O_2 = 292 g/m^3 O_2. Hence, in theory, every m^3 of sewage percolating through the bed nearly requires the oxygen supplied by 1 m^3 of air (292 g of O_2

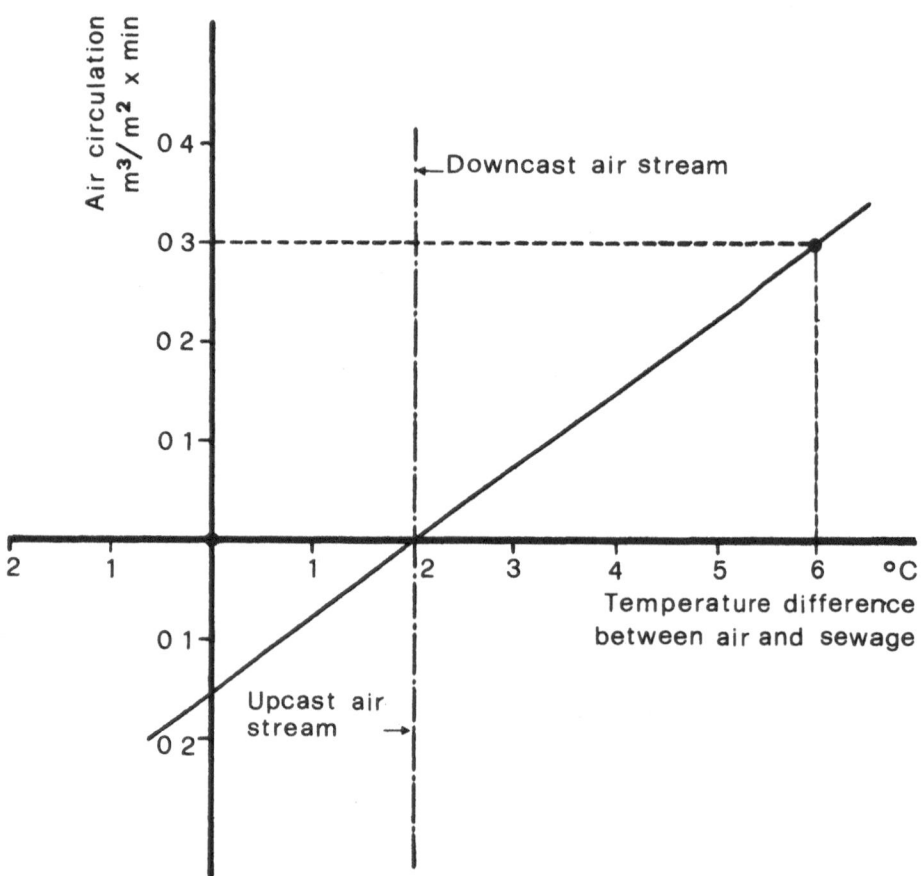

Fig. 14: Conditions of Natural Air Circulation in a
Trickling Filter as a Function of the Difference in Temperature between the External
and Internal Temperature.

necessary for 1 m³ of sewage against 280 g of O_2 supplied by 1 m³ of air).

In the practice, air circulation is far more intense than that theoretically required by the amount of sewage passing through the percolating filter. In fact, let us suppose that the hydraulic load of the percolating filter is 0.8 m³/m² x h (equal to 19,3 m³/m² x day; this is a quite high hydraulic load and corresponds to intensive filter capacity); in the case of a 6°C temperature difference between the outer athmosphere and the sewage *(Fig. 8)* the air amount would be 0.3 m/min = 18 m³/m² x h, with an air/sewage circulation ratio = 18/0.8 = 22,5. This means that 22 m³ of air circulate in each m³ of sewage and that the oxygen amount is 22,5 times higher than theoretically required (hence the oxygen amount used in only 5% the available one). Such conditions represent a considerable safety coefficient and in any case guarantee a sufficient air stream even when the temperature difference is very low.

This example shows how even with very high loads (high-rate percolators) the natural ventilation is enough to supply to oxygen required.

Forced ventilation is justified only under particular conditions (very deep filters, exceptionally high loads). However, it is quite seldom applied. More frequently, it is adopted when filters are covered because of a number of reasons (fly control and bad odour elimination). In quite cold climates, forced ventilation with heated air may be convenient to increase the trickling filter temperature or prevent to present ice formation.

Forced ventilation is carried out by fans that either blow in or suck air in the trickling filter. In the case of covered filters, air is generally blown in the upper part.

Fans must be dimensioned in such a way as to guarantee an air stream through the filter of 18 m/h [17] (this value has already been reported in the above example). Such a capacity guarantees excess air (oxygen) even in the case of high-rate filters with a particularly high load.

The need of such a high amount is justified by the need of guaranteeing air stream distribution over the whole bed section and this involved the need of foreseeing bed covering (blowing in from the top). Hence the increase in the percolator construction cost is not negligible. However, such a covering is also useful for fly control and bad odour elimination.

Reversal fans should be preferably applied (either with a plenum or vacuum effect depending upon the cases). In the case of simple-action fans, the conditions would periodically occur of forced

184

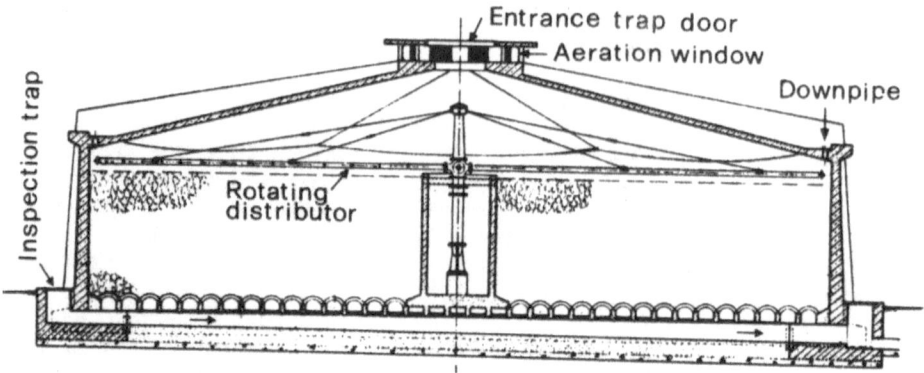

Fig. 15: Example of a Covered Trickling Filter
with Natural Air Draft; section.

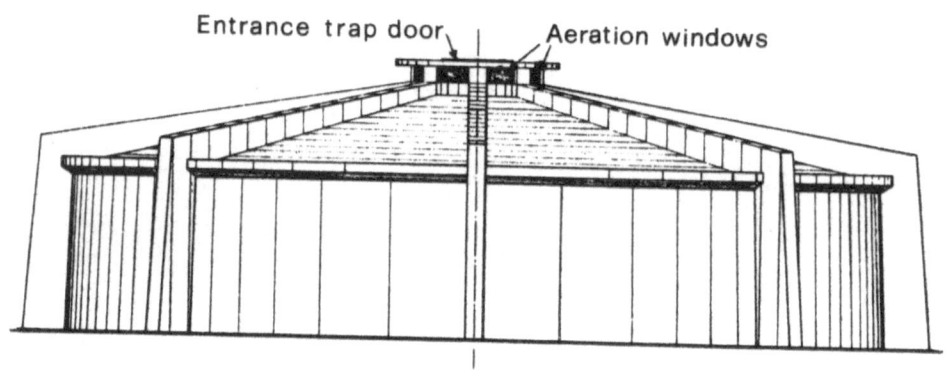

Fig. 16 : Example of a Covered Trickling Filter
with Natural Air Draft; view.

air flow with a direction opposite to that of natural ventilation.

In any case, forced ventilation is seldom applied; plants of this type are quite few in the U.S. and are being adopted ever less frequently; instead, natural ventilation is extensively adopted in very many percolating filters presently in operation. On the contrary, covered percolating filters equipped with forced ventilation have been applied to the recent Berne plant, (Switzerland). These trickling filters have height of 4.50 m and diameter of 36 m (a surface of ca. 1000 m^2) with a maximum hydraulic load in dry weather of 40 m^3/m^2 x day. Hence they have a considerable size and loads and consequently particular exigencies.

7. THE PERFORMANCE of TRICKLING FILTERS

The performance of trickling filters is extremely simple; this characteristic makes them particularly fit for small plants, where a convenient control of the quite complex mechanical equipments can be hardly done owing to the lack of specialized staff. The hydraulic winch is the only element in motion, which does not require a particular maintenance in addition to internal cleaning and lubrication.

However, trickling filters may cause some inconveniences (clogging, bad odours, flies): hence, they must be continually and conveniently controlled.

Clogging of a trickling filter occurs when the spaces among the various filtering media are completely filled up with the biological material constituting the film. Two may be the reasons for such an inconvenience:

- too small or insufficiently uniform filtering media
- a too high organic load on the filter in respect of the hydraulic load (too high BOD).

In order to obviate such an inconvenience, various possibilities exist: [18]

- to rake the bed surface, by turning the upper layer as much as possible;

- to wash the bed surface with water jets under pressure;

- to stop the winch so that one of the rotating distributor arms stops in correspondence of the obstructed area and to abundantly let the sewage flow down in order to remove the biological film;

- to introduce high chlorine concentration into the bed (5 mg/ℓ)

free chlorine in the percolator influent) for several hours
at weekly intervals. Such an operation should be carried
out at the minimum inflow times (possibly at night) in order
to reduce chlorine consumption;

- to leave the filter out of use for one day or more in order
  to dry it;

- to submerge the bed for 24 hrs at least. Such an operation
  may be carried out if the holes for air passage and drainage
  canals are equipped with convenient sluice gates; such a device is quite expensive and generally little applied;

- to replace the filtering media if the previous methods do not
  give positive results. It is usually cheaper to replace the
  filtering media rather than to completely clean the old one.

Another cause of clogging may be the presence of trees around
the filters; the leaves falling on the bed may obstruct the surface
layer. This inconvenience may be definitively overcome by avoiding
caducous-leaved trees.

Bad Odour may be caused by the anaerobic decomposition of both
sewage and biological film that may take place in the bed. In order to avoid such an inconvenience, the sewage must be kept under
non septic conditions; this problem may arise even in the sewer
line and not only in the various treatment plant units, the primary
sedimentation inclusive. This inconvenience may be corrected by
preventing an excessive growth of the biological film; that is, the
organic load will be reduced by increasing the hydraulic load by
recirculation. The phenomenon may also be controlled by chlorinating the percolator influent, preferably at minimum flow times, due
to the above mentioned reasons.

A further inconvenience frequently occurring in trickling filters consists in the development of winged insects, in particular
the midges *Psychoda,* which may abundantly develop on trickling filters, especially in summer. The superficial biological film of the
filter shelters the larvae, which then develop favoured as they are
by the alternance of wet and dry conditions. Hence the phenomenon
occurs more easily in the intermittently fed filters and it is less
frequent in high-rate filters, at least compared with its occurrence
in low-rate filters. Covered filters avoid this phenomenon.

When larvae, such metazoa are useful since they pierce the biological membrane thus giving their contribution to make aeration
easier. When insects, they constitute a troublesome inconvenience
both for the plant operators and for the possible inhabitants in
the neighbourhood. Their range of action is rather limited; according to the experience acquired at the Foggia experimental station,
(Italy) midges depart from the bed only by a few meters. However,

the wind may convey them even farther away. Their living cycle ranges from 22 days at 15°C, to 7 days at 40°C. They may be controlled:

- by feeding the filter continuously rather than intermittently;

- by removing the excessive development of the superficial biological film by the same methods as those indicated to control clogging phenomena;

- by submerging the bed for 24 hours at weekly or biweekly intervals;such a measure may be adopted when the lower bed openings are fitted with water-tight seal devices; such a measure is quite expensive and generally little applied;

- by chlorinating the sewage (0.5 - 1 mh/ℓ) for some hours, at one or two week intervals, in order to avoid that the insects may complete their cycle between one treatment and the other;

- by using insecticides.

In some exceptional cases, the *atmospheric temperature* may affect the filter performance by freezing the outerbed surface and the sewage in the distributors. Such a phenomenon may occur in very cold climates and in case of exceptional events. Such inconveniences may be controlled by different methods:

- by reducing the residence time in primary sedimentation tanks; this may be done when tanks are more than one. Some of them are eliminated and thus the exposure time of the sewage to the atmosphere before its immission into the filter is reduced;

- if it is a recirculation system, by reducing or completely eliminating the recirculation capacity, even with the purpose of reducing the detention time of the sewage contacted with the atmosphere and hence its gradual cooling

- by operating filters in parallel rather than in series (due to the same reason as in the previous item);

- by covering filters *(Fig.'s. 15 and 16)*: tests have shown that even at temperature of -47°C, ice formation is avoided if filters are covered;

- by continuously and not intermittently, distributing the sewage on the filter;

- by frequently removing ice formations;

- by adopting forced circulation with pre-heated air.

REFERENCES

1. McKinney R.E.: *Microbiology for Sanitary Engineers,* McGraw Hill Book Co., 1962.

2. Klein, L.: *Aspects of River Pollution,* Butterworths Scientific Publications, London, 1957.

3. Imhoff, K. and K.: *der Stadtentwässerung*; R. Oldenburg Verlag, 1972.

4. Ippolito, G.: Fognature in Cremonese, *Manuale dell'Ingegnere Civile,* Ed. Perrella.

5. Eckenfelder, W.W., O'Connor, P.J.: *Biological Waste Treatment* Pergamon Press, Inc. 1961.

6. Fair, G.M., Geyer, J.C.: *Water Supply and Wastewater Disposal* John Wiley & Sons, New York, 1956.

7. Federation of Sewage and Industrial Wastes Association: *Units of Expression for Wastes and Waste Treatment*; Manual of Practice N° 6, Sewage and Industrial Wastes, Vol. 30, p. 706.

8. Texas Water and Sewage Works Association: *Manual for Sewage Plant Operators,* Lancaster Press, 1957.

9. Upper Mississippi River Board of Public Engineers – Great Lakes Board of Public Health Engineers: *Standards for Sewage Works,* 1964.

10. Tonolli, V.: *Introduzione allo Studio della Limnologia,* Edizioni dell'Institute Italiano di Idrobiologia, Verbania Pallanza, 1964.

11. Eden, G.E., Progress and New Development in the Field of Biological Filtration, *Schweizerische Zeirschrift für Hydrologie,* XXVI, 2.

12. Burrows, W., *Textbook of Microbiology,* Ed. W.B. Saunders Co. 1959.

13. Subcommittee Report on Toxicity, Committee on Research FSIWA: Review of Literature on Toxic Material Affecting Sewage Treatment Processes, Streams and BOD Determinations, *Sewage and Industrial Wastes,* 22, pp. 1157 + 1191.

14. American Society of Civil Engineers, Filtering Materials for Sewage Treatment Plants, *Manuals of Engineering Practice,* New York, 1937.

15. Federation of Sewage and Industrial Wastes Association: *Sewage Treatment Plant Design,* Washington. 1959.

16. British Standards Institution, *Media for Biological Percolating Filters,* British Standards, 1438÷1948.

17. Halvorson, H.P., Savage, G.M., Piret, E.L.: Some Fundamental Factors Concerned with the Operation of Trickling Filters, *Sewage Works J.*, VIII, pp. 888+9]4.

18. Federation of Sewage and Industrial Wastes Association: *Operation of Wastewater Treatment Plants*, Washington, 1959.

OXYGEN TRANSFER AND AERATION EQUIPMENT SELECTION

Davis L. Ford, Ph.D., P.E.

Senior Vice President,
Engineering-Science, Inc.

DISCUSSION OF THEORY AND CONCEPTS

The supply of oxygen to an aerobic biological treatment system is one of the critical aspects of proper design and operation. Oxygen, which is a sparingly soluble gas in water, is transferred from the gas phase to the liquid phase by diffusion and convection to a concentration in accordance with Henry's Law,

$$C_s = k_s P \qquad \dots (1)$$

where

C_s = saturation concentration of oxygen in water,

k_s = proportionality constant,

P = partial pressure of oxygen in the gas phase.

Under turbulent flow conditions assuming that the resistance of the liquid film controls oxygen transfer rate, transfer of oxygen from the gas to the liquid phase is a function of the overall transfer coefficient, $K_L a$, and the oxygen deficit:

$$\frac{dC}{dt} = K_L a (C_s - C) \qquad \dots (2)$$

The rearranged and integrated form, converted to $\log_{10}$ yields:

$$K_La = \frac{2.303 \ log_{10}(\frac{C_s - C_1}{C_s - C_2})}{t_2 - t_1} \quad \quad \dots (3)$$

where

    C =   concentration of D.O. at any time t

    K_La =   overall mass transfer coefficient (time^{-1})

    C_s =   saturation D.O. concentration of the liquid at a specific temperature, barometric pressure, and salinity

    C_1 =   D.O. concentration at time t_1

    C_2 =   D.O. concentration at time t_2

The overall mass transfer coefficient, K_La, can be physically measured as described in subsequent sections of this treatise, and includes the effects of changes in the liquid film coefficient, K_L, and the interfacial area per unit volume, A/V. The liquid film coefficient as defined by Danckwerts [1] and Higbee [2] is the square root product of the molecular diffusion coefficient, D_L, and the rate of surface renewal:

$$K_L = \sqrt{D_L r} \quad \quad \dots (4)$$

The surface renewal rate, r, is the average frequency with which the interfacial film is replaced with liquid from the body of solution. In the turbulent regime which prevails in most aerobic biological systems, the oxygen transfer rate is a function of surface renewal. The overall mass transfer coefficient is therefore equal to:

$$K_La = K_L A/V \quad \quad \dots (5)$$

and can be determined in the laboratory or field using Equation (3). From Equation (5), it is obvious that K_La is a function of K_L, which is dependent on the surface tension and molecular characteristics which prevail at the gas-liquid interface of a given fluidized system, and A/V , which depends on the turbulence and bubble patterns in an aeration system.

One of the most significant factors which effect K_La based on diffusivity and viscosity is temperature. This temperature effect can be defined by the relationship:

$$K_La_{(t)} = K_La_{(20°)}\theta^{(T-20)}$$

The θ value has been reported to vary from 1.016 to 1.037 [3,4]. Values of 1.020 to 1.028 are normally used for bubble systems, while a value to 1.024 is considered to be correct for mechanical aeration systems [5].

When oxygen is supplied to fluidized systems treating wastewaters via aerobic biological oxidation, it is necessary to define a correction factor which relates the oxygen transfer to the nature of the waste. Using the transfer of oxygen to tap water as the datum, the parameter, alpha, serves as this correction factor. Specifically, it related the overall mass transfer coefficent (K_La) of the wastewater to that of tap water:

$$\alpha = \frac{K_La \text{ wastewater}}{K_La \text{ tap water}}$$

Although α is represented in the mathematical relationships describing diffused, turbine, and surface aeration system performance, there are many variables which affect its magnitude. These include:

1. temperature of the mixed liquor,
2. nature of dissolved organic and mineral constituents,
3. characteristics of the aeration equipment (diffused or mechanical),
4. the liquid mixing intensity which affects the surface renewal rate, and
5. the liquid depth and geometry of the aeration basin.

The temperature effect is attributable to the temperature dependence of the liquid film coefficient (K_L). Alpha (α) might be expected to increase, decrease, or approach unity during the course of biological oxidation since dissolved organic material affecting the transfer rate is being removed in the biological process. Alpha generally increases with the degree of treatment as shown in Figure 1.

The mixing intensity affects the magnitude of alpha, particularly using surface aeration systems as indicated in *Fig. 2* [6]. Under quiescent or moderately turbulent conditions, the presence of surface active agents inhibit molecular diffusion of oxygen through the gas-liquid interface and there is a decrease in surface

renewal. However, α may approach or even exceed unity at high
mixing intensities as high surface renewal and increased surface
contact areas are associated with turbulent conditions. Determination of α based on bench scale experiments is only an approximation because of the difficulty of accurately representing field
mixing conditions in the laboratory reactor. It is difficult to
scale up diffused or mechanical aeration equipment operating
characteristics from bench scale to full scale. Recognizing
these constraints, a bench or pilot-scale determination of α
does give an indication of its magnitude and can be used to
estimate diffused, turbine, or surface aeration requirements
for an aerobic biological system.

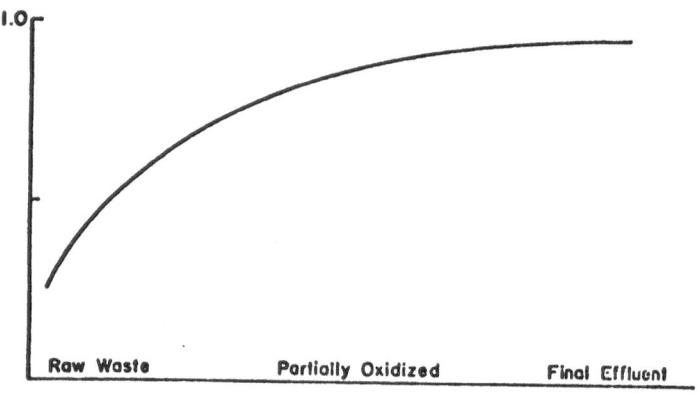

Fig. 1 : Degree of Biological Stabilization

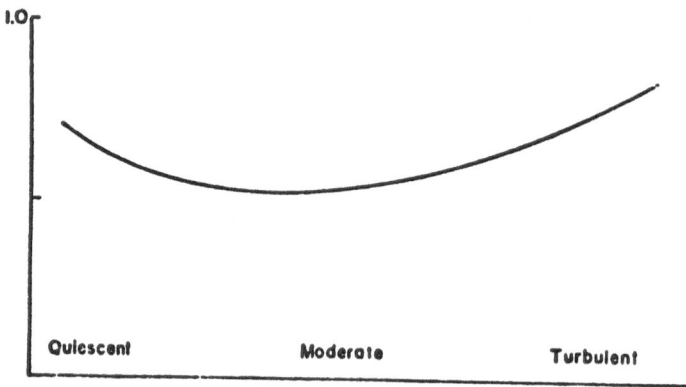

Fig. 2 : Liquid Mixing Intensity

AERATION SYSTEMS

For the purpose of this discussion, aeration equipment is categorized into three systems, namely: diffused aeration, turbine aeration, and mechanical aeration. These systems are illustrated schematically in Figure 3 and discussed as follows.

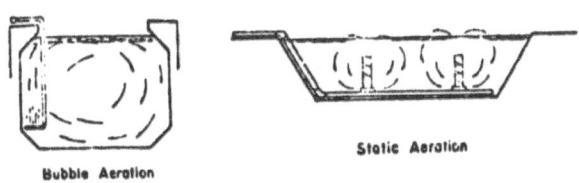

Bubble Aeration

Static Aeration

(A) Diffused Aeration Systems

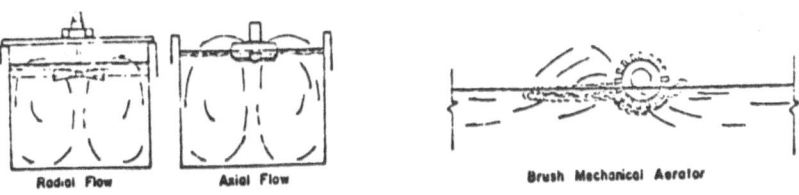

Radial Flow　　**Axial Flow**　　　**Brush Mechanical Aerator**

(C) Mechanical Aeration Systems

Fig. 3 : Aeration Systems

DIFFUSED AERATION SYSTEMS

Bubbler Aeration: Diffused aeration devices are commercially available in several basic types. Small orifice devices are constructed of silicon dioxide or aluminum oxide held in a porous mass with a ceramic binder, or in tubes or bags wrapped with Saran or nylon material. The size of bubbles released from this type of diffuser ranges from 2.0 to 2.5 mm. The absorption efficiency (oxygen absorbed/oxygen supplied) depends on the size of the air bubbles released and the turbulence generated in the system. Another type of air diffuser uses a large orifice device such as the sparjer. The sparjer contains four short tube orifices at 90° centers from which the air is emitted at high velocity. Tank turbulence tends to redivide large bubbles into smaller bubbles. The orifice diameters normally range from 0.508 to 0.635 cm. Other commercial units in this category include the hydraulic shear diffuser, the Venturi diffuser, and the INKA diffuser.

There are several factors which affect the transfer characteristics of diffusion units. The transfer efficiency is first a function of the gas flow as indicated in the following equation:

$$N = CG_s^n \qquad\qquad \dots (8)$$

where

N = oxygen transfer efficiency, Kg O_2/hr/unit

G_s = air flow rate, standard cm^3/sec

C, n = constants which are a function of the diffuser equipment design

The effect of the liquid depth above the diffusers and the width of the tank on transfer efficiency has been previously assessed and is applied to Equation (8):

$$N_o = CG_s^n \; (H^m/W^P) \qquad\qquad \dots (9)$$

where

H = depth of aeration tank

W = width of aeration tank

P = width correction exponent ($\simeq 0.36$)

m = depth correction exponent ($\simeq 0.72-0.88$)

Incorporating the oxygen deficit or driving force, the temperature correction, and the transfer correction coefficient, Equation (9) becomes:

$$N = CG_s^n (H^m/W^P) \; (\frac{C_{sm} - C_L}{C_s}) \; \theta^{T-20} \; \alpha \qquad\qquad \dots (10)$$

where

C_{sm} = saturation concentration of dissolved oxygen at tank mid-depth, mg/ℓ

C_L = operating dissolved oxygen concentration in aeration tank, mg/ℓ

c_s = saturation dissolved oxygen concentration at 20°C, one atmosphere and pure water, mg/ℓ

The c_{sm} value can be calculated according to:

$$c_{sm} = \frac{c_{sw}}{2} \left(\frac{P_b}{P_o} + \frac{O_t}{O_o} \right) \qquad \dots \ (11)$$

where

P_o = pressure at sea level = 1 atm

O_o = % O_2 in air = 21%

c_{sw} = saturation concentration of dissolved oxygen in wastewater, mg/ℓ

P_b = absolute pressure at the depth of air release, atm

O_t = oxygen in the exit gas, percent

Reported oxygen transfer efficiencies of bubbler diffused aeration systems range from 0.55 to 0.91 Kg O_2/Kw-hr while levels of 1.22 to 1.83 Kg O_2/Kw-hr have been reported for static diffuser systems.

Static Aeration: Static aeration systems consist of vertical cylindrical tubes placed at specified intervals in an aeration basin and containing fixed internal elements. A central compressor supplied an air source through a sparjer at the bottom of the tubes, and an air-water mixture is forced through the cylinder. There is an air-water contact as the mixture travels up through the elements within the static mixer cylinder. Most of the oxygen transfer occurs within the cylinder, although there is additional transfer at the surface turbulent area where the air-water mixture is discharged from the cylinder. The transfer efficiency of static aeration systems is reported to be higher than that obtained from the conventional diffused air-sparjer or orifice system 7 . The exact transfer obtained using a static aeration system is a function of several design features, namely:

1. the bottom sparjer design;
2. diameter of the cylinder;
3. length of the cylinder;
4. the air flow rate;
5. the liquid submergence;
6. the imported liquid velocity and mixing level; and,
7. the design of the fixed elements and associated pressure drop through the cylinder.

The air flow per mixer cylinder normally varies from 5.70 to 14.20 ℓ/sec with a delivery air pressure in the range of 0.70 to 1.05 Kg/cm².

To date, static aeration systems have been used primarily in aerated lagoons carrying relatively low concentrations of biological suspended solids. They have several advantages in this application, including low annual costs and relatively high transfer efficiencies. They have also been applied as mixers in small mechanical or neutralization rapid mix tanks.

TURBINE AERATION SYSTEMS

In turbine aeration, air is discharged from a pipe or sparge ring beneath the rotating blades of an impeller. The air is broken into bubbles and dispersed throughout the tank contents. Present commercial units employ one or more submerged impellers and may utilize an additional impeller near the liquid surface for oxygenation from induced surface aeration. In addition to air flow the diameter and speed of the impeller will affect the bubble size and velocity, thus influencing the overall transfer coefficient, $K_L a$. The transfer equation for turbine aeration systems is:

$$N = C G_s^n \, d_t^y \, R^x (C_{sw} - C_L) \theta^{(T-20)} \alpha \qquad \dots \dots (12)$$

where

d_t = impeller diameter, meters

R = impeller peripheral speed, meters

y, x = exponents

A significant correlation between oxygen transfer efficiency and the power supplied to the system from the rotor (Kw_R) and the compressor (Kw_C) has been demonstrated [8];specifically:

$$N = C \, P_d^n \qquad \dots \dots (13)$$

where

P_d = power split, Kw_R/Kw_C

C, n = constants

It was further shown that oxygenation efficiency can be related to P_d by differentiating the oxygenation efficiency with respect to P_d, equating the differential to zero, and solving for the optimum power distribution. The value is a function of the exponent "n" in Equation (13).

$$P_d^* = (\frac{n}{1 - n}) \qquad \qquad \text{.... (14)}$$

in which P_d^* is the power distribution for the optimum oxygenation efficiency.

In most cases, P_d^* occurs near 1.0. (This implies an equal power expenditure by the turbine and the compressor.) At extremely high air rates ($P_d \ll 1.0$), large bubbles and flooding of the impeller yield poor oxygenation efficiencies, while at very low rates ($P_d \gg 1.0$), too much turbine horsepower is being expended in fluid mixing without associated oxygen transfer.

Variation in oxygen demand in the system can most easily be adjusted by varying the air rate under the impeller. This in turn will change P_d. The anticipated range of operation should cover the maximum range of oxygenation efficiency as related to P_d.

Available data indicate that the oxygenation efficiency of turbine aerators in water should vary from 0.91 to 1.83 Kg O_2/Kw-hr (including motor and reducer losses).

MECHANICAL AERATION SYSTEMS

Mechanical aerators have become increasingly popular in recent years, particularly in industrial treatment applications. There are three basic classifications of mechanical aerators; specifically, the radial flow slow speed aerators, the axial flow high speed unit, and the brush mechanical aerator. A brief discussion of each follows.

The radial flow, slow speed aerator is essentially a low head high volume pump. Assuming an exit liquid velocity of 2.44 m/sec and impeller submergence of 0.15 m, for example, the total dynamic head would be only 0.46 m. The volume pumped per unit kilowatt (nameplate) is a function of the motor size as indicated in Figure 4, and ranges from 76 to 570 ℓ/sec/Kw [9]. Oxygen transfer is accomplished primarily through entrainment associated with an induced hydraulic jump. The primary component of these units are the motor, gear reducer, and impeller. The impellers vary in diameter from 0.90 to 3.66 m, and motor sizes ranges from 2.24 to 112 Kw.

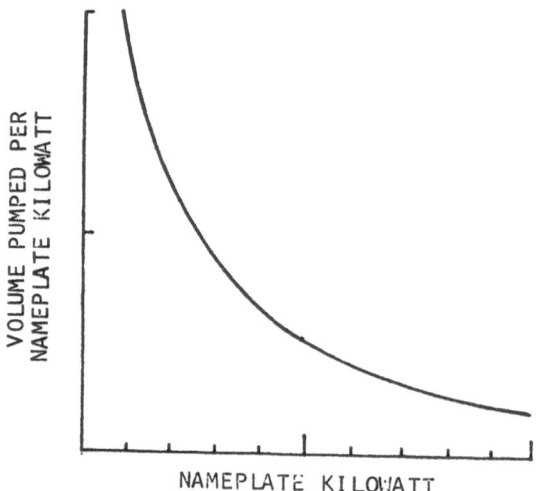

Fig. 4 : Relation Between Volume Pumped per Unit
Kilowatt and Motor Size in Radial Flow
Slow Speed Aerators

The output speed ranges from 30 to 60 rpm. These aerators are
generally capable of mixing the contents of large aeration basins
varying in depth from 0.90 to 5.50 m. This depth can be increased
by utilizing draft tubes or increasing the impeller submergence
depth. These units usually are affixed to bridges or platforms.

The axial flow, high speed aerator is widely used and consists
of a motor and propeller assembly which is usually float-mounted.
The major components are a non-submersible motor, riser tube,
fiberglass or stainless steel float, and propeller deflector.
They have a lower pumping capacity per unit horsepower than the
radial flow units, but impart a higher velocity to the liquid
(3.66 to 5.50 m/sec). As the radial flow speed units, they are
low head, high volume pumps working at a total dynamic head of
0.90 to 1.80 m. Most of the oxygen transfer occurs in the spray
pattern, although transfer also occurs in the turbulent area in
the outer peripheral area of the spray. These units range in size
from one to 112 Kw, and the rotational speed is in the 900 to
1,400 rpm range. The liquid depths required for proper operation
vary from 0.90 to 4.60 m, depending on the aerator size.

The brush aerators, which are the mechanical aeration devices
used in aeration ditches, transfer oxygen from the gas to the
liquid phase in a manner similar to that of the aforementioned
mechanical aerators; namely, through spray contact and air entrainment. These units rotate in the 30 to 60 rpm range with a wide
range of reported oxygen transfer efficiencies. The exact efficiency depends on the type of brush, rotational speed, submergence,

and aeration conditions.

The oxygen transfer efficiencies of mechanical aeration systems will depend on the inherent design of the equipment, such as the impeller diameter and configuration, deflector plate, and speed and submergence depth of the rotating element. The reported efficiencies have ranged from 1.5 to 2.5 Kg O_2/Kw-hr, athough 1.52 to 1.83 Kg O_2/Kw-hr is probably a more accurate range.

The traditional design equation for mechanical aerators is:

$$N = N_o \left(\frac{C_{sw} - C_L}{C_s}\right)(1.024^{(T-20)})\alpha \qquad \dots \quad (15)$$

where

N = oxygen transfer efficiency, field conditions, Kg O_2/Kw-hr

N_o = oxygen transfer efficiency, test tank conditions, Kg O_2/Kw-hr

If the power level (Kw/1000 m³) is shown to affect N_o during the test series, as is the case in Figure 5, then:

$$N_o = K + S(P.L.) \qquad \dots \quad (16)$$

where

K = intercept (N_o at an infinitely low power level)

S = slope

$P.L.$ = power level, usually expressed as Kw/1000 cubic meters

and Equation (15) must be modified accordingly. This, of course, requires a trial and error solution by assuming a power level and using the corresponding N_o value to calculate N. If the resultant power level is different than the first assumed, then the calculation must be repeated until the two values coincide [10]. It should be noted that Figure 5 applies to one aerator type only and cannot be applied for all designs. The influence of power level on transfer efficiency will depend on aerator spacing, tank geometry, and overall mixing and circulating patterns.

A generalized summary of the characteristics and applications for the various types of aeration equipment as discussed herein is given in Table 1 [11].

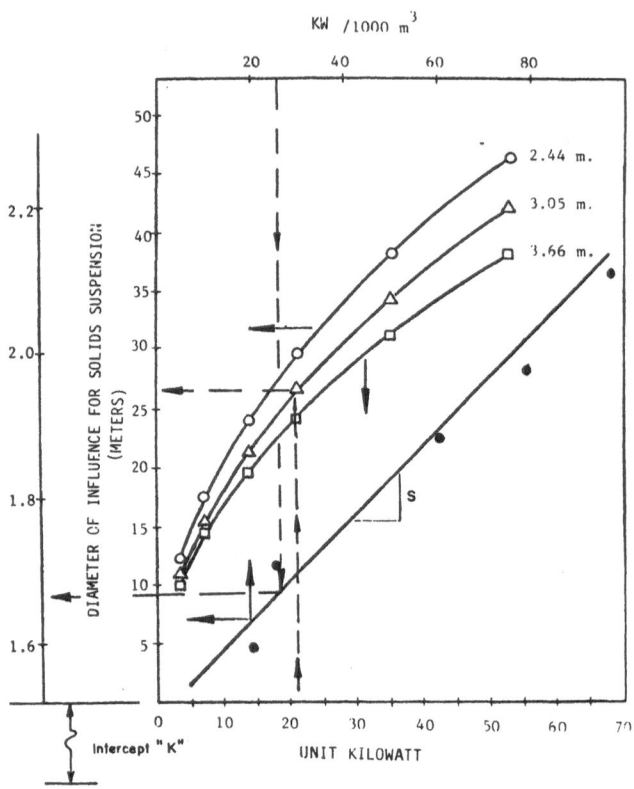

Fig. 5 : Surface Aerator Characteristics

EVALUATION AND TESTING OF AERATION SYSTEMS

The need for accurately evaluating aeration equipment in terms of oxygen transfer capability is well recognized both by manu-facturers and by design engineers. The three most commonly accep-ted techniques used to determine transfer efficiency of aeration equipment are:

1. the non-steady state reaeration of deoxygenated "pure" water in a test tank or actual aeration basin;
2. the steady state evaluation in an activated sludge basin; and,
3. the non-steady state evaluation in an activated sludge basin.

Each of these procedures and their limitations are discussed as follows.

TABLE 1 : Characteristics of Available Aeration Equipment

Equipment Type	Equipment Characteristics	Processed Where Used	Advantages	Disadvantages	Reported Transfer Efficiency[*] (Kg O_2/Kw-hr)
Diffused Aeration: (Bubbler)					
Porous Diffusers	Produce fine or small bubbles. Made of ceramic plates or tubes, plastic-wrapped or plastic-cloth tube or bag.	Large, conventional, activated sludge process.	High Oxygen transfer efficiency; good mixing; maintain high liquid temperature. Varying air flow provides good operational flexibility.	High initial and maintenance costs; tendency to clog; not suitable for complete mixing.	0.30-0.91
Nonporous Diffusers	Made in nozzle, valve, orifice or shear types, they produce coarse or large bubbles. Some made of plastic with check-valve design.	All sizes of conventional activated sludge process.	Nonclogging; maintain high liquid temperature; low maintenance cost.	High initial cost; oxygen trasfer efficiency; high power cost. Clogging occurs.	
(Static)	Produces high shear and entrainment as water-air mixture is forced through vertical cylinder containing static mixing elements. Cylinder construction is metal, plastic, or polyethylene.	Primarily aerated lagoon applications.	Economically attractive; low maintenance; high transfer efficiencies for diffused air systems. Well suited for aerated lagoon application.	Ability to adequately mix aeration basin contents is questionable. Application for use in high rate biological systems unconfirmed.	1.22-1.83
Mechanical Aeration: Radial Flow	Low output speed; large diameter turbine, usually fixed-bridge or platform mounted. Used with gear reducer.	All sizes of conventional, activated sludge and aerated lagoon processes.	High Oxygen transfer efficiency; tank design flexibility; good transfer efficiency. High pumping capacity.	Some icing in cold climates. Initial cost higher than axial flow aerators. Gear reducer often causes maintenance problems.	
Axial High Speed	High output speed. Small diameter propeller. They are direct, motor-driven units mounted on floating structure.	Aerated lagoons and activated sludge processes.	Low initial cost; simple to install and operate; good transfer efficiency; adjust to varying water level. Flexible operation.	Some icing in cold climates; poor maintenance accessibility.	1.14-2.13
Brush Aeration	Low output speed, with gear reducer.	Oxidation ditch applied either as an aerated lagoon or as an activated sludge process.	Relatively low initial cost, easy to install and operate; good maintenance accessibility, moderate transfer eff.	Subject to operational variables which may affect efficiency.	
Turbine Aeration:	Units contain a low speed turbine and provide compressed air on sparge ring. Fixed-bridge application.	Conventional, activated sludge process.	Good mixing; high capacity input per unit volume; deep tank application; moderate efficiency, wide oxygen input range; operational flexibility.	Require both gear reducer and compressor; tendency to foam; high total power requirements.	0.77-1.46

[*]Test conditions and procedures not documented.

NON-STEADY STATE REAERATION OF DEOXYGENATED WATER

This is the most common method of evaluating aerator transfer performance and has generally been adopted as a standard. It involves reaeration of "pure" water which has been deoxygenated by the addition of sodium sulfite and a cobalt catalyst. This reaction stoichometrically requires 7.9 ppm of Na_2SO_3 per ppm of dissolved oxygen although a 20 to 30 percent excess is normally used. Concentrations of the cobalt ion should be a minimum of 0.5 ppm and should not exceed 1.5 ppm. The predissolved chemicals are added to the test basin and the aeration equipment is turned on to mix the chemical-water mixture until the dissolved oxygen is completely depleted. Once this condition prevails, the test for reaeration can be initiated. The aeration equipment is started and dissolved oxygen levels at various points in the basin are monitored. Both an oxygen probe and a Winkler dissolved oxygen analysis should be used. If a discrepancy between the Probe and Winkler data exists, then the Winkler analysis can be considered to be the more accurate.

Reaeration should continue until 75 to 90 percent of saturation is observed. The oxygen deficit is plotted with respect to time on a semi-log plot as shown in the example in Figure 6. The C_s value for the measured temperature of the "pure" test water can be taken from the standard oxygen saturation value shown in Table 2.

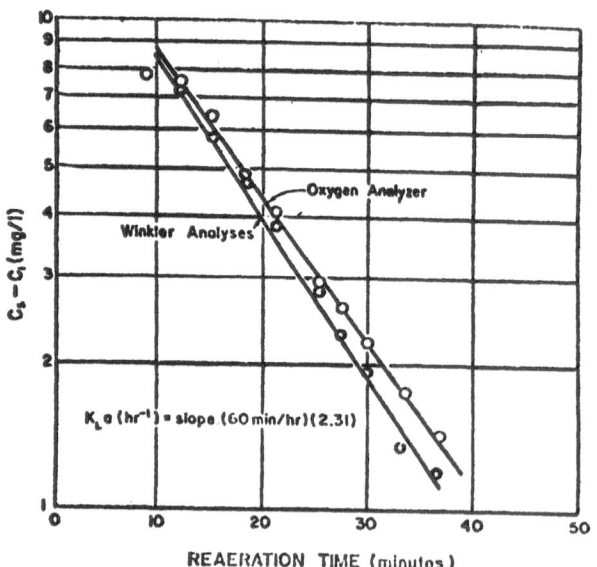

Fig. 6 : Variation of Oxygen Deficit with Time

TABLE 2: Oxygen Saturation Values (C_s) for Distilled
Water at Standard Conditions*

T° (F)	T° (C)	mg/ℓ O_2	T° (F)	T° (C)	mg/ℓ O_2
32.0	0	14.6			
33.8	1	14.2	78.8	26	8.2
35.6	2	13.8	80.6	27	8.1
37.4	3	13.5	82.4	28	7.9
39.2	4	13.1	84.2	29	7.8
41.0	5	12.8	86.0	30	7.6
42.8	6	12.5	87.8	31	7.5
44.6	7	12.2	89.6	32	7.4
46.4	8	11.9	91.4	33	7.3
48.2	9	11.6	93.2	34	7.2
50.0	10	11.3	95.0	35	7.1
51.8	11	11.1	96.8	36	7.0
53.6	12	10.8	98.6	37	6.9
55.4	13	10.6	100.4	38	6.8
57.2	14	10.4	102.2	39	6.7
59.0	15	10.2	104.0	40	6.6
60.8	16	10.0	105.8	41	6.5
62.6	17	9.7	107.6	42	6.4
64.4	18	9.5	109.4	43	6.3
66.2	19	9.4	111.2	44	6.2
68.0	20	9.2	113.0	45	6.1
69.8	21	9.0	114.8	46	6.0
71.6	22	8.8	116.6	47	5.9
73.4	23	8.7	118.4	48	5.8
75.2	24	8.5	120.2	49	5.7
77.0	25	8.4	122.0	50	5.6

*From *Standard Methods for the Examination of Water and Wastewater* (1965).

This value can be corrected for atmospheric pressure and test
water quality differences from "pure" water "as required" according to Equation (17):

$$C_s = C'_s (P/P_o) (\beta) \qquad \qquad \text{.... (17)}$$

where

β = saturation of test water/saturation of pure water

C'_s = uncorrected D.O. saturation volume

P = water pressure

P_o = atmospheric pressure at sea level = 1.0 atm

The overall mass transfer coefficient can then be calculated as
the slope of the line in Figure 6 according to Equation (3). As
the test standard is 20°C, K_La can be corrected to this temperature using Equation (6). The oxygenation capacity (O.C.) can
then be calculated as follows:

$$\text{O.C.} = K_La_{(20°C)} (C_s) (\text{vol of test tank, m}^3) \qquad \text{.... (18)}$$

The transfer efficiency (T.E.) in terms of name plate Kilowatt
(NPKw) is then:

$$\text{T.E.} = (\text{O.C./NPKw}) \qquad \qquad \text{.... (19)}$$

The T.E. can be expressed in terms of wire to water Kw_s ($Kw_{w/w}$)
by the following calculation:

$$(Kw_{w/w}) = [\text{line voltage}][\text{line amperage}][\text{power factor}]$$

$$[\text{conversion Kw/watts}][(\text{motor eff.})(\text{gear eff.})] \qquad (20)$$

This non-steady state reaeration approach is valid only when the
entire liquid content of the test basin is completely mixed and
K_La is constant and time-independent. There are several limitations to this method of evaluation, the most obvious of which is
power level. As indicated in Figure 5, N_o is a function of power
level and because of the lack of a "standard" power level in the
aeration test, it is misleading to compare aeration equipment
rated under different conditions. The test should be conducted

at the design power level. Another effect which influences the transfer efficiency of an aerator is the basin geometry of the test tank. As the transfer efficiency is dependent on flow patterns, the exact T.E. value can be related to various geometrical parameters such as a diameter/depth ratio, wetted area/basin volume, and/or length to width ratio. This aspect limits precise translation of T.E. values determined from test tanks to application in a field basin of different configuration, even though power levels, temperature, and alpha correction factors are properly considered. The effect of increasing sodium sulfate and cobalt concentrations inherent in the non-steady state tests on T.E. values also has been reported [5, 12]. Strong evidence has indicated that sodium sulfate values above 2,000 ppm increases the T.E., probably due to a reduction in surface tension with an increasing interfacial area and $K_L a$ value. Moreover, an increase in T.E. with cobalt concentrations exceeding 1 to 1.5 ppm has been implied [13].

It is apparent that the non-steady state test has many short-comings and idiosyncrasies. However, if more standard test conditions can be established and testing procedures clearly defined and followed, it does offer an acceptable method for rating and comparing aeration equipment [14].

STEADY STATE REAERATION OF ACTIVATED SLUDGE

There are obvious incentives to field test aerators in the activated sludge basin. As the equipment is performing under actual field conditions, the problem of translating "clean water" test results to the "field situation" is circumvented. This does force, however, a delay in aerator evaluation until the plant has been put "on line" and is working properly. Nevertheless, aeration equipment can be assessed "in-situ" once a quasi-steady state condition has been established in the aeration basin. Once this is obtained, Equation (2) can be modified:

$$\frac{dc}{dt} = K_L a (C_{sw} - C) - r_r \qquad \ldots\ldots (21)$$

where

r_r = average oxygen uptake rate in the basin by the activated sludge, mg/ℓ-hr

Under quasi-steady state conditions, dc/dt approaches zero, and $K_L a$ can be determined as follows:

$$K_L a = r_r / (C_{sw} - C) \qquad \ldots\ldots (22)$$

The oxygen uptake rate should be measured at various points within the basin to insure uniformity and statistical accuracy. The $K_L a$ value obtained includes the effects of alpha, temperature, and solids concentration and is specific only to the particular wastewater being treated. It is important that the wastewater treatment system be stabilized with respect to a constant organic loading, feed composition, and basin dissolved oxygen level during the test.

NON-STEADY STATE REAERATION OF ACTIVATED SLUDGE

Another approach in evaluating $K_L a$ under process conditions is the non-steady state reaeration of activated sludge. This involves stopping wastewater flow to the system and assuming the oxygen uptake rate will become relatively constant during the ensuing 0.5 to one hour of aeration time. Within this period, the aeration is stopped for a sufficiently long period of time for the activated sludge to deplete the liquid contents of dissolved oxygen, then the system is reaerated and the oxygen level monitored. The mass transfer coefficient, $K_L a$, can be evaluated without actually measuring the oxygen uptake rate using the following approach: (Equation (21) is reexpressed as follows)

$$\frac{dc}{dt} = K_L a \, C_{sw} - r_r - K_L a \, C \qquad \dots \ (23)$$

If the dissolved oxygen level is monitored following aeration start-up as described above, then the profile would approximate that shown in Figure 7. By taking tangents at various points on the curve, dc/dt values for corresponding C levels (at the point of tangency) can be determined. These data can be replotted as shown in Figure 8, and according to Equation (23), $K_L a$, is the slope of the line and can be determined accordingly. One advantage of this approach is the fact that the oxygen uptake rate, which is difficult to accurately determine, is included in the intercept and does not have to be physically measured.

FIELD PERFORMANCE

Manufacturers of aeration equipment are normally required to furnish a guarantee of transfer efficiency. Although this guarantee, based on standard test conditions in a clean water test tank, can be easily defined, it may or may not relate to field oxygen supply requirements. For this reason, some vendors are basing performance guarantees on BOD removal and oxygen driving force in the actual basin. It is first shown that the T.E. is a function

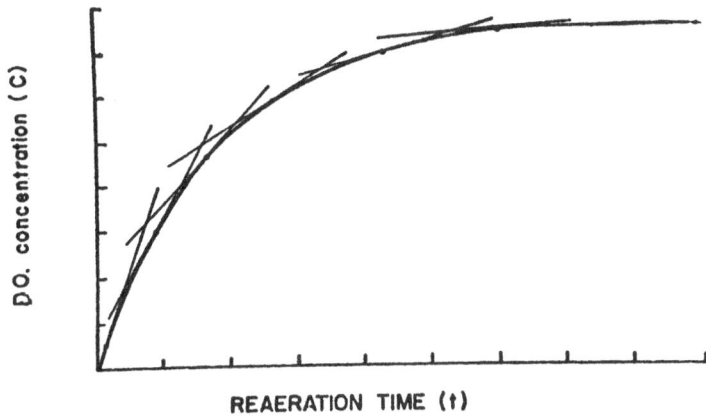

Fig. 7 : Variation of Dissolved Oxygen
Concentration with Reaeration
Time.

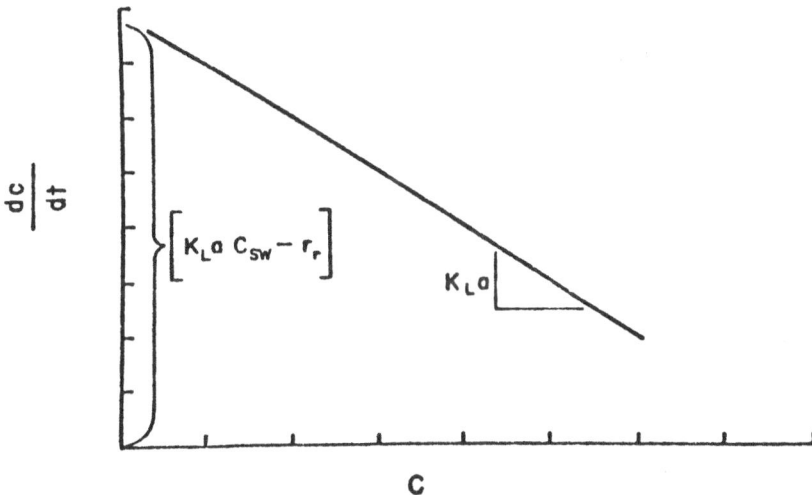

Fig. 8 : Variation of dc/dt with
Dissolved Oxygen.

of the D.O. in the aerator incoming water as depicted in *Fig. 9*
[9]. Secondly, if one assumes that the oxygen utilized is proportional to the BOD removed and constant, then:

$$X = y'N = K\Delta C \qquad \qquad \dots \ (24)$$

where

X = Kg BOD removed/hr

y' = Kg BOD removed/Kg O_2

K = constant

This relationship is shown in Figure 10 and would serve as a basis for guaranteeing process performance based on oxygen residuals. This is reasonable because in the final analysis, the residual D.O. in the aeration basin is the important process consideration. For example, a guarantee of mechanical aeration performance in terms of oxygen transfer could be specified based on the concept shown in Figure 11. However, it should be recognized that even this approach has some limitations. From a process point of view, the oxygen utilized per pound of BOD removed in fact may not be constant due to varying substrate characteristics and oxygen demanding inorganic constituents. If the operating D.O. level falls below one mg/ℓ (i.e., $\Delta c = c_{sw} - 1$), there will most probably be a drop in BOD removal efficiency because of a diminution of biochemical reaction rates at low oxygen tension. Alpha values which deviate from the design prediction would also affect this approach.

PROBLEMS IN HYDRAULIC AND GEOMETRIC SIMILITUDE

The problems of transferring aerator performance from test conditions to field conditions have been discussed previously. To truly develop similitude, one should apply the principles of fluid mechanics; specifically, the Reynolds Number, the Froude Number, and the Weber Number as defined below:

$$Re = (V_p d\rho/\mu) \quad \text{or} \quad (ND^2\rho/\mu) \quad \dots (25)$$

$$Fr = V_p^2/gd \quad \text{or} \quad N^2 d/g \quad \dots (26)$$

$$We = V_p^2 d\rho/\sigma \quad \text{or} \quad N^2 d^3\rho/\sigma \quad \dots (27)$$

where

V_p = peripheral velocity of aerator

d = aerator diameter

N   =   rpm of aerator

g   =   gravity constant

ρ =   fluid density

μ =   fluid viscosity

σ =   fluid surface tension

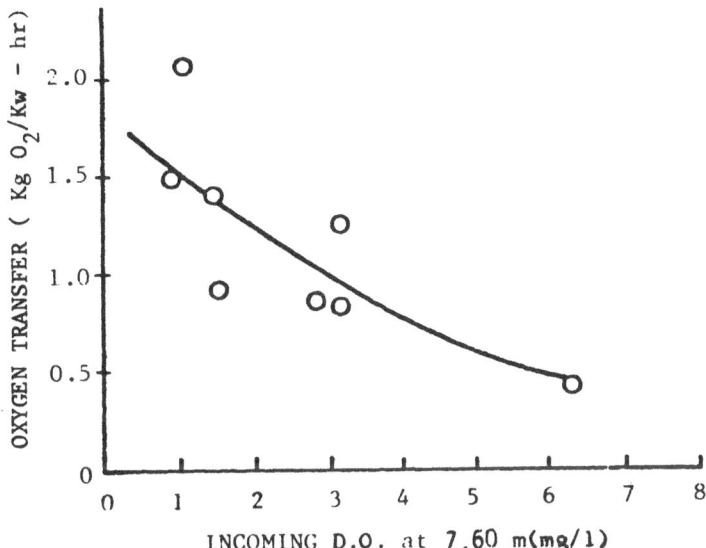

Fig. 9 : The Relation Between
Oxygen Transfer and
Incoming Dissolved Oxygen

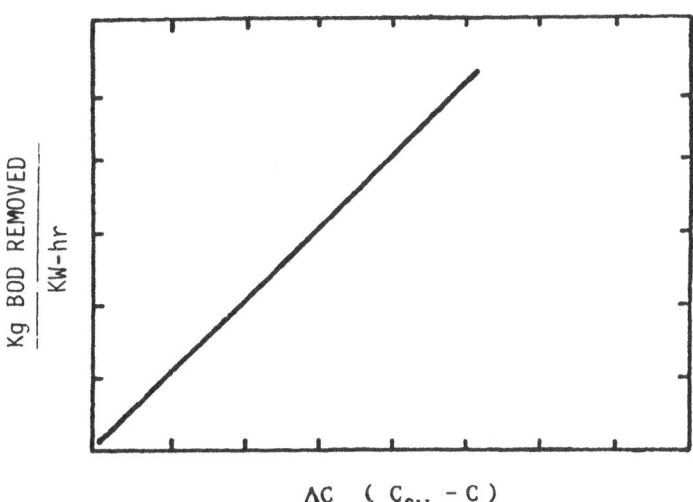

Fig. 10 : The Relation Between
$\Delta C(C_{sw}-C)$ and BOD Removed.

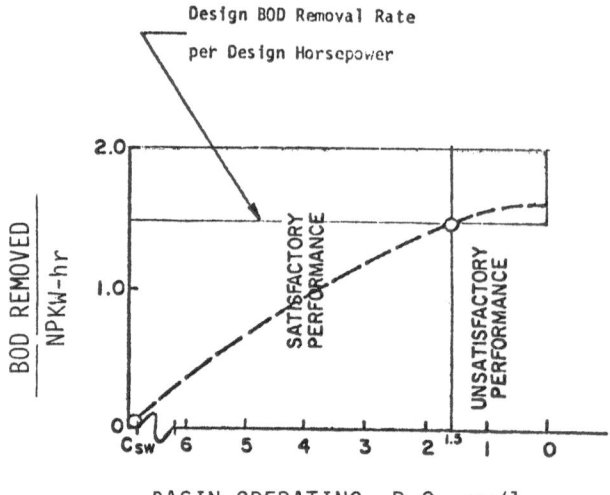

Fig. 11 : Mechanical Aeration
Performance in terms
of Oxygen Transfer

The attainment of true dynamic similarity would require that all
three numbers be the same for all aerator sizes, and geometric
similarity must prevail 12 . As this is difficult, if not impossible, to establish, empirical data such as flow patterns,
velocity magnitudes and vectors, oxygen dispersion, etc., must be
established to evaluate the similarity or dissimilarity of test
and field conditions.

REFERENCES

1. Danckwerts, P.V., Significance of liquid film coefficients
    in gas adsorption, *Ind. Engr. Chem.*, 43, 1460, 1951.

2. Higbee, Ralph, The rate of adsorption of a pure gas into a
    still liquid during short periods of exposure, *Trans. Amer.
    Inst. Chem. Eng.*, 31, 365, 1935.

3. Eckenfelder, W.W., and Barnhart, E.B., Paper presented
    Amer. Inst. of Chem. Engrs., Atlanta, Georgia, February 1960.

4. Eckenfelder, W.W., and Ford, D.L., New concepts in oxygen
    transfer and aeration, *Advances in Water Quality Improvement*,
    Univ. of Texas Press, Austin, Texas, 1968.

5. Crocker, J.D., Analysis of relative mixing abilities of
    various mechanical aerator types. Unpublished Report. 1972.

6. Eckenfelder, W.W., and Ford, D.L., Water Pollution Control -
    Experimental Procedures for Design, *Pemberton Press*, Austin,
    Texas, 1971.

7. Epstein, A.C., and Glover, C., Full scale aeration studies
    of static aeration systems, *Permutit Corporation Report of
    Kenics Corp.*, 1971.

8. Quirk, T.P., Optimization of gas-liquid contacting systems,
    Unpublished Report, 1962.

9. National Council of the Paper Ind. for Air and Stream Improve-
    ment, A study of mixing characteristics of aerated stabili-
    zation basins, *Tech. Bulletin* No. 245, 1971.

10. Eckenfelder, W.W., and Ford, D.L., Engineering aspects of
    surface aeration design, 22nd Industrial Waste Conference,
    Purdue University, 1967.

11. Nogaj, R.J., Selecting wastewater aeration equipment,
    Chemical Engineering, April 17, 1972.

12. Kalinski, A.A., Hydraulics of localized mechanical aerators
    in waste treatment basins and flowing streams, *WPCF Confe-
    rence*, Bal Harbour, 1964.

13. Conway, R.A., and Kumke, G.W., Field techniques for evalua-
    ting aerators, *ASCE Journal*, April 1966.

14. Ferrel, J.F., and Ford, D.L., Select aerators carefully,
    Hydrocarbon Processing, October 1972.

ATTACHMENT A: LABORATORY PROCEDURE FOR DETERMINATION OF THE
                OXYGEN TRANSFER COEFFICIENT (Alpha)

EQUIPMENT REQUIRED (Laboratory Evaluation of Alpha)

1. Galvanic Cell Oxygen Analyzer and Probe, properly calib-
    rated (chemicals and equipment necessary for performing
    Winkler analysis of dissolved oxygen may be used as an
    alternative).

2. Aeration source, either mechanical aeration device or
    compressed air line with diffuser. A tachometer is
    required with mechanical aerator and a rotameter or wet
    test meter would be required for diffused air system.

3. Water receptacle, with volume of two liters to 208 liters depending on desired scale of alpha evaluation test.

4. Deoxygenation equipment – nitrogen source with feed line and diffuser, or deoxygenating chemicals (Na_2SO_3 + cobalt catalyst).

LABORATORY PROCEDURE

The following procedure can be used in the bench scale determination of the oxygen transfer coefficient, α.

1. Vessel similar to those shown in Figure 12 can be used. Mechanical or diffused aeration devices can be employed as desired. If a diffused system is used, an air measuring rotometer should be installed for control. When a mechanical surface aerator is used, the mixing intensity can be controlled with a variable speed motor and a rheostat.

2. The container is filled with a defined volume of tap water and the temperature recorded. The water is then deoxygenated chemically using sodium sulfite and a cobalt catalyst (approximately 8 to 12 mg/ℓ of sodium sulfite per mg/ℓ of dissolved oxygen) or stripping the oxygen from solution with an inert gas such as nitrogen. In the laboratory unit, the latter is preferred in order to eliminate any possible chemical interferences.

3. Once the contents have been deoxygenated, the water is reaerated at a controlled diffused air flow or mechanical rotational speed. The oxygen concentration for various time intervals is recorded at the predetermined level of liquid turbulence. The oxygen deficit at various aeration times is then plotted on semi-log paper, the slope taken as the coefficient K_La. This step can be repeated for various mixing intensities and temperatures. The test using tap water can be repated, substituting a mechanical aeration device for the diffused air system:

   Mechanical Aeration: Rotation Speed = as desired
   Rotation Speed = as desired
   Temperature = 20°C, 33°C
   Deoxygenation = nitrogen stripping
   Liquid contents = tap water

4. Following the deoxygenation and reaeration of tap water
    for the various test conditions, steps 2 and 3 should be
    repeated, using the same volume of wastewater. As the
    transfer coefficient, α, is a function of the degree of
    biological stabilization as previously indicated, every
    effort should be made to use a wastewater similar to the
    mixed liquor anticipated in the prototype aeration basin.
    For example, the effluent from the bench scale biological
    reactors should be used in the alpha test when a completely
    mixed system is contemplated.

    There may be some problem with interferences of the waste-
    water with the Winkler Method (Standard Methods) of deter-
    mining dissolved oxygen. A properly calibrated galvanic
    cell oxygen analyzer and probe is therefore a practical
    and reliable method of determining the oxygen concentra-
    tion throughout the testing period.

5. The $K_L a$ values as determined in step 4 can then be com-
    pared to the $K_L a$ values for tap water at the same tempe-
    rature and mixing conditions and α can be calculated in
    accordance with Equation (7).

    Once the α values for a range of temperature and mixing
    conditions are determined, the controlling magnitude for
    the most critical design condition can be used as a basis
    for sizing aeration equipment.

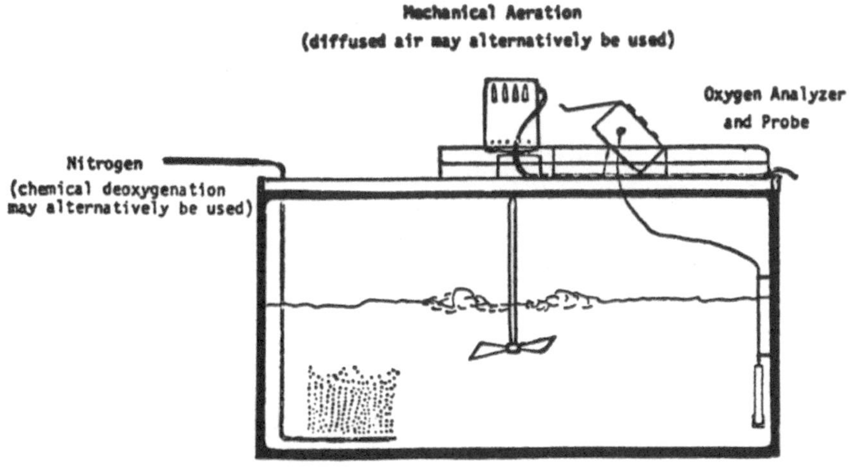

Fig. 12 : Laboratory Apparatus for Determination of α

AERATED LAGOON

Carl E. Adams, Jr., Ph.D., P.E.

President, Associated Water
and Air Resources Engineers, Inc., Nashville, Tennessee

W. Wesley Eckenfelder, Jr. P.E.

Distinguished Professor, Vanderbilt University,
Nashville, Tennessee

DISCUSSION of PRINCIPLES

An aerated lagoon is any basin utilizing artificial aeration in which biological organisms are allowed to grow and proliferate. Oxygen may be supplied to the basin by either mechanical or diffused aeration units. There are two types of aerated lagoons: the aerobic lagoon, in which dissolved oxygen is maintained throughout the basin; and the aerobic-anaerobic or facultative lagoon, in which oxygen is maintained in the upper layer of liquid in the basin only. These basins are depicted in *Fig. 1*.

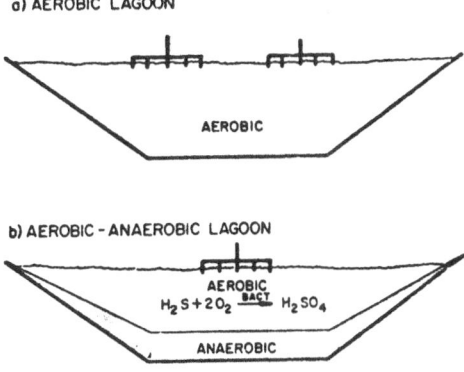

Fig. 1 : Comparison of Alternative
Aerated Lagoon Systems

In the aerobic lagoon, all solids are maintained in suspension, and this system may be thought of as a "flow-through" activated sludge system without solids recycle. Thus, the effluent suspended solids concentration will be equal to the aeration basin solids concentration. In most cases, separate sludge settling and disposal facilities are required. The aerobic lagoon can readily be modified to the activated sludge process.

In the facultative lagoon, a portion of the suspended solids settle to the bottom of the basin where they undergo anaerobic decomposition. The anaerobic by-products are subsequently oxidized in the upper aerobic layers of the basin. The facultative lagoon can also be modified to yield a more highly clarified effluent by the inclusion of a separate or baffled settling compartment.

BOD REMOVAL CHARACTERISTICS

As in the completely-mixed activated sludge process, organic removal is primarily a function of detention time, biological solids concentration, temperature, and the nature of the waste. Organic removal in completely-mixed lagoons can be determined using the following relationship:

$$\frac{S_o - S_e}{X_v t} = kS_e \qquad \qquad \dots \quad (1)$$

where:

S_o = influent total BOD, COD, or TOC, mg/ℓ

S_e = effluent soluble BOD, COD or TOC, mg/ℓ

X_v = average or equilibrium concentration of VSS in aeration basin, mg/ℓ

t = detention time, days

k = specific organic removal rate coefficient, ℓ/mg-day

Thus, at a constant basin detention time, the equilibrium biological solids concentration and overall rate of organic removal can be expected to increase as the influent organic concentration increases. For a soluble waste, the equilibrium biological solids concentration, X_v, can be predicted from the following relationship:

$$X_v = \frac{X_{ov} + aS_r}{1 + bt} \qquad \qquad \dots \quad (2)$$

where:

X_{ov} = influent biological volatile suspended solids concentration, mg/ℓ

a = sludge synthesis coefficient, mg VSS produced/mg organics removed or Kg VSS produced/Kg organics removed

S_r = organics (BOD, COD, or TOC) removed (S_o-S_e), mg/ℓ

b = sludge auto-oxidation coefficient, mg VSS destroyed/day-mg VSS in the aeration basin or Kg VSS destroyed/day-Kg VSS in the aeration basin.

Assuming no influent biological volatile suspended solids (X_{ov}=0), Equations 1 and 2 can be combined and the effluent soluble organic concentration can be defined as follows:

$$S_e = \frac{1 + bt}{akt} \qquad \qquad .... \quad (3)$$

Equation 3 can be linearized as:

$$S_e = \frac{1}{akt} + \frac{b}{ak} \qquad \qquad .... \quad (3a)$$

The specific organic reaction rate coefficient, k, can thus be determined from a plot of S_e versus 1/t. From Equation 3, it may be concluded that the effluent soluble organic concentration is independent of the influent organic concentration. For lagoons with fixed detention times, this conclusion is justified because higher influent organic concentrations will result in higher equilibrium biological solids levels, and, therefore, higher overall BOD removal rates. It should be emphasized, however, that the unit organic removal rate remains proportional to the soluble organic concentration in the lagoons as shown in Equation 1. The minimum detention time for the use of Equation 3 occurs as S_e approaches S_o. This minimum detention time can be determined using the following relationship:

$$t_m = \frac{1}{ak-b} \qquad \qquad .... \quad (4)$$

where:

t_m = minimum detention time for organic removal, days

The specific organic reaction rate coefficient, k, is highly temperature dependent. The following relationship can be used to correct k for temperature:

$$k_2 = k_1 \; \theta^{(T_1 - T_2)} \qquad \qquad .... \quad (5)$$

where:

k_2 = specific organic reaction rate coefficient at T_2, °C, ℓ/mg-day

k_1 = specific organic reaction rate coefficient at T_1, °C, ℓ/mg-day.

θ = temperature correction coefficient, assumed to be 1.0639 for aerated lagoons.

Depending on the aeration power level and the nature of the solids, a portion of the solids may deposit on the bottom of the basin. At low organic and suspended solids levels, the biological growth is dispersed and tends to remain in suspension even at very low power levels. For example, all solids have been maintained in suspension at power levels as low as 0.80 Kw/1000 m^3 in pulp and paper mill aerated lagoons operated at organic loadings of less than 2246 Kg/10000 m^2-day. Even at higher BOD loadings, data from the operation of pulp and paper mill aerated lagoons indicate that 50 mg/ℓ of solids can be maintained in suspension at this same power level. The solids in suspension as a function of power level may be approximated as shown in *Fig. 2*.

Solids deposited on the bottom of facultative lagoons will undergo anaerobic degradation. This degradation results in a feedback of soluble organics to the upper aerobic layer. For these conditions, Equation 1 should be modified as follows:

$$S_e = (\frac{S_o}{1 + kX_v t}) \; F \qquad \qquad .... \quad (6)$$

where:

F = organic feedback coefficient due to benthal activity, dimensionless.

The degree of anaerobic activity is highly temperature dependent and the coefficient, F, may be estimated to vary from 1.0 to 1.4 under winter and summer conditions, respectively.

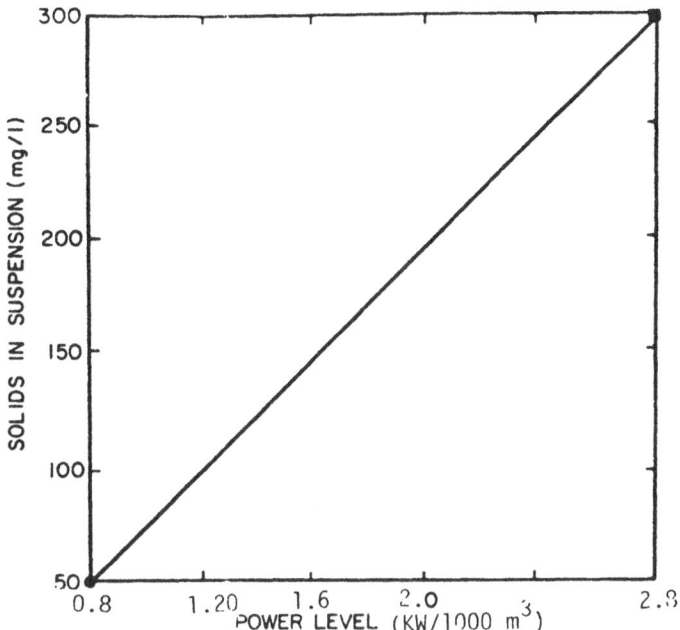

Fig. 2 : Power Level vs Solids
Concentration in Suspension

OXYGEN REQUIREMENTS

For completely-mixed aerobic lagoons, oxygen requirements can be computed in a manner similar to that for activated sludge. The relationship is:

$$R_r = a'S_r + b'X_v \qquad \qquad \dots \dots (7)$$

where:

R_r = total oxygen utilization, Kg O_2/day

a' = oxygen utilization coefficient for synthesis, Kg O_2 utilized/Kg organics removed or mg O_2 utilized/mg organics removed.

S_r = organics (BOD, COD or TOC) removed (S_o-S_e), Kg/day

b' = oxygen utilization coefficient for endogenous respiration, Kg O_2 utilized/day-Kg VSS in the aeration basin or mg O_2 utilized/day-mg VSS.

X_v = average of equilibrium concentration of VSS in aeration basin, Kg/day.

In facultative aerobic-anaerobic lagoons, biological solids are maintained at a lower level and soluble organics are fed back to the liquid as anaerobic degradation of the settled solids proceeds. In this instance, the oxygen requirements can be related to organic removal and estimated from the following relationship:

$$R_r = F'S_r \qquad \qquad \text{.... (8)}$$

where:

F' = overall oxygen utilization coefficient for facultative lagoons, dimensionless.

The results obtained for various industrial wastes indicate that the value of F' is a function of the degree of organic feedback. This feedback depends on the basin temperature, and F' can be estimated to vary from 0.8 to 1.1 during winter operation when anaerobic activity in the bottom of the basin is low and from 1.2 to 1.5 during summer operation when anaerobic activity in the bottom of the basin is at a maximum. Oxygen requirements should generally be designed for summer operation since the rates of organic removal and benthal feedback will be the greatest during the warmer months of the year.

Multiple basins may be effectively used in aerated lagoon systems under the proper conditions. Although it can be shown from Equation 1 that there is no advantage to series basin operation at a given temperature in terms of effluent organic concentration, there are two considerations which may favor series operation. First, series operation may be desired when land availability is a concern. A minimum total basin volume can be obtained by employing two basins. The first basin volume is minimized to maintain a high temperature and a resulting high BOD reaction rate. Second, series operation may be favored when stringent effluent solids standards must be met. In the case where a low solids effluent is required, a primary, completely-mixed basin is followed by a secondary facultative basin. The facultative basin at low power (mixing) levels will permit solids settling. An optimization procedure can be utilized to determine the smallest total basin volume and the lowest aeration horsepower for a specified effluent quality.

TEMPERATURE EFFECTS

The performance of aerated lagoons is significantly influenced by changes in basin temperature. In turn, the basin temperature is influenced by the temperature of the influent wastewater and the ambient air temperature. Although heat is lost through evaporation, convection and radiation, it is gained from solar radiation.

In order to accurately predict the water temperatures in aerated biological systems, heat transfer theory is used. Formulas associated with heat transfer theory take into account water temperature changes due to wind velocity, solar radiation, and the heat transfer associated with aeration.

The aeration basin temperature can be calculated from the following equation:

$$T_w = \frac{T_i + (\frac{AK}{Q})\,(T_o)}{1 + (\frac{AK}{Q})} \qquad \ldots \ (9)$$

where:

T_w = water temperature in completely-mixed aeration basin, °C

T_i = influent raw waste temperature, °C

A = surface area of basin, sq m

Q = hydraulic flow, cu m/day

K = overall heat transfer coefficient, m/day

T_o = equilibrium temperature, °C

The overall heat transfer coefficient can be computed from the following formula:

$$K = 0.114 + 0.0326\ x^{-0.1} \cdot V_w + 0.432\ \frac{Q_A}{A}$$
$$+ e^{0.06\,T_a}(1.9\ 10^{-2} \cdot x^{-0.1} \cdot V_w + 0.2525\ \frac{Q_A}{A}) \qquad \ldots \ (10)$$

where:

T_a = average air temperature, °C

x = characteristic dimension of the basin, $\sqrt{A}$, m

V_w = wind velocity, Km/hr

Q_A = air flow through aeration spray, m^3/min

e = $8.34 \cdot N \cdot E \cdot V_w$

N = number of aerators

E = aerator spray area, sq m (this area represents the vertical cross-section of the aerator spray).

The equilibrium temperature can then be computed from the following formula:

$$T_o = T_a + \frac{1}{K} [0.02396\ H_s + 6.94267\ (\beta - 0.874)$$

$$- e^{0.06\ T_a} (1 - \frac{f_a}{100})\ (0.35258 \cdot X^{-0.1} \cdot V_w + 0.8432\ \frac{Q_A}{A})] \ \dots\ (11)$$

where:

H_s = solar radiation, Kg-cal/sq m-hr

β = Raphael's long wave radiation coefficient, dimensionless.

f_a = relative humidity, percent

Knowing the equilibrium temperature, the basin temperature can then be computed from Equation 9.

EXCESS SLUDGE PRODUCTION

During the process of organic removal in a conventional aerated lagoon, organisms grow and reproduce, thereby resulting in an accumulation of aeration suspended solids. In order to obtain the highest effluent quality from an aerated lagoon system, the biological solids should be separated from the effluent. In completely-mixed single-stage systems where aerobic and facultative lagoons are operated in series, solids separation occurs due to settling of the solids on the bottom of the facultative lagoon. The rate of solids build-up on the bottom of facultative lagoons must be estimated so that the lagoons can be cleaned out before the treatment efficiency is impaired.

The amount of excess biological volatile sludge produced is a function of the sludge produced through organic removal and that sludge which is destroyed endogenously. Excess biological volatile sludge production can be determined using the following equation:

$$\Delta X_v = aS_r - bX_v \qquad \qquad \cdots\cdot (12)$$

where:

ΔX_v = excess biological volatile sludge production, Kg VSS/day

S_r = organics (BOD, COD or TOC) removed $(S_o - S_e)$, Kg/day

X_v = average or equilibrium concentration of VSS in aeration basin, Kg/day

The total excess sludge production is a function of the influent suspended solids concentration, the biodegradable fraction of the influent suspended solids concentration, the excess biological sludge produced, and the effluent suspended solids concentration. Total excess sludge production can be determined by using the following equation:

$$\Delta X = fX_o + \frac{\Delta X_v}{f_v} - X_e \qquad \qquad \cdots\cdot (13)$$

where:

ΔX = total excess sludge production, Kg SS/day

f = non-biodegradable fraction of influent suspended solids

X_o = influent suspended solids, Kg SS/day

X_e = effluent suspended solids, Kg SS/day

f_v = volatile fraction of equilibrium basin SS (VSS/SS)

FINAL SETTLING BASIN

The final settling basin should be designed with four major constraints in mind:

1. Sufficiently long detention time to effect the desired suspended solids removal;
2. Adequate volume for sludge storage;
3. Minimization of development of algal colonies;
4. Minimization of odor development from anaerobic benthal activity.

Unfortunately, these objectives in design are not always compatible. At times the short detention times required to inhibit algal growth are too short to obtain proper settling. Also, adequate volume must remain above the sludge deposits at all times to prevent the escape of odorous gases of decomposition. Thus, sludge storage volume is taken out of service. Different locales will dictate requirements, e.g., algal colonies and high degrees of anaerobic feedback are more predominant in warmer climates than colder climates. In most cases, the following guidelines are applicable:

1. A minimum detention time of one day is required to settle the majority of settleable suspended solids. Thus, the settled sludge should not be allowed to accumulate to depths which would result in less than a one-day detention time of the supernatant liquid. A calculated one-day detention time may be too short for proper settling if uneven distribution of solids has occurred and channelling has developed.

2. If algal growth possess potential problems, a maximum detention time of one to two days is recommended. This short detention time should prevent the algal colonies from proliferating thus causing blooms, etc. Algal growth may lead to a diminished effluent quality during certain periods of the year. Consequently, when the basin is empty, the maximum detention time should be two days for algal control.

3. For odor control, it is recommended that a minimum water level of 0.90 m be maintained above the sludge deposits at all times. In most climates this layer is deep enough to oxidize any hydrogen sulfides to sulfates, thereby destroying the odorous compounds. In addition, ninety centimeter water layer assists in oxidizing soluble organics generated during anaerobic decomposition of the sludge deposits. In cases of high summer temperatures or excessive sulfide concentrations, it may be necessary to maintain water layers of up to 1.80 m to prevent odorous conditions from developing.

The rate of anaerobic decomposition of the sludge which is deposited in the settling basin or in a low power level aerated lagoon can be estimated by the following decay equation:

$$\frac{W_t}{W_o} = e^{-D_k t_d} \qquad \qquad \dots \dots (14)$$

where:

W_o = initial concentration of deposited VSS, Kg

W_t = concentration of deposited VSS at time, t, which have not degraded, Kg.

D_k = rate of VSS degradation, $time^{-1}$

t_d = duration of degradation, time

Integradation of the above equation yields:

$$W_t = \frac{W_o}{D_k} (1 - e^{-D_k t_d}) \qquad \qquad \dots \ (15)$$

A certain percentage of the deposited volatile solids will degrade each year and may be estimated in the range of 40 to 60 percent degradation per year. The sludge accumulation will be equal to the deposited volatile solids which have not degraded plus the deposited inert or non-volatile suspended solids.

NUTRIENT REQUIREMENTS

Biological volatile solids in aerated lagoon systems usually contain up to 12.3 percent nitrogen and 2.6 percent phosphorus, and each nutrient must be present in quantities sufficient for organism growth and synthesis. These nutrients are contributed to the system by the influent wastewater or by cell lysis accompanying endogenous respiration. Wastewaters are often deficient in one or both nutrients; therefore, the addition of each nutrient may be required. Nitrogen and phosphorus requirements for aerated lagoon systems may be calculated using the following equations:

$$Kg \ N/day \ = \ 0.123 \ \Delta X_v \qquad \dots \ (16)$$
$$Kg \ P/day \ = \ 0.026 \ \Delta X_v \qquad \dots \ (17)$$

DESIGN EQUATIONS and NOMENCLATURE

EQUATIONS

BOD Removal Relationships

Completely Mixed

$$S_e = \frac{S_o - S_e}{X_v tk}$$

$$S_e = \frac{1 + bt}{akt}$$

Facultative

$$S_e = \frac{S_o}{1 + kX_v t} F$$

$$k_2 = k_1 \, \theta^{(T_1 - T_2)}$$

Oxygen Requirements

Completely Mixed

$$R_r = a'S_r + b'X_v$$

Facultative

$$R_r = F'S_r$$

$$N = N_o \left(\frac{\beta C_{sw} - C_L}{C_s}\right) \alpha \, \theta^{T-20}$$

Temperature Considerations

$$K = 0.114 + 0.0326 \cdot X^{-0.1} \cdot V_w + 0.432 \frac{Q_A}{A}$$

$$+ e^{0.06\,Ta}(1.9 \times 10^{-2} \cdot X^{-0.1} \cdot V_w + 0.2525 \frac{Q_A}{A})$$

$$T_o = Ta + \frac{1}{K} [0.02396 \cdot H_s + 6.94267(\beta - 0.874)$$

$$- (1 - \frac{f_a}{100}) e^{0.06\,Ta} \, (0.35258 \cdot X^{-0.1} \cdot V_w + 0.8432 \frac{Q_A}{A})]$$

$$T_w = \frac{T_i + (\frac{AK}{Q})(T_o)}{1 + (\frac{AK}{Q})}$$

Sludge Production

$$X_v = \frac{X_{ov} + aS_r}{1 + bt}$$

$$\Delta X_v = aS_r - bX_v$$

$$\Delta X = fX_o + \frac{\Delta X_v}{f_v} - X_e$$

Sludge Accumulation

$$W_t = \frac{W_o}{D_k}(1 - e^{-D_k t_d})$$

Nutrient Requirements

$$Kg\ N/day = 0.123\Delta X_v$$

$$Kg\ P/day = 0.026\Delta X_v$$

REQUIRED DESIGN INFORMATION

KINETIC PARAMETERS

1. Specific BOD reaction rate coefficient, k, ℓ/mg-day
2. Oxygen coefficients, a' and b', Kg O_2 utilized/Kg organics removed and Kg O_2 utilized/day-Kg VSS in basin
3. Sludge coefficients, a and b, Kg VSS produced/Kg organics removed and Kg VSS destroyed/day-Kg VSS in basin
4. Oxygen transfer coefficient, α

230

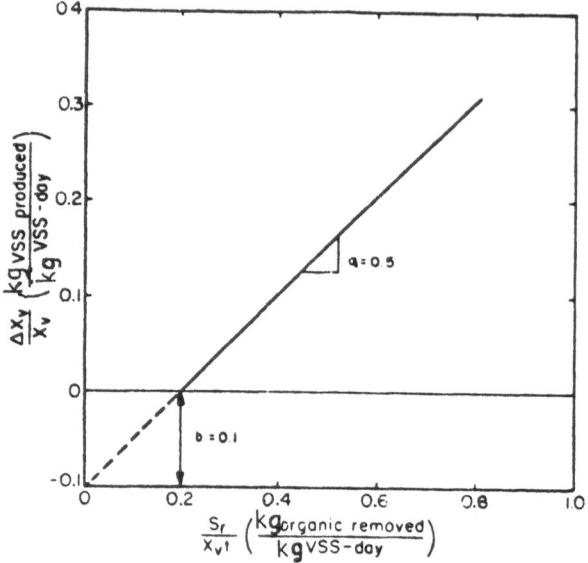

Fig. 3 : Determination of Sludge
Production Coefficients

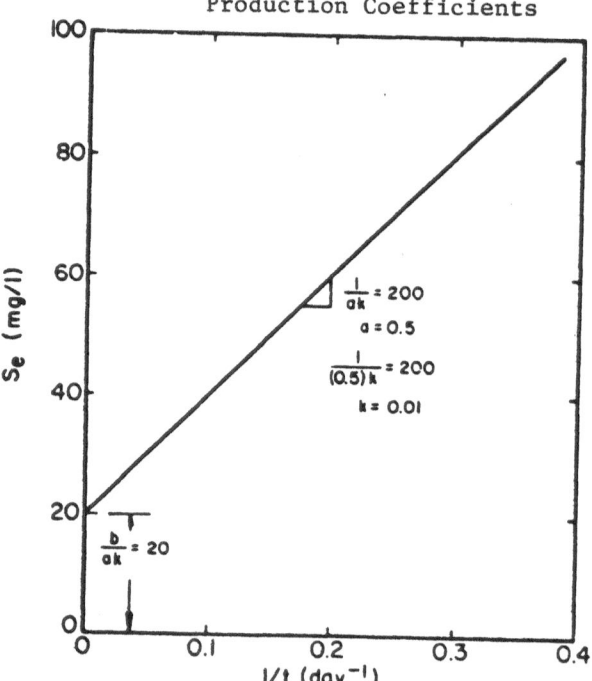

Fig. 4 : Determination of
Removal Rate Coefficient

5. Oxygen transfer efficiency, standard conditions, N_o
6. Oxygen saturation coefficient, β
7. Temperature correction coefficient for specific BOD reaction rate coefficient and oxygen transfer efficiency, Θ.

DESIGN CONDITIONS

1. Average and maximum total wastewater flow, Q, cu m/day
2. Influent wastewater temperature, T_i, °C
3. Average extreme ambient temperatures, summer and winter, T_a, °C
4. Average and maximum influent total BOD, S_o, mg/ℓ
5. Influent suspended solids, X_o, mg/ℓ
6. Influent nitrogen, mg/ℓ
7. Influent phosphorus, mg/ℓ

DESIGN CRITERIA TO BE MET

1. Average and maximum effluent soluble BOD, S_e, mg/ℓ
2. Effluent suspended solids, X_e, mg/ℓ

DEVELOPMENT of DESIGN PARAMETERS

The required design parameters, especially the kinetic parameters, must be estimated by the designer or developed from comprehensive bench and pilot-scale studies. Generally, several aerated lagoon systems are operated in parallel at various detention times. Influent and effluent streams are monitored for total and soluble BOD, total and volatile suspended solids, nitrogen, phosphorus, pH and other characteristics, such as phenols, cyanides, etc. which may be unique to the wastewater being studied. The aeration basin contents are monitored for specific oxygen uptake rate expressed in g oxygen utilized/day-g VSS in the basin.

The specific design parameters can be developed by correlating the data obtained during the studies by the procedures outlined in the following sections.

DETERMINATION of SLUDGE PRODUCTION COEFFICIENTS, a and b

Correlate the excess biological volatile sludge production with the BOD removal.

Plot $\Delta X_v/X_v$ versus $S_r/X_v t$ on the Y and X axes, respectively, as shown in *Fig. 3*. The slope of the resulting straight line is equal to a, and the Y axis intercept is equal to b.

DETERMINATION of the
SPECIFIC BOD REACTION RATE COEFFICIENT, k

Correlate the data according to Equation 3a:

$$S_e = \frac{1}{akt} + \frac{b}{ak}$$

Plot S_e versus ℓ/t. The slope of this plot is equal to $1/ak$ and the Y axis intercept is equal to b/ak (See *Fig. 4*).

The specific BOD reaction rate coefficient, k, can be obtained by substituting the value for a obtained from Fig. 3 into the equation for the slope obtained from *Fig. 4*.

DETERMINATION of
OXYGEN COEFFICIENTS, a' and b'

Correlate the total oxygen utilization rate with the BOD reaction rate. Plot R_r/X_v versus $S_r/X_v t$ as shown in *Fig. 5*. The slope of the resulting straight line is equal to a' and the intercept equal to b'.

DESIGN PROCEDURE

A straight-forward calculation of the aeration basin detention time is not possible because of the dependence of the BOD removal rate coefficient on temperature. The detention time is calculated from Equation 1 or Equation 3 after correcting the coefficient, k, for the aeration basin temperature. However, the aeration basin temperature cannot be calculated unless the detention time is known. Consequently, a trial-and-error procedure is required. The design procedures presented herein will yield optimized designs based upon either:

1. The minimum aeration basin volumes and associated horse-power necessary to meet effluent criteria, or
2. The minimum horsepower and associated aeration basin volumes necessary to meet effluent criteria.

Capital and operational economic considerations will dictate which design is chosen for implementation. A two-stage design is presented herein.

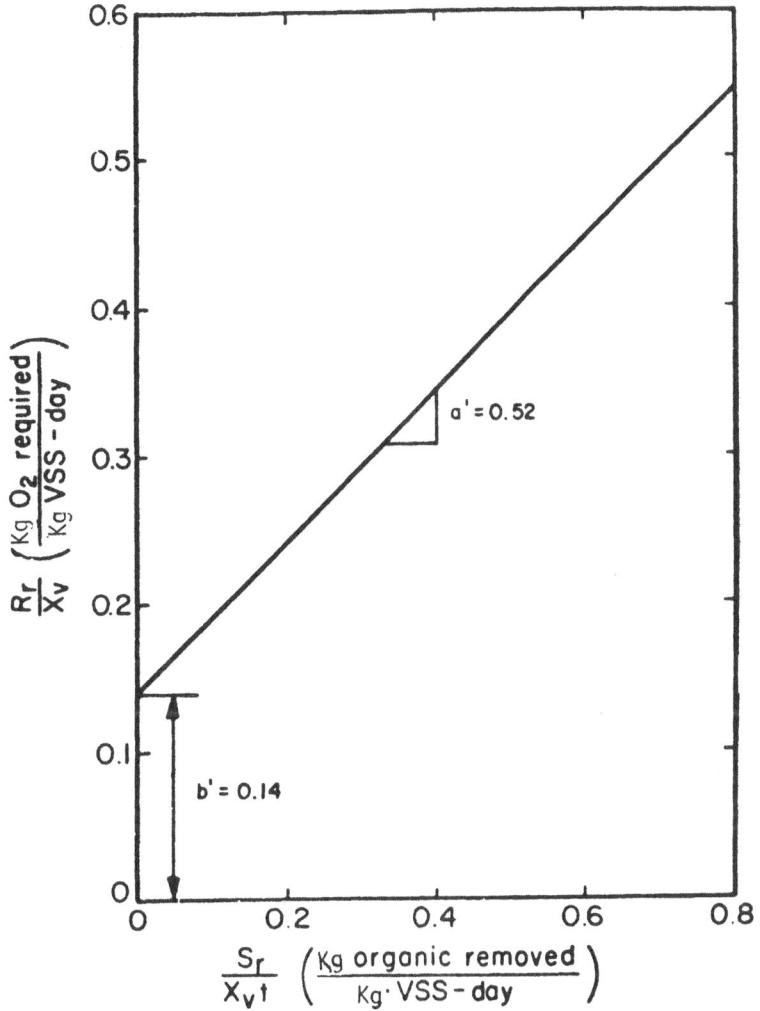

Fig. 5 : Determination of Oxygen
Utilization Coefficients

DESIGN of BASIN I
BASED on MINIMUM DETENTION TIME

The first of a two-stage system is designed by assuming a detention time and calculating the effluent BOD which will be generated
at the selected detention time. The results can be tabulated as
shown in *Table 1*. The procedures for developing *Table 1* are outlined below.

1. Assume a Detention Time

   Initial assumption usually, to 5 days.

2. Compute Basin Volume

$$V = Qt$$

where:

V = volume, cu m

Q = flow, cu m/day

t = detention time, days

3. Compute Basin Area

$$A = V/d$$

where:

A = area, sq m

V = volume, cu m

d = depth, m (Selected by designer)

4. Estimate Horsepower and Number of Aerators

   Assume a power level of at least 2.75 Kw/1000 m^3. Multiply this power level by the volume obtained in Step 2 to obtain the total horsepower. Determine the number of aerators required.

5. Calculate Winter Water Temperature in Basin

   Using Equation 9 to determine the aeration basin temperature:

$$T_w = \frac{T_i + (\frac{AK}{Q})}{1 + (\frac{AK}{Q})}$$

6. Correct the Biological Reaction Rate Coefficient for Winter Conditions

$$k_2 = k_1 \theta^{(T_1 - T_2)}$$

TABLE 1 : Design Summary of Basin I
            (Aerobic Lagoon)

A. Summer Conditions

Detention Time (days)	T_w (°C)	k (ℓ/mg-day)	S_e (mg/ℓ)	X_v (mg/ℓ)	O_2 (Kg/day)	Total Kw	Power Level Kw/ 1000 m³
0.5	36.3	0.0175	240	100	10403	308	6.5
1.0	35.6	0.0171	128	146	17785	534	5.7
2.0	35.1	0.0168	71	158	22843	686	3.5
2.5	34.9	0.0167	60	156	24365	736	3.2
3.0	34.7	0.0166	52	153	25673	773	2.75
4.0	34.6	0.0165	42	146	27822	1043 (894)	2.75 (2.36)
5.0	34.4	0.0164	37	138	29473	1310 (931)	2.75 (1.97)

B. Winter Conditions

Detention Time (days)	T_w (°C)	k (ℓ/mg-day)	S_e (mg/ℓ)	X_v (mg/ℓ)	O_2 (Kg/day)
0.5	32.6	0.0154	273	84	9269
1.0	30.9	0.0146	150	136	16570
2.0	28.7	0.0135	89	150	21745
2.5	27.8	0.0131	76	150	23378
3.0	26.9	0.0127	68	147	24647
4.0	25.4	0.0120	58	140	26716
5.0	24.1	0.0115	52	133	28404

7. Calculate the Winter Effluent Soluble BOD

Using Equation 3 assuming no influent biological volatile suspended solids:

$$S_e = \frac{1 + bt}{akt}$$

8. Calculate the Summer Water Temperature in the Basin as illustrated in Step 5

9. Correct the Biological Reaction Rate for Summer Conditions as illustrated in Step 6

10. Calculate the Summer Effluent Soluble BOD as illustrated in Step 7

11. Calculate the Summer Equilibrium or Average Biological Volatile Suspended Solids Concentration

    Using Equation 2:

    $$X_v = \frac{X_{ov} + aS_r}{1 + bt}$$

12. Calculate Oxygen Requirements

    Using Equation 7:

    $$R_r = a'S_r + b'X_v$$

13. Calculate Oxygen Transfer Efficiency of Aerators

    Aeration equipment is normally rated for standard conditions. Standard conditions are: $\alpha = 1.0$, $\beta = 1.0$, one atmosphere, 0.0 mg/ℓ dissolved oxygen and 20°C. The following equation is used to correct the rated transfer efficiency to design field conditions:

    $$N = N_0 [(\beta C_{sw} - C_L)/C_s)]\alpha\Theta^{T_w-20}$$

14. Compute Horsepower Requirements

    $$Kw = \frac{Kg \ O_2/day \ required}{(N)(24 \ hr/day)}$$

15. Check Power Level

    The power level is checked to ensure that sufficient mixing is provided to maintain all suspended solids in suspension. For a conservative design, the minimum power level should be 2.75 Kw/1000 m^3

    $$Power \ level = Kw/volume$$

    If the power level is significantly less than 2.75 Kw/1000 m^3, then 2.75 Kw/1000 m^3 should be used.
    If the calculated Kw requirements differ by more than 5 percent from the assumed Kw in Step 5, a new Kw must be

assumed and the temperature of the basin must be recalculated. Thus, a trial and error procedure is required.

16. Repeat Steps 2 through 13 Using Different Detention Times

17. Correlate Effluent Soluble BOD to Detention Time
(See Figure 6).

DESIGN of BASIN II
BASED on MINIMUM DETENTION TIME

The design of the second basin is based on selecting an effluent
BOD concentration from the first basin which would, of course, be
in the influent to the second basin. Based on this influent BOD
and the desired effluent level, set by regulatory constraints, a
detention time is calculated. Since the second basin will probably
be operated at a low power level with solids sedimentation, certain
considerations are necessary.

1. Assume the Detention Time Required to Reduce the Soluble
BOD to the Prescribed Level

   Higher influent BOD values require longer detention times,
   generally less than 25 days.

2. Calculate Volume and Surface Area Corresponding to the
Assumed Detention Time

   (See Figure 2).

3. Select a Kw Level and the Corresponding Solids Concentration

   Using the selected power level and the volume from Step 2,
   Basin II, the total Kw for the second stage can be calcu-
   lated. The number of aerators can then be determined.
   (See Figure 2).

4. Calculate the Temperature in Basin II and Adjust the Reaction Rate

   Using the method shown in Steps 5 and 6 for Basin I.

5. Compute the Detention Time Required to Reduce the Soluble
BOD to the Prescribed Level (defined by regulatory requirements)

   Using Equation 6:

   $$S_e = (\frac{S_0}{1 + kX_v t}) F$$

6. Check the Computed Detention Time Against the Assumed Detention Time

   If necessary, recompute Steps 1 through 4 for Basin II.

7. Calculate the Summer Water Temperature in the Basin Using the Method shown in Step 5 for Basin I.

8. Calculate the Biological Reaction Rate Coefficient for Summer Conditions

   Using the method shown in Step 6 for Basin I.

9. Calculate the Summer Effluent Soluble BOD

   Using the method shown in Step 5 for Basin II.

10. Calculate the Summer Equilibrium or Average Volatile Suspended Solids Concentration

    Using the method shown in Step 11 for Basin I.

11. Calculate Oxygen Requirements for the Facultative Basin

    Using Equation 8:

    $$R_r = F'S_r$$

12. Calculate Oxygen Transfer Efficiency of Aerators

    The Oxygen transfer efficiency for Basin II is calculated the same as shown in Step 13 for Basin I.

13. Compute Kw Requirements

    Using the methods shown in Step 14 for Basin I.

14. Check Power Level

    In the facultative lagoon, the power level is checked to ensure that sufficient mixing is provided to distribute oxygen throughout the liquid layer. The minimum power level should be 0.78 Kw/1000 m^3.

    $$\text{Power Level} = \text{Kw/volume}$$

    If the power level is significantly less than 0.78 Kw/1000m^3, then 0.78 Kw/1000 m^3 should be used. If the Kw requirements are significantly different from the assumed Kw in Step 4 for Basin II, a new detention time must be assumed in Step 1 for Basin II and the temperature of the basin recalculated, etc. Thus, a trial and error procedure is again required.

CALCULATION of SLUDGE PRODUCTION and ACCUMULATION BASED UPON the MAXIMUM INFLUENT BOD

1. Calculate the Excess Biological Volatile Sludge Production

   Using Equation 12:

$$\Delta X_v = aS_r - bX_v$$

2. Determine Influent SS and VSS to the Basin and Assume Effluent SS and VSS from the Basin

   Influent SS and VSS are equal to effluent SS and VSS from Basin I. Effluent SS and VSS from Basin II are usually 50–100 mg/ℓ.

3. Calculate SS and VSS Deposition, Q_0

   Non-Volatile
   SS Deposition = (influent SS - influent VSS)(days/month)
   $\qquad\qquad$ - (effluent SS - influent VSS)(days/month)

   SS Deposition = $X_{ov} - X_{ev}$ = (influent VSS)(days/month)
   $\qquad\qquad\qquad\qquad$ - (effluent VSS)(days/month)

   where:

   SS, VSS deposition = Kg/month
   SS, VSS $\qquad\qquad$ = Kg/day

4. Calculate Rate of Volatile Suspended Solids Degradation, D_k

   Using Equation 14:

   $$W_t/W_o = e^{-D_k t_d}$$

   $$\therefore \quad D_k = -\frac{\ln(W_t/W_o)}{t_d}$$

   where:
   $\qquad$ W_t = quantity of VSS after degradation for time, t, Kg/month
   $\qquad$ W_o = quantity of VSS before degradation $(X_{ov}-X_{ev})$, Kg/month
   $\qquad$ t_d = time of degradation, 12 months/yr

5. Calculate Deposited Volatile Solids that have not Degraded Using Equation 15:

   $$W_t = \frac{W_o}{D_k} (1 - e^{-D_k t})$$

6.   Calculate Total Sludge Accumulation

Total Accumulation = Non-volatile SS deposition + W_t

Where total accumulation and non-volatile ss deposition are expressed in Kg SS/month, and W_t, in Kg VSS/month

7.   Calculate Rate of Total Sludge Accumulation

Assume the initial solids will compact to a specific percentage (usually 4 percent).

$$\text{Rate of Accumulation} = \frac{\text{Total Accumulation}}{(\%)(10,000)}$$

where:

% = percent solids

CALCULATION of NUTRIENT REQUIREMENTS

1.   Calculate Nitrogen Requirement

Assume that the biological solids contain 12.3 percent nitrogen.

$$\text{Nitrogen} = 0.123\Delta x_v$$

2.   Calculate Phosphorus Requirement

Assume that the biological solids contain 2.6 percent phosphorus.

$$\text{Phosphorus} = 0.026\Delta x_v$$

EXAMPLE PROBLEM

Given the following information, design a two-stage lagoon system that will yield:

1.   An average effluent soluble BOD = 18 mg/ℓ
2.   A maximum effluent soluble BOD = 30 mg/ℓ
3.   An average effluent suspended
     solids = 50 mg/ℓ at 80 percent volatile content

KINETIC PARAMETERS

1.   BOD reaction rate coefficent
     k = 0.01 day at 20°C

2.   Sludge synthesis coefficient
     a = 0.5 mg VSS/mg BOD removed

3. Sludge auto–oxidation coefficient
    b = 0.1 mg VSS/day-mg VSS

4. Oxygen utilization coefficient
    a'= 0.52 mg O_2/mg BOD removed

5. Endogenous oxygen utilization rate
    b'= 0.14 mg O_2/day-mg VSS

6. Temperature correction coefficient
    Θ = 1.035 for BOD reaction rate coefficient
    Θ = 1.028 for oxygen transfer efficiency correction for
        mechanical surface aeration

7. Coefficient to account for organic feedback due to benthal
    activity
    Summer, F = 1.4
    Winter, F = 1.0

8. Overall oxygen utilization coefficient for facultative
    lagoon
    F'= 1.5

9. Oxygen transfer efficiency
    N_0= 1.95 Kg O_2/KW-hr for low-speed aerators

10. Oxygens transfer coefficient
    α = 0.80

11. Saturation Coefficient
    β = 0.90

12. BOD contribution due to effluent volatile suspended solids
    = 0.3 mg BOD/mg VSS

DESIGN CONDITIONS

1. Total wastewater flow
    Q = 94625 m^3/day

2. Influent waste temperature
    Summer, T_i = 37.7°C
    Winter, T_i = 35°C

3. Average extreme ambient temperature
    Summer, T_a = 26.6°C
    Winter, T_a = 0°C

4. Average wind speed
    Summer, V_w = 10.8 Km/hr
    Winter, V_w = 8.5 Km/hr

5. Solar radiation
    Summer, H_s = 298 Kcal/m^2-hr
    Winter, H_s = 5.4 Kcal/m^2-hr

6. Average relative humidity
    Summer, f_a = 85 percent
    Winter, f_a = 50 percent

7. Pond depth, d= 4.6 m

8. Wastewater BOD
    Influent, S_0
    50 percentile value = 37864 Kg/day
    ....= 400 mg/ℓ
    90 percentile value = 42596 Kg/day
    ....= 450 mg/ℓ

9. Wastewater suspended solids
    Influent, X_o ......= 0 Kg/day
    ....= 0 mg/ℓ

COMMENTS on DESIGN of BASIN I
(Aerobic Lagoon)

The BOD removal rate will be lower in winter than in summer.
Therefore, the optimum design, as far as required detention time
for reaching a terminal BOD is concerned, will be controlled by
winter temperatures. On the other hand, oxygen requirements and
power level will usually be controlled by summer conditions in the
basin system.

The results of the calculations for summer and winter conditions
for the example problem are summarized in *Table 1*. Corrections are
shown in the table for instances where the calculated oxygen requirements resulted in power levels less than the 14 hp/mil gal
level required for complete mixing. In these instances, the calculated hp and power level requirements are shown in parentheses
beside the values that should be used for design.

COMMENTS on DESIGN of BASIN II
(Facultative Lagoon)

The design of the facultative lagoon requires a series of trial
and error calculations. The hp level in a facultative lagoon is
generally lower than in an aerobic lagoon and normally not sufficient to maintain all solids in suspension. The minimum hp level
is about 0.78 Kw/1000 m^3 with a suspended solids concentration of
50 mg/ℓ. For a hp level of 2.75 Kw/1000 m^3 or greater, all solids
can be assumed to be in suspension. The solids concentration is
important because the required detention time to meet a certain
effluent standard decreases with increasing solids concentration.

The calculations for the required detention time in the second
basin are based on winter conditions. Subsequently, the required

oxygen and corresponding power level are calculated for summer conditions. The results of these calculations are shown in *Table 2*. Corrections are shown in the table for instances where the calculated oxygen requirements resulted in power levels less than the 0.78 Kw/1000 m³ level required for complete oxygen dissolution. In these instances, the calculated Kw and power level requirements are enclosed in parentheses beside the values that should be used for design. To reduce the high solids concentrations in the Basin II effluent, one can include a settling compartment in the basin or add a separate settling basin. The results in Table 2 do not include such a settling compartment.

The total effluent BOD is the sum of the soluble BOD and that BOD contributed by the carryover suspended solids:

$$BOD_T = BOD + (mg\ BOD/mg\ VSS)(effluent\ VSS)$$

For the sludge age in most aerated lagoon systems, about 0.3 mg BOD/mg VSS will be contributed. Since the effluent suspended solids are estimated to be 50 mg/ℓ with an 80 percent volatile content and the average effluent soluble BOD concentration is calculated to be 18 mg/ℓ, the effluent total BOD can be computed as follows:

$$BOD_T = 18 + 0.3\ (0.8)(50) = 30\ mg/\ell$$

OPTIMUM LAGOON SYSTEM

This two-stage lagoon system can be optimized to minimize either the detention time or the total hp to be installed.

In *Fig. 6*, the effluent BOD of the first basin is plotted versus the detention time. This figure also shows the detention time required in Basin II to achieve an effluent BOD of 18 mg/ℓ with any given influent BOD. The total required detention time of the lagoon system necessary to meet the effluent standards is determined by totalling the detention time of the two lagoons and plotting this versus the detention time in Basin I. This is illustrated in *Fig. 7*.

The required power for Basin I and II versus detention time is plotted in *Fig. 8*, and in *Fig. 9* the total required power for Basins I and II is plotted versus the required detention time for Basin I. By choosing from *Fig. 7*, a basin system which gives the desired effluent quality, the total Kw required for such a system can be determined in *Fig. 9*.

The minimum required detention time is 10.1 days, with a corresponding total installed Kw of 1497. The minimum total Kw is

TABLE 2

DESIGN SUMMARY of BASIN II

(Facultative Lagoon)

Basin I T_w (°C)	S_o (mg/ℓ)	Basin II T_w (°C)	k (ℓ/mg-day)	t (days)	S_e (mg/ℓ)	X_v (mg/ℓ)	O_2 (Kg/day)	Kw required	Power level Kw/1000 m³	Total Kw Required for Basin I & II
32.6	273*					40			0.78	
36.3	240**					40			0.78	
30.9	150	11.6	0.0075	23.5	18	40	18955	1818(667)	0.78 (0.2)	2352
35.6	128	33.4	0.0159	23.5	13	40	16784	1818(564)	0.78 (0.2)	
28.7	89	15.6	0.0086	11.5	18	40	9590	857(328)	0.78 (0.4)	1543
35.1	71	33.4	0.0159	11.5	13	40	8236	857(281)	0.78 (0.2)	
27.8	76	16.5	0.0089	9.1	18	40	7679	678(264)	0.78 (0.4)	1414
34.9	60	33.4	0.0159	9.1	13	40	6674	678(226)	0.78 (0.2)	
26.9	68	17.1	0.0091	7.6	18	40	6490	566(224)	0.78 (0.4)	1340
34.7	52	33.4	0.0159	7.6	13	40	5539	566(171)	0.78 (0.2)	
25.4	58	17.3	0.0091	6.1	18	40	5202	454(180)	0.78 (0.4)	1497
34.6	42	33.4	0.0159	6.1	13	40	4118	454(149)	0.78 (0.2)	
24.1	52	17.1	0.0091	5.2	18	40	4430	387(153)	0.78 (0.4)	1691
34.4	57	33.4	0.0159	5.2	13	40	3410	387(127)	0.78 (0.2)	

*winter
**summer

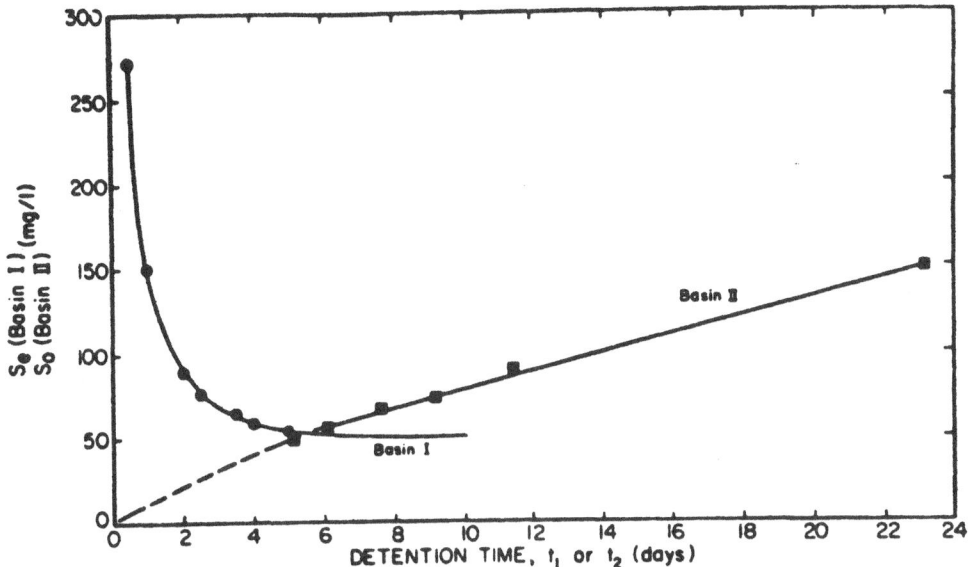

Fig. 6 : Total Required Detention Time in
Basins I and II vs Time in Basin I

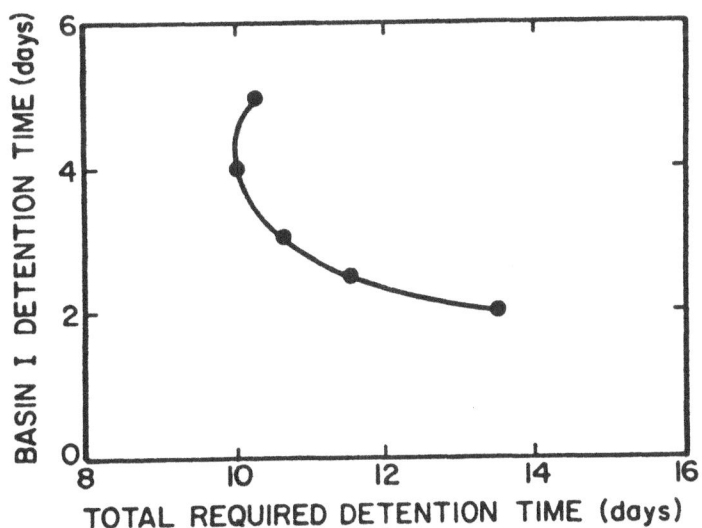

Fig. 7 : BOD Reduction in Lagoon System
for Winter Conditions

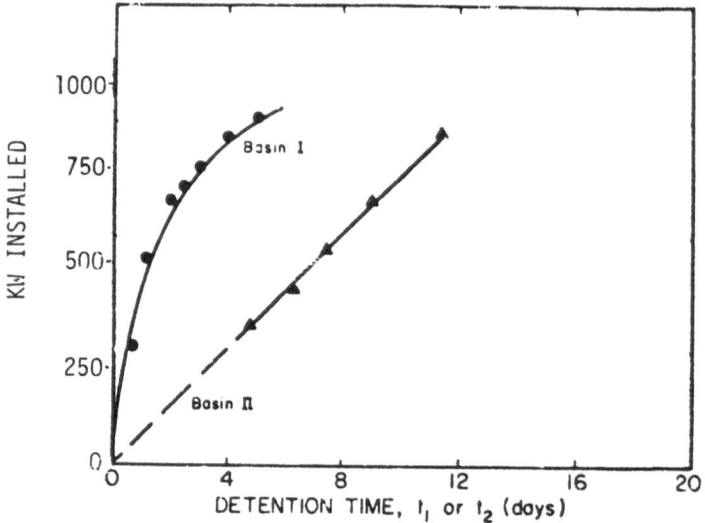

Fig. 8 :   Total Required KW for
             Lagoon System for
             Summer Conditions

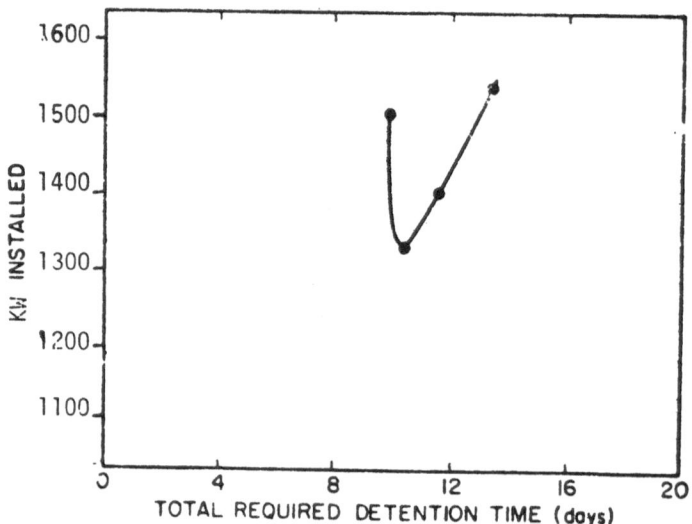

Fig. 9 :   Total Required KW vs Detention
             Time in Basins I and II

1,798 with a total detention time of 10.6 days. The second alternative is the optimum system since the increased basin size will be justified by an economic gain realized by operating with fewer aerators. In this alternative, the detention time in Basin I is 3 days and the detention time in Basin II is 7.6 days. Sample calculations for the optimum system are presented in the following section.

SAMPLE CALCULATIONS

First Aeration Basin

Design for a detention time of 3 days

Basin Volume

$$V = Qt$$
$$= 94625 \times 3 = x$$
$$= 283875 \ m^3$$

Basin Area

At a 4.6 m depth

$$A = V/d$$
$$= 283875/4.6 = x$$
$$= 61712 \ m^2$$

Basin Temperature

For a detention time of 3 days, assume a power level of 2.75 KW/1000 m^3. 782 KW or 14-56 KW aerators are required.

Characteristic Dimension of the Basin

$$X = (A)^{0.5}$$
$$= (61712)^{0.5}$$
$$= 248 \ m$$

Air Flow Through Aerator Spray

$$Q_A = (8.34)(N)(\text{Spray area})(\text{windspeed})$$

where:

N = number of aerators
Winter:

$$Q_A = (8.34)(14)(9.5)(8.5)$$
$$= 9428 \ m^3/min$$

Summer:

$$Q_A = (8.34)(14)(9.5)(10.8)$$
$$= 11980 \ m^3/min$$

Heat Transfer Coefficient

$$R = 0.114 + 0.326 \cdot X^{-0.1} \cdot V_w + 0.432 \frac{Q_A}{A} + e^{0.06 \ T_a} \cdot$$
$$\cdot (1.9 \cdot 10^{-2} \cdot X^{-0.1} \cdot V_w + 0.2525 \frac{Q_A}{A})$$

Winter:

$$K = 0.114 + 0.0326 \cdot (248)^{-0.1} \cdot 8.5 + 0.432 \cdot \frac{9428}{61712} + e^{0.06 \cdot 0} \cdot$$
$$\cdot [1.9 \cdot 10^{-2} \cdot (248)^{-0.1} \cdot 8.5 + 0.2525 \frac{9428}{61712}] =$$
$$= 0.47 \ m/min$$

Summer:

$$K = 0.114 + 0.0326 \cdot (248)^{-0.1} \cdot 10.8 + 0.432 \frac{9428}{61712} + e^{0.06 \ 26.67} \cdot$$
$$\cdot [1.9 \cdot 10^{-2} \cdot (248)^{-0.1} \cdot 10.8 + 0.2525 \frac{9428}{61712}] =$$
$$= 1.23 \ m/min$$

Equilibrium Temperature

$$T_o = T_a + \frac{1}{K} [0.02396 \ H_s + 6.94267 (\beta - 0.874) - e^{0.06 \ T_a} (1 - f_a/100) \cdot$$
$$\cdot (0.35258 \cdot X^{-0.1} \cdot V_w + 0.8432) \frac{Q_A}{A}]$$

Winter:

$$T_o = 0 + \frac{1}{0.47} [0.02396 \cdot 54.25 + 6.94267 (0.9 - 0.874) - e^{0.06 \cdot 0} (1 - \frac{50}{100}) \cdot$$
$$\cdot (0.35258 \cdot (248)^{-0.1} \cdot 8.5 + 0.8432 \cdot \frac{9428}{61712})]$$
$$= 1 °C$$

Summer:

$$T_O = 26.67 + \frac{1}{1.23} \, [0.02396 \cdot 298.35 + 6.94267(0.9 - 0.874) - e^{0.06 \cdot 26.57} \cdot$$

$$\cdot \, (1 - \frac{85}{100}) \cdot (0.35258 \cdot (248)^{-0.1} \cdot 8.5 + 0.8432 \, \frac{9428}{61712} \,)]$$

$$= 31°C$$

Basin Temperature

$$T_w = \frac{T_i + (\frac{AK}{Q})(T_O)}{1 + (\frac{AK}{Q})}$$

Winter:

$$T_w = \frac{35 + \frac{(61712)(0.47)(1)}{(94625)}}{1 + \frac{(61712)(0.47)}{(94625)}}$$

$$T_w = 27°C$$

Summer:

$$T_w = \frac{37.78 + \frac{(61712)(1.23)(31)}{94625}}{1 + \frac{(61712)(1.23)}{94625}}$$

$$T_w = 34.7°C$$

Temperature Correction of Reaction Rate Coefficient

$$k_2 = k_1 \, \Theta^{(T_i - T_2)}$$

Winter:

$$k_2 = (0.01)(1.0639)^{-2.834 + 6.67}$$

$$= 0.0127 \; \ell/mg\text{-day}$$

Summer:

$$k_2 = (0.01)(1.0639)^{1.5 + 6.67}$$

$$= 0.0166 \; \ell/mg\text{-day}$$

Organic Removal Rate

$$S_e = \frac{1 + bt}{akt}$$

Winter:

$$S_e = \frac{1 + (0.01)(3)}{(0.5)(0.0127)(3)}$$

$$= 68 \text{ mg/}\ell$$

Summer:

$$S_e = \frac{1 + (0.1)(3)}{(0.5)(0.0166)(3)}$$

$$= 52 \text{ mg/}\ell$$

Oxygen Requirements

The oxygen requirements are based on summer conditions and the 90 percentile BOD value.

$$X_v = \frac{X_{ov} + aS_r}{1 + bt}$$

$$= \frac{0 + (0.5)(450-52)}{1 + (0.1)(3)}$$

$$= 153 \text{ mg/}\ell$$

O_2 Required

$$R_r = a'S_r + b'X_vV$$

$$= (0.52)(450-52)(\frac{94625}{1000}) + (0.14)(153)(\frac{283875}{1000})$$

$$= 19584 + 6081$$

$$= 25665 \text{ Kg/day}$$

Aeration Requirements

$$N = N_o[(\frac{\beta C_{sw} - C_L}{C_s})\alpha \, 1.028^{T-20}]$$

$$= 1.95 \left[\frac{(0.9)(7.1) - (1.6)}{9.2} (0.90)(1.028)^{34.7-20}\right]$$

$$= 1.37 \text{ Kg } O_2/\text{Kw-hr}$$

Power Requirements

$$\text{Kw} = \frac{\text{Kg } O_2 \text{ required}}{N \times 24}$$

$$= \frac{25655}{1.37 \cdot 24}$$

$$= 780 \text{ Kw}$$

Power Level

$$PL = \text{Kw}/V$$

$$= \frac{780}{283875} \cdot 1000$$

$$= 2.75 \text{ Kw}/1000 \text{ m}^3$$

A summary of the characteristics for Basin I for various detention times is presented in Table 1.

SECOND BASIN

Assume a power level of 4 hp/mil gal. This degree of mixing will maintain 40 mg/ℓ of volatile suspended solids in suspension. Assume a detention time of 7.6 days.

Basin Volume

$$V = Q_t$$
$$= 94625 \cdot 7.6 = 719150 \text{ m}^3$$

Basin Area

At a 4.6 m depth
$$A = V/d$$
$$= \frac{719150}{4.6} = 156337 \text{ m}^2$$

Basin Temperature

For a power level of 0.78 Kw/1000 m^3, 566 Kw are required for or 10-56 Kw aerators.

Characteristic Dimension of the Basin

$$X = (A)^{0.5}$$
$$= (156337)^{0.5}$$
$$= 395.4 \text{ m}$$

Aerator Flow Through Aerator Spray

$$Q_A = (8.34)(N)(\text{Spray area})(\text{windspeed})$$

where:

N = number of aerators

Winter:

$$Q_A = 8.34 \cdot 10 \cdot 9.5 \cdot 8.5$$
$$= 6735 \text{ m}^3/\text{min}$$

Summer:

$$Q_A = 8.34 \cdot 10 \cdot 9.5 \cdot 10.8$$
$$= 8557 \text{ m}^3/\text{min}$$

Heat Transfer Coefficient

$$K = 0.114 + 0.326 \cdot x^{-0.1} \cdot V_w + 0.432 \frac{Q_A}{A} + e^{0.06 \, Ta} (1.9 \cdot 10^{-2} \cdot x^{-0.1} \cdot V_w + 0.2525 \frac{Q_A}{A})$$

Winter:

$$K = 0.114 + 0.0326 \cdot (395.4)^{-0.1} \cdot 8.5 + 0.432 \frac{8557}{156337} + e^{0.06 \cdot 0} (1.9 \cdot 10^{-2} \cdot (395.4)^{-0.1} \cdot 8.5 + 0.2525 \frac{6735}{156337})$$

$$K = 0.375 \text{ m/min}$$

Summer:

$$K = 0.114 + 0.0326 \cdot (395.4)^{-0.1} \cdot 10.8 + 0.432 \cdot \frac{8557}{156337} + e^{0.06 \cdot 26.67} \cdot$$

$$\cdot 1.9 \cdot 10^{-2} \cdot (395.4)^{-0.1} \cdot 10.8 + 0.2525 \frac{8557}{156337}) =$$

$$= 0.943 \text{ m/min}$$

Equilibrium Temperature

$$T_O = T_a + \frac{1}{K} [0.02396 \ H_s + 6.94267 (\beta - 0.874) - e^{0.06} \ T_a (1 - \frac{f_a}{100}) \cdot$$

$$\cdot (0.35258 \cdot X^{-0.1} \cdot V_w + 0.8432) \frac{Q_A}{A}]$$

Winter:

$$T_O = 0 + \frac{1}{0.375} [0.02396 \cdot 54.25 + 6.94267 (0.9 - 0.874) - e^{0.06 \cdot 0} (1 - \frac{50}{100}) \cdot$$

$$\cdot (0.35258 \cdot (395.4)^{-0.1} \cdot 8.5 + 0.8432) \frac{6735}{156337}]$$

$$T_O = 1.7°C$$

Summer:

$$T_O = 26.67 + \frac{1}{0.943} [0.02396 \cdot 298.35 + 6.94267 (0.9 - 0.874) - e^{0.06 \cdot 26.67} \cdot$$

$$\cdot (1 - \frac{85}{100}) \cdot (0.35258) \cdot (395.4)^{-0.1} \cdot 10.8 + 0.8432 \frac{8557}{156337}]$$

$$T_O = 32.75°C$$

Basin Temperature

$$T_w = \frac{T_i + \frac{AK}{Q} T_O}{1 + \frac{AK}{Q}}$$

Winter:

$$T_W = \frac{26.9 + \dfrac{(156337)(.375)(1.67)}{(94625)}}{1 + \dfrac{(156337)(.375)}{(94625)}}$$

$T_W = 17.24\,°C$

Summer:

$$T_W = \frac{34.7 + \dfrac{(156337)(.943)(32.75)}{(94625)}}{1 + \dfrac{(156337)(.943)}{(94625)}}$$

$T_W = 33.5\,°C$

Temperature Correction of Reaction Rate Coefficient

$$k_2 = k_1 \Theta^{(T_1 - T_2)}$$

Winter:

$k_2 = 0.01(1.0639)^{-8.28 + 6.67}$

$\quad = 0.0091 \; \ell/mg\text{-}day$

Summer:

$k_2 = 0.01(1.0639)^{.78 + 6.67}$

$\quad = 0.0159 \; \ell/mg\text{-}day$

Organic Removal Rate

$$S_e = \left(\frac{S_o}{1 + k\,X_v t}\right) F$$

Winter:

$$S_e = \frac{68(1.0)}{1 + (0.0091)(40)(7.6)}$$

$$= 18 \text{ mg}/\ell$$

Summer:

$$S_e = \frac{52(1.4)}{1 + (0.0159)(40)(7.6)}$$

$$= 13 \text{ mg}/\ell$$

Oxygen Requirements

The oxygen requirements are based on summer conditions.

$$O_2 \text{ Required} = F'S_r$$

$$= (1.5)(52-13) \cdot \frac{94625}{1000} = 5536 \text{ Kg/day}$$

Aeration Requirements

$$N = N_o[(\frac{\beta C_{sw} - C_L}{C_s})\alpha 1.028^{T-20}]$$

$$= 1.95[\frac{(0.9)(7.3) - (1.6)}{(9.2)} (0.90)(1.028)^{33.4-20}]$$

$$= 1.365 \text{ Kg } O_2/\text{Kw-hr}$$

Power Requirements

$$Kw = \frac{Kg \; O_2 \text{ required}}{N \times 24}$$

$$= \frac{5536}{(1.365)(24)}$$

$$= 169 \text{ Kw}$$

Power Level

$$PL = Kw/V = \frac{169}{719150} \cdot 1000 = 0.235 \frac{Kw}{1000 \ m^3}$$

This is insufficient to maintain complete oxygen dissolution, therefore, use 0.78 $Kw/1000 \ m^3$.

Power Required = 0.78·719.150

= 566 Kw

A summary of characteristics for Basin II is presented in *Table 2*.

Biological Sludge Production

First Basin

$$X_{v_1} = \frac{0 + aS_{r1}}{1 + bt_1}$$

$$= \frac{0 + (0.5)(450-52)}{1 + (0.1)(7.6)}$$

$$= 153 \ mg/\ell$$

Second Basin

$$X_{v2} = \frac{X_{v1} + aS_{r2}}{1 + bt_2}$$

$$= \frac{153 + .(0.5)(52 - 9)}{1 + (0.1)(7.6)}$$

$$= 99 \ mg/\ell$$

$$\Delta X_v = aS_r - bX_v V$$

$$= (0.5)(450-9) \cdot \frac{94625}{1000} - (0.1)(99) \cdot \frac{719150}{1000}$$

$$= 20865 - 7120$$

$$= 13745 \ Kg/day$$

Nutrient Requirements

Nitrogen

$N = 0.123 \ \Delta x_v$

$\quad = (0.123)(13745)$

$\quad = 1691$ Kg/day

Phosphorus

$P = 0.026 \ \Delta x_v$

$\quad = (0.026)(13745)$

$\quad = 357$ Kg/day

Sludge Accumulation

Influent VSS = 13745 Kg/day
$\qquad\qquad = 145$ mg/ℓ

Influent SS = Assume 80 percent volatile content
$\qquad\quad x_i = (13745)/(0.80)$
$\qquad\qquad = 17181$ Kg/day
$\qquad\qquad = 214$ mg/ℓ

Effluent SS = 4731 Kg/day
$\qquad\qquad = 50$ mg/ℓ

Effluent VSS = 3785 Kg/day
$\qquad\qquad = 40$ mg/ℓ

Amount of
Deposited SS = $(17181 - 13745)(31) - (4731 - 3785)(31)$
$\qquad\qquad = 106516 - 29326$
$\qquad\qquad = 77190$ Kg/month

Amount of
Deposited VSS=
$\qquad\quad W_o = (13745)(31) - (3785)(31)$
$\qquad\qquad = 426,095 - 117,335$
$\qquad\qquad = 308,760$ Kg/month

Volatile Solids Accumulated

$$W_t = \frac{W_o}{D_k} (1 - e^{-D_k t})$$

Assume a 60 percent degradation/year

$= 5.0$ percent/month

$$\frac{W_t}{W_o} = e^{-D_k t}$$

$\ell n\ 0.40 = -D_k t$

$D_k = 0.076$/month

$$W_t = (\frac{308760}{0.076}) (1 - e^{-(0.076)}$$

$= 297,318$ Kg/month

Total Accumulation

$= 77,190 + 297,318$

$= 374508$ Kg/month

At 4 percent compaction $= \frac{374508}{(40.000)} = 9362$ m^3/month

Surface area $= 156337$ m^2

Accumulation $= \frac{9362}{156337} = 0.059$ m/month

PROCESS DESIGN SUMMARY

	Basin I	Basin II
Detention time, days	3	7.6
Basin Volume, m^3	283875	719150
Depth, m	4.6	4.6
Oxygen Requirements, Kg/day	25665	5536
Power Requirements, Kw	780	169
Power Level, Kw/1000 m^3	2.75	0.235
Basin Temperature		
Summer, T_w, °C	34.7	33.44
Winter, T_w, °C	27	17
Oxygen Transfer, N, Kg O_2/Kw-h	1.37	1.365
Nutrient Requirements (maximum)		
Nitrogen, Kg/day	1691	-
Phosphorus, Kg/day	357	-
Excess Sludge Production, Kg/day	-	13745
Sludge Accumulation, m/month	-	0.059

NOMENCLATURE

BOD Removal Relationships

S_o	=	influent BOD, COD or TOC, mg/ℓ
S_e	=	effluent BOD, COD, or TOC, mg/ℓ
X_v	=	average VSS concentration in an aeration basin, mg/ℓ
t	=	detention time, days
k	=	removal rate coefficient, ℓ/mg-day
k_2	=	substrate removal rate coefficient at T°C, ℓ/mg-day
k_1	=	substrate removal rate coefficient at 20°C, ℓ/mg-day
a	=	Kg VSS produced per Kg BOD removed
b	=	Kg VSS destroyed endogeneously per Kg VSS in the system per day, day^{-1}
F	=	organic feedback coefficient due to benthal activity, dimensionless
Θ	=	temperature coefficient
T_1, T_2 =		temperature, °C

Oxygen Requirements

R_r	=	total oxygen utilization, Kg O_2/day
a'	=	Kg oxygen required per Kg BOD synthesized
S_r	=	Kg BOD removed per day
b'	=	Kg oxygen required endogenously per Kg VSS
X_v	=	average concentration of VSS in aeration basin, Kg/day
F'	=	oxygen utilization coefficient for facultative lagoons, dimensionless
N	=	oxygen transfer efficiency, field conditions, Kg O_2/Kw-hr
N_o	=	oxygen transfer efficiency, standard conditions, Kg O_2/Kw-hr
β	=	ratio of oxygen saturation in waste to that in water
C_{sw}	=	saturation oxygen concentration (depending on temperature and elevation), mg/ℓ
C_L	=	residual oxygen concentration in prototype system, mg/ℓ
C_s	=	saturation oxygen concentration at 20°C, mg/ℓ
α	=	oxygen transfer coefficient
T	=	temperature, °C

Temperature Considerations

X	=	characteristic dimension of basin, m
A	=	basin surface area, sq m

K	=	surface heat exchange coefficient, m/day
T_a	=	ambient air temperature, °C
V_w	=	wind velocity, Km/hr
Q_A	=	air flow rate through aerator spray, m^3/sec
T_o	=	equilibrium temperature, °C
H_s	=	solar radiation, Kg-cal/sq m-hr
β	=	Raphael's constant, dimensionless
f_a	=	relative humidity, percent
T_i	=	influent temperature, °C
T_w	=	aeration basin temperature, °C
Q	=	hydraulic flow, cu m/day

Sludge Production

X_v	=	average quantity of VSS in the system, Kg/day or mg/ℓ
X_{ov}	=	influent biological volatile suspended solids, Kg/day
a	=	Kg VSS produced per Kg BOD removed
S_r	=	substrate removed, Kg/day
b	=	Kg VSS destroyed endogenously per Kg VSS in system, day^{-1}
ΔX_v	=	excess sludge produced, Kg/day
ΔX	=	total excess sludge production, mg SS/day
f	=	non-biodegradable fraction of influent suspended solids
X_o	=	influent suspended solids, Kg SS/day
X_e	=	effluent suspended solids, Kg/day
f_v	=	volatile fraction of equilibrium basin SS, VSS/SS

Sludge Accumulation

W_t	=	concentration of deposited VSS at time, t, which have not degraded, Kg
W_o	=	initial concentration of deposited VSS, Kg
D_k	=	rate of VSS degradation, $time^{-1}$
t_d	=	duration of degradation, time

USE OF DISPERSED FLOW MODEL IN DESIGNING WASTEWATER TREATMENT UNITS

S. J. Arceivala

Professor, Sanitary Engineer
WHO, Regional Office for Europe, Copenhagen

It is well known that the performance of any waste treatment unit depends essentially on two factors: the treatability of the waste at the reactor conditions, and the mixing pattern in the reactor insofar as it affects the time of exposure. Removal of organic matter in waste treatment generally follows apparent first-order kinetics, and the rate of substrate removal can be stated as

$$\frac{dS}{dt} = - KS \qquad \qquad \dots\ (1)$$

in which, S = substrate concentration, (mass/volume)
K = overall substrate removal
     rate constant for the reactor, (t^{-1})
t = time of exposure to treatment in
     reactor, (t)

MIXING PATTERN

The time of exposure to treatment in a reactor depends on the actual mixing conditions in the reactor. In effect, each element of incoming flow resides in the reactor for a different length of time and the system can be described in a manner analogous to Fick's second law. Taking a steady-state materials balance at any cross-section in a reactor, Levenspiel [1] gave a differential equation which for first-order reactions can be written as follows:

$$D \cdot \frac{\delta^2 S}{\delta x^2} - U \cdot \frac{\delta S}{\delta x} - KS = 0 \qquad \qquad \dots\ (2)$$

in which,   D = axial dispersion coefficient, $(L^2)(t^{-1})$
            U = mean displacement velocity
                along reactor length, $(L)(t^{-1})$
            x = distance in the direction of mean flow, (L)

Wehner and Wilhem [2] analytically solved the above equation
nearly 20 years ago. Their solution is now finding useful applications in the wastewater treatment field where many removal processes follow first-order kinetics. Their equation can be written as

$$\frac{S}{S_o} = \frac{4a \cdot e^{\frac{1}{2}d}}{(1 + a)^2 \cdot e^{a/2d} - (1 - a)^2 \cdot e^{-a/2d}} \quad \cdots \cdots \quad (3)$$

in which,

    S & S_o = initial and final substrate concentration
              (mass/volume)
        a = $\sqrt{1 + 4Kt \cdot d}$, (dimensionless)
        d = $\frac{D}{UL}$, the reactor dispersion number
              (dimensionless)
        L = length of axial travel path of a typical
            particle in the reactor
        t = theoretical detention time in the reactor

and other terms have been defined earlier.

The term S/S_o denotes the fraction of the initial substrate concentration that remains after treatment. The fraction removed is
$(1 - S/S_o)$.

The term $\frac{D}{UL}$, the reactor dispersion number, characterizes the
mixing conditions in the reactor. It depends directly on the axial
dispersion coefficient D, and inversely on U and L. The basic
mechanism causing dispersion is the mixing that occurs due to differences in velocity of flow in different parts of a reactor.
Some sections are faster flowing and some slower. Materials are
also carried laterally and vertically by turbulent diffusion. Equation 2 is only a one-dimensional model for the sake of simplicity
in which D indicates the overall effect of the mixing process.

By definition, the value of D (and, therefore, of D/UL) is equal
to zero for ideal plug flow conditions, and infinity for ideal
completely mixed conditions. In between these two boundary conditions, the values of D and D/UL can vary widely from reactor to

reactor depending on the scale of the mixing phenomenon, the geometry of the reactor, the type of inlet and outlet, the types of baffles, projections and surfaces, inflow velocity and its fluctuations, wind, mechanical or pneumatic power input per unit volume, and such other factors.

Often, for the sake of simplicity, the mixing conditions in reactors have been assumed to follow one of the two ideal conditions stated above. Thus, some of the more commonly used models to describe substrate removal have been the following:

For ideal plug flow conditions:

$$S = S_o \, e^{-Kt} \qquad\qquad \cdots\cdots \quad (4)$$

For ideal complete-mixing, single-cell reactors:

$$S = \frac{S_o}{1 + Kt} \qquad\qquad \cdots\cdots \quad (5)$$

For ideal complete-mixing, equal cells in series:

$$S = \frac{S_o}{(1 + Kt')^n} \qquad\qquad \cdots\cdots \quad (6)$$

where,

n = number of equal-sized cells in series
t' = detention time per cell

USE of DISPERSED FLOW MODEL

In reality, waste treatment reactors are neither ideal plug flow type nor complete-mixing. They all lie somewhere in between. This is particularly true for the lagoons and ponds used in waste treatment. Thus, use of the dispersed flow model (Eq. 3), given earlier, helps to better characterize the treatment process by taking into account both the actual mixing conditions in the reactor through the use of the term D/UL *and* the treatability of the wastewater as reflected by its K value.

For reactors more or less approaching plug flow conditions ($D/UL{\to}0$), use of the ideal condition equation (Eq. 4) gives prac-

tically the same result as obtained by using the dispersed flow
model with very low values of D/UL. Similarly for reactors approaching completely mixed conditions, use of the ideal condition
equation (Eq. 5) gives practically the same result as obtained by
using the dispersed flow model with high values of D/UL. But, with
all other mixing conditions, only the dispersed flow model gives
appropriate results while both the "ideal" models tend to give er-
roneous results, the extent of error depending on how far the actual mixing conditions depart from the ideal ones. Thus, for first-
order type reactions, the dispersed flow model is more universal
in its applications to reactor design than the "ideal" models.

Murphy has shown the application of the dispersed flow model to
the design of activated sludge systems [3] and aerated lagoons [4].
Thirumurti [5] has advocated its application to stabilization ponds,
and Arceivala [7][8] to both ponds and lagoons.

Use of the dispersed flow model is aided by the development of
curves to enable graphical solutions to be readily obtained. *Fig.1*
shows a plot of Kt versus S/S_0 for a few values of D/UL. It faci-
litates application of the Wehner-Wilhem equation over a wide range
or removal efficiencies as may be required for BOD removal, bacterial die-off, etc., in waste treatment [6]. For a given efficiency and D/UL value, the product Kt is found from the figure. If
either K or t is known, the other can be readily found.

For a given substrate removal rate, K, and a desired efficiency,
the required value of the detention time, t, varies considerably,
depending on the dispersion number, D/UL, for the reactor. Thus,
D/UL is as much as a design parameter as K. This is particularly
important when high efficiencies of removal are required as, for
example, bacterial removal in stabilization ponds. From the curve
shown in *Fig. 1* it is evident that high bacterial removal efficiencies can only be obtained if D/UL is about 0.2 or less.

In the case of both laboratory models and prototypes, conservative tracer substances (e.g., dyes) can be used to find the values
of D/UL using methods suggested by Levenspiel [1] and Murphy and
Timpany [9].

The approximate range of values of the dispersion number D/UL
for different waste treatment units given by Arceivala [6] are
shown in *Table 1*. Relatively limited data are as yet available.
However, it is evident from the values listed in the table that
considerable variations in mixing occur due to various factors enumerated earlier. None of the units listed have ideal plug flow
or ideal complete-mixing conditions.

Values of D/UL are considerably affected by the geometry of the
reactor vessel, the type and disposition of the inlet-outlet system,

TABLE 1:   Likely Range of Values of the Dimensionless
           Dispersion Number D/UL for Some Waste Treatment
           Units [6].

UNIT	D/UL
Rectangular Sedimentation Tanks:	0.2 - 2.0
Activated Sludge Aeration Tanks:	
long, plug flow type	0.1 - 1.0
complete-mixing type	3.0 - 4.0 and over
Pasveer & Carrousel ditches	3.0 - 4.0 and over
Waste Stabilization Ponds:	
single cells	1.0 - 4.0 and over
cells in series	0.1 - 1.0
Mechanically Aerated Lagoons:	
square-shaped	2.0 - 4.0 and over
rectangular, long	0.2 - 1.0

and other factors as stated earlier. The statistical nature of
D/UL values must also be borne in mind as they depend on the occurrence of a number of events [6]. Thus, it is not a constant at
all times for a given reactor. In this regard, cells placed in series tend to perform more uniformly over varying conditions than
single-celled units. In a laboratory model study, no significant
effect of water temperature ranging from 12° to 30°C was found on
the observed value of D/UL [9].

For design purposes, a specific value of D/UL may need to be
aimed at depending on process requirements. For example, process
conditions may demand that complete-mixing be striven for. For
practical considerations, a value of D/UL = 4 or more may then be
selected. Conversely, where plug flow conditions are to be preferred, a value of D/UL = 0.2 or less may be sought. In each case,
the nature of the process and degree of accuracy required in predicting the efficiency will determine the choice of D/UL value. In
the past, model studies have often been recommended to ensure that
the desired value of D/UL will be obtained in the full scale reactor. This has evidently been cumbersome to use, and efforts have
been made to develop analytical procedures to facilitate estimation
of D/UL values. If one has some idea of the likely value of D in
different reactors, the other two variables U and L can be varied
at will to achieve the desired value of D/UL. A few typical values
of D are shown in Table 2.

The rather wide range of values of D shown above for aerated
lagoons and waste stabilization ponds is due to the fact that the

266

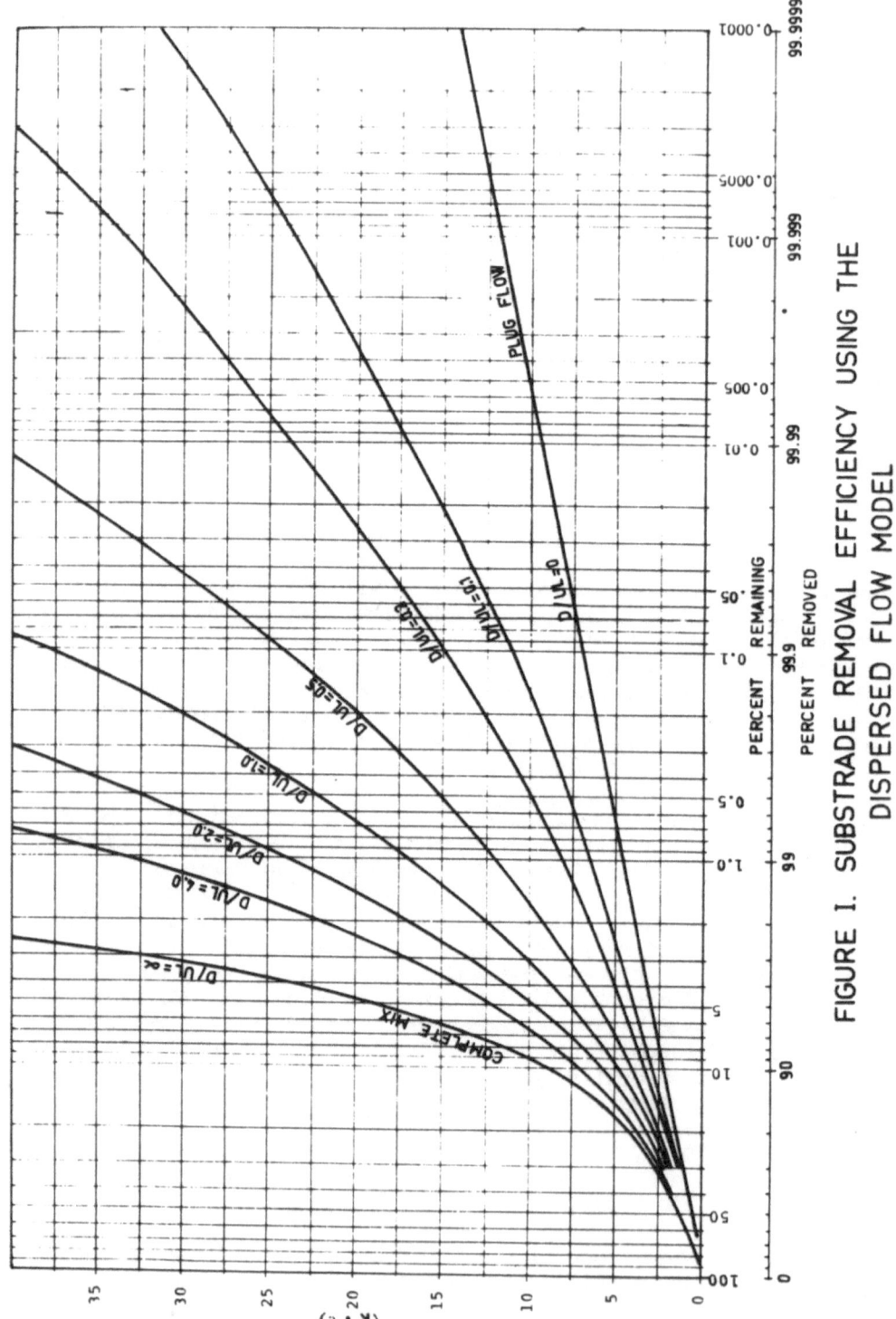

FIGURE I. SUBSTRADE REMOVAL EFFICIENCY USING THE DISPERSED FLOW MODEL

TABLE 2: Typical Values of Dispersion Coefficient,
D m^3/hour, in Some Reactors

Unit	D, m^2/hour	Reference
Activated Sludge Aeration Tanks	186 – 560	Murphy {3}
3 Aerated Lagoons (large)	2880	Murphy & Wilson {4}
Aerated Lagoons and Stabilization Ponds (various widths)	500 – 3000	Arceivala {7}

value of D depends on various factors enumerated above. Large sized, single-celled units show higher values of D while units with a narrow width or having cells in series show lower values. Arceivala [7] has developed an empirical method for estimating the D value in lagoons and ponds.

The provisions of just two cells in series can generally give *overall* values of D/UL equal to 0.2 or 0.7. Thus, where it is desired to approximate plug flow conditions the provision of cells in series is called for, the more the cells in series the surer is one of having a low value of D/UL. Analytical methods for estimating overall D/UL values in the case of cells in series are not yet available. They have to be estimated separately for each cell.

SUBSTRATE REMOVAL RATE K

The overall substrate removal rate K per unit time for a process is either determined from laboratory studies or field data. In either case, the actual mixing conditions must be known at the time that the substrate removal (eg., BOD, coliforms) data are obtained. This does not present much difficulty in laboratory studies where the mixing conditions can be manipulated to be either plug flow (batch type) or reasonably completely mixed (continuous flow type). But when field data are used, often K is calculated assuming either ideal complete-mixing or ideal plug flow conditions (Eg. 4, 5, or 6). The value of K so obtained are specific for the prevailing mixing conditions in the reactor at the time of study and cannot be extrapolated to other reactors with different mixing conditions. This neglect of actual mixing conditions when field data are used to calculate K values is one reason for the wide range of values reported from different studies.

For example, Murphy [3] studied four conventional activated sludge plants in Canada and found that the mixing conditions in the aeration tanks varied widely, eg., D/UL varied from 0.10 to 3.36 in the four plants. If the ideal complete-mixing model was applied to all cases, equally widely varying K values would have been erroneously obtained. But by using the dispersed flow model and taking into account the actual mixing conditions in each case, the K values obtained showed less variation and a value of K = 19.2 per day appeared to give a reasonable representation of the overall performance in all the four cases.

In the case of the activated sludge process and its modifications, K values can be found from batch tests (which are akin to ideal plug flow conditions) in the laboratory, using the method given by Bhatla, Stack and Weston [10].

Some typical K values observed in practice for different processes treating domestic sewage are given in Table 3. These values

TABLE 3: Typical K Values for BOD Removal After Taking Actual Mixing Conditions into Account [6].

Process	Substrate Removal Rate K/day (20°C)
Activated Sludge (conventional)	13 – 21
Aerated Lagoons (facultative)	0.6 – 0.8
Waste Stabilization Ponds	0.04 – 0.16

are no doubt considerably affected by various process conditions, thus still showing a variation in values. Process loading itself may be a factor affecting the removal rate.

APPLICATIONS to DESIGN

In order to illustrate the use of the dispersed flow model, a stabilization pond design is given (see design example) using two methods: (a) Gloyna's well-known equation based on the assumption of ideal complete-mixing conditions [11], and (b) the dispersed flow model just discussed.

The results are interesting to observe. Use of Gloyna's equation gives a pond requirement that is nearly three times larger than that obtained using the dispersed flow model and designing for more plug flow type conditions. The pond geometry is so manipulated by the provision of around-the-end type baffles or compartments series that a low value of the dispersion number D/UL is possible to use for design purposes.

Use of Gloyna's equation would be acceptable where the process would benefit from complete-mixing and pond geometry is selected to promote complete-mixing. The irony lies in the fact that sometimes lagoons and ponds are designed using the ideal complete-mixing model and their dimensions are then selected almost arbitrarily with even one or more baffles thrown in for good measure! A more economic design can be prepared in such cases if the pond geometry and the resulting mixing condition is made a design parameter besides the K value and the dispersed flow model is used in arriving at the required detention time in the unit.

DESIGN EXAMPLE

Estimate the detention time required in a waste stabilization pond to obtain 90% efficiency at 20°C in treating domestic sewage at 160 ℓ/cap-day and BOD_5 at 54 g/cap-day (BOD_u = 1.4 BOD_5). For use in the dispersed flow model assume K = 0.158 per day taking into account the temperature and likely loading conditions [7].

1. Using Gloyna's equation

   Pond Volume,

   $$V \, m^3/cap = (3.5 \times 10^{-5})(\ell/cap-day)(BOD_u \, mg/\ell)(1.085)^{(T-35)}$$

   $$= (3.5 \times 10^{-5})(160)(472)(1.085)^{20-35}$$

   $$= 9 \, m^3/cap$$

   Hence, detention time,

   $$= \frac{9 \times 10^3}{160} = 56 \text{ days}$$

2. Using the dispersed flow model

   From *Fig. 1*, the calculated detention times to achieve 90% removal (S/S_o= 0.10) at different mixing conditions are given below:

Assumed Mixing Condition	D/UL	Value of Kt (Fig.1)	Calculated Value of t days (= Kt/0.158)
Ideal plug flow	0	2.3	14.6
Cells in series or around-the-end type baffles	0.2	3.25	20.6
Single cell	4.0	6.9	43.7
Ideal complete-mixing	∞	9.0	57.0

Most ponds in the field will fall within the mixing condition given by D/UL equal to 0.2 to 4.0. Thus, ponds built with a detention time ranging from 20 to 43.7 days are all likely to show same efficiency at the given conditions, only depending on their actual mixing characteristics. The result obtained by using Gloyna's equation (56 days) gives an identical value of detention time as obtained by using the dispersed flow model with assumed complete-mixing conditions (57 days).

In this example, where complete-mixing condition is not essential for the successful operation of the process, the detention time can be reduced from 57 days to about 40 days if a typical single-celled pond is provided, and to about 20 days if the pond is adequately baffled or constructed in the form of at least two or three cells in series. This illustrates the considerable economy in design that can be achieved by paying greater attention to mixing conditions.

ESTIMATION OF D/UL

As an illustration, the value D/UL has been computed below for a pond of, say, 40 days' detention time, and serving a population of 10,000 people:

Pond Volume required = 40 x 160 ℓ/c/d x 10,000 = 64,000 m^3

Letting depth = 1.6 m,

area = 40,000 m^2

Provide two cells in parallel (*Fig. 2a*), each of 50 m width and 400 m length.

$$\text{Velocity, } U = \frac{1600 \text{ m}^3/\text{day}}{(2)(50\text{m})(1.6)} = 0.41 \text{ m/hour}$$

From *Table 2*, assume D = 900 m^2/hour, and

$$D/UL = \frac{900}{(0.41)(400)} = 5.5$$

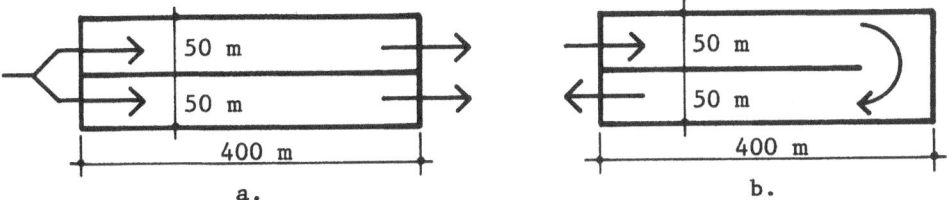

Fig. 2 :   Two Different Flow Patterns
Through Ponds of the Same
Overall Size

The pond configuration can be revised if required to give D/UL=4.0 or any other desired value. If a low value of D/UL is required, the pond configuration shown in *Fig. 2b* may be followed. This will give U = 2 x 0.41 m/hr, and L = 2 x 400 m. Thus, assuming the same value of D as above, we get

$$\frac{D}{UL} = \frac{900}{(0.82)(800)} = 1.4$$

If a yet longer value of D/UL is desired, we may provide two longitudinal baffles instead of one. As explained earlier, when D/UL is reduced there is often scope for reducing the detention time itself without affecting performance.

REFERENCES

1. Levenspiel, O., *Chemical Reaction Engineering,* John Wiley and Sons, New York, 1962.

2. Wehner, J. and Wilhem, R., "Boundary Conditions of Flow Reactor", Chem. Eng. Sc., Vol. 6, p. 89, 1956.

3. Murphy, K., "Significance of Flow Patterns and Mixing", published in *Biological Waste Treatment,* R. Canale, (ed.), Interscience Publishers, John Wiley and Sons, New York, 1971.

4. Murphy, K., Wilson, A., "Characterisation of Mixing in Aerated Lagoons", Jour. Env. Eng. Div., ASCE, p. 1105, October 1974.

5. Thirumurthi, D., "Design of Stabilization Ponds", Jour. San. Eng. Div., ASCE, April 1969.

6. Arceivala, S., *Wastewater Disposal Technology,* under publication, Marcel Dekker, New York.

7. Arceivala, S., "Design of Stabilization Ponds Using Dispersed Flow Model", under publication, Jour. ASCE.

8. Arceivala, S., *Simple Waste Treatment Methods,* Middle East Technical University, Ankara, Turkey, 1973.

9. Murphy, K., Timpany, P., "Design and Analysis of Mixing for an Aeration Tank", Jour. San. Eng. Div., ASCE, p. 1, October 1967.

10. Bhatla, M., Stack, V., Weston, R., "Design of Wastewater Treatment Plants from Laboratory Data", Jour. Wat. Pol. Con. Fed., Vol. 38, p. 601, 1966.

11. Gloyna, E., "Waste Stabilization Ponds", W.H.O., Monograph No. 60, 1971.

AEROBIC DIGESTION

W. Wesley Eckenfelder, Jr.

Vanderbilt University, Nashville, Tennessee

Carl E. Adams, Jr.

Associated Water & Air Resources Engineers, Inc.,
Nashville, Tennessee

DISCUSSION of PRINCIPLES

Aerobic digestion may be defined as the destruction of degradable organic sludges by aerobic, biological mechanisms. Generally, aerobic digestion is most applicable to excess biological sludges, such as those generated by the activated sludge and trickling filter processes. In the absence of an external substrate, micro-organisms enter the "endogenous" phase of the life cycle and deplete internal cellular carbon sources. Due to the presence of a heterogenous population of microorganisms and an extremely complex ecosystem, various microbial species may serve as food sources for other members of the population. Eventually, some organisms undergo cellular lysis releasing protoplasm into the environment which is then utilized by other bacteria. The result is a net decrease in the degradable portion of the microbial population or sludge mass. A simplified summary of the aerobic conversion of organic materials into cellular materials and the subsequent breakdown into by-products and residue is shown in *Fig. 1*.

The factors which control the aerobic digestion process are:

1. Concentration of biodegradable volatile solids
2. Temperature
3. Mixing
4. Oxygen requirements
5. Characteristics of the Solids
6. Nutrient Concentration
7. Physiological condition of the Microorganisms
8. Detention time

274

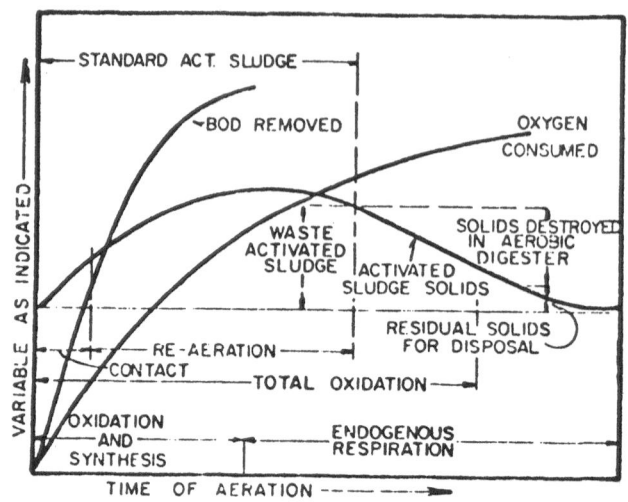

Fig. 1 :   Schematic Depiction
            of Aerobic Biological
            Treatment

The aerobic digestion kinetics are considerably lower under winter conditions, with approximately a two-fold rate of change for a
10°c temperature change.   Nutrients, such as nitrogen and phosphorous, are required for endogenous respiration; however, these requirements are usually satisfied by nutrient release during degradation of protoplasm.   It has been shown that power levels of 1.50-
2.04 m^3/min/100m^3 using diffused air or 0.02 Kw/m^3 with surface mechanical aerators are adequate for providing both mixing and oxygen
requirements.   The characteristics of the solids influence the aerobic digestion processes and the system is most feasible where the
volatile fraction is greater than 60 percent.   The physiological
state of the microorganisms, e.g., high rate activated sludge or extended aeration, will control the performance of the organisms present.   Uusallly there is an increase in stabilization with increased
detention time although a major portion of the digestion is completed in 10-15 days.

Several kinetic models have been proposed to estimate the required detention time for a desired destruction of organic sludge.
However, many aerobic digester systems have been underdesigned due
to the application of first-order kinetics to a continuous-flow
system [1].

In batch-fed reactors or continuously-fed plus flow systems, the
destruction of volatile degradable organics may be approximated by
first order kinetics as follows [2].

$$\frac{(x_d)_e}{(x_d)_o} = e^{-k_b t} \qquad \dots \text{(1a)}$$

where;

$(x_d)e$ = degradable VSS remaining after batch aeration time, t, mg/ℓ

$(x_d)_o$ = initial degradable VSS, mg/ℓ

k_b = batch reaction rate for degradable VSS destruction, day^{-1}

t = time of aeration, days

The degradable VSS, x_d, may be expressed as a function of total VSS, x_e, by incorporating the non-degradable residue, x_n:

$$(x_d)_e = (x_e - x_n) \qquad \dots \text{(2a)}$$

$$(x_d)_o = (x_o - x_n) \qquad \dots \text{(2b)}$$

where;

x_e = effluent total VSS remaining at time, t, mg/ℓ

x_o = influent total VSS, mg/ℓ

x_n = non-degradable portion of VSS, assumed constant throughout aeration period, mg/ℓ

Equation (1a) may therefore be modified to consider total VSS rather than degradable VSS by incorporating Equations (2a) and (2b).

$$\frac{(x_e - x_n)}{(x_o - x_n)} = e^{-k_b t} \qquad \dots \text{(1b)}$$

Most conventional digesters are operated on a "flow through" basis where raw excess sludge enters the digester directly from the final clarifier of an activated sludge system. The digested sludge is aerated for a sufficient period to obtain the desired VSS destruction and passes from the digester to subsequent thickening and dewatering processes. As such the digester is a completely-mixed reactor with no solids recirculation. Generally, design procedures have been comprised of determining a rate coefficient, k_b, by cor-relating degradable vss with aeration time on a batch basis with a semi-log correlation. The rate coefficient is then employed in Equation (1b) along with the desired removal efficiency to calcu-late the required detention time, t.

Unfortunately, this approach often results in an under-designed system because the aeration basin is not operated as a batch or plug-flow reactor, but as a flow-through, completely-mixed tank. A mass balance around a completely-mixed digester is depicted in *Fig. 2* and described below:

Degradables In - Degradables Out = Degradables Destroyed

$$Q(X_d)_o - Q(X_d)_e = \frac{d(X_d)_e}{dt} V \qquad \dots (3a)$$

$$(X_d)_o - (X_d)_e = k(X_d)_e t$$

where:

Q = flow rate, m^3/sec
V = digester volume, m^3

Substituting Equations (2a) and (2b) and solving for detention time:

$$t = \frac{X_o - X_e}{k_b(X_e - X_n)} \qquad \dots (3b)$$

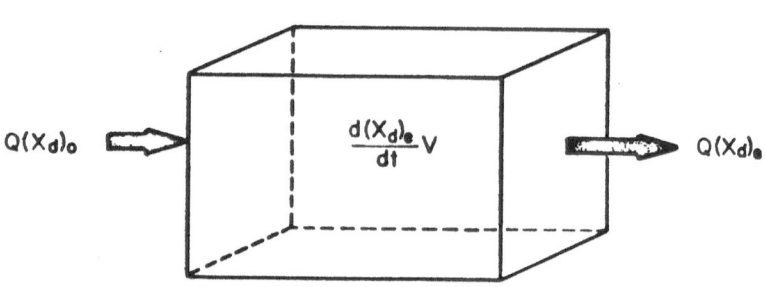

Fig. 2 :   Balance Through a
Completely Mixed
Aerobic Digester

Based on the above sequence of equations, it has been shown that a batch experimental procedure may be utilized to establish the reaction rate coefficient, k_b, but Equations (1a) and (1b) cannot be used to predict a completely-mixed flow-through system. Instead, the batch coefficient can be utilized in Equation (3c) to predict the performance of the completely-mixed basin (1). The rate coefficient, k_b, must be corrected for temperature as shown below:

$$(k_b)_T = (k_b)_{20} \, \theta^{t-20}$$

where:

$$\theta = \text{temperature coefficient (it has been found to range from 1.02 to values as high as 1.11)}$$

DESIGN EQUATIONS and NOMENCLATURE

EQUATIONS

Detention Time for a Continuous Digester

$$t = \frac{X_o - X_e}{k_b (X_e - X_n)}$$

$$k_b \text{ (base e)} = k_b \text{ (base 10) (2.305)}$$

$$(k_b)_T = (k_b)_{20} \, \theta^{T-20}$$

$$(X_d)_o = X_o - X_n$$

$$(X_d)_e = X_e - X_n$$

Oxygen Requirements

$$\left. \begin{array}{c} \text{Average daily} \\ O_2 \quad \text{used} \end{array} \right\} = \frac{\text{Area under batch curve}}{\text{time}}$$

NOMENCLATURE

t	=	detention time, days
x_e	=	effluent VSS, mg/ℓ
x_o	=	influent, VSS, mg/ℓ
x_n	=	non-degradable VSS, mg/ℓ
k_b	=	batch reaction rate coefficient, day^{-1}
$(x_d)_o$	=	influent degradable VSS, mg/ℓ
$(x_d)_e$	=	effluent degradable VSS, mg/ℓ
Θ	=	temperature coefficient

REQUIRED DESIGN INFORMATION

1. Influent flow and solids loading
2. Influent volatile suspended solids concentration, x_o
3. Reaction rate coefficient, k_b
4. Temperature coefficient, Θ
5. Non-degradable fraction of VSS, x_n
6. Oxygen requirements for VSS destruction, R_r
7. Warm and cold weather operating temperatures

DEVELOPMENT of DESIGN PARAMETERS

The laboratory or pilot studies are designed specifically to determine the rate of volatile suspended solids destruction, the maximum percent reduction of volatile suspended solids which can be expected, and the oxygen requirements for various degrees of volatile solids destruction. Either batch or continuous systems can be initiated, however, it is possible to accurately predict the continuous digester performance by correctly incorporating the batch digester results into Equation (3b).

BATCH DIGESTERS

The batch reactors should be set up to cover anticipated ranges in suspended solids and temperatures.

1. Initiate three or four batch units with approximately 4 to 20 liters of excess activated sludge in each unit. The suspended solids concentration in the batch units should be varied in order to cover the entire range of anticipated concentrations in the proposed digester.

2. Aerate the units, and, after they are completely mixed, immediately perform the following analyses in each system:

a.   Suspended solids (SS), mg/ℓ
b.   Volatile suspended solids (VSS), mg/ℓ
c.   Oxygen uptake, mg/ℓ-hr

3. Continue to aerate the systems for about 25–30 days and perform the above analyses every three days on each unit.

4. The above sequence of steps should also be performed with at least one reactor in a refrigerator at anticipated cold weather temperatures.

5. Correlate the data as shown in *Figures* (3)(4) and (5).

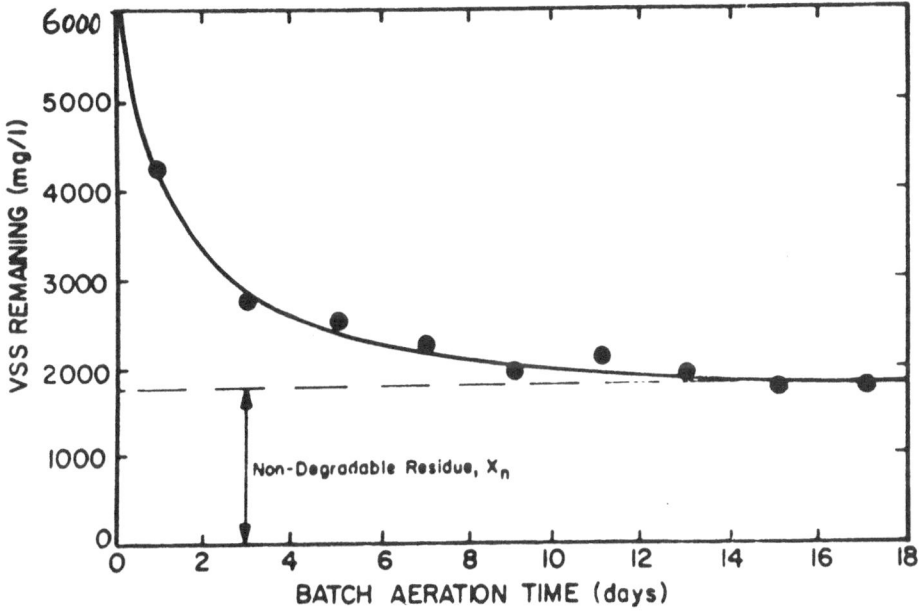

Fig. 3 :   Chronological Destruction
of VSS in Batch Reactor

DESIGN PROCEDURE

1.   Correct reaction rate coefficient for temperature

$$k_T = k_{20} \Theta^{t-20}$$

2.   Calculate non-volatile content of solids.
Use the experimental data to establish the percentage of the influent VSS which is nondegradable

$$\% \text{ non-degradable} = \frac{X_n}{X} (100)$$

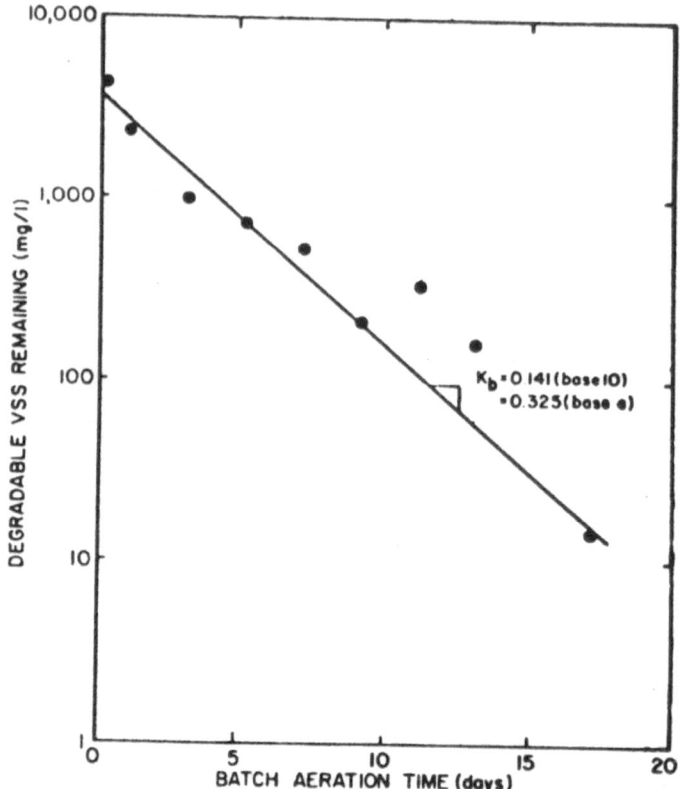

Fig. 4 : Correlation of Degradable
VSS with Detention Time

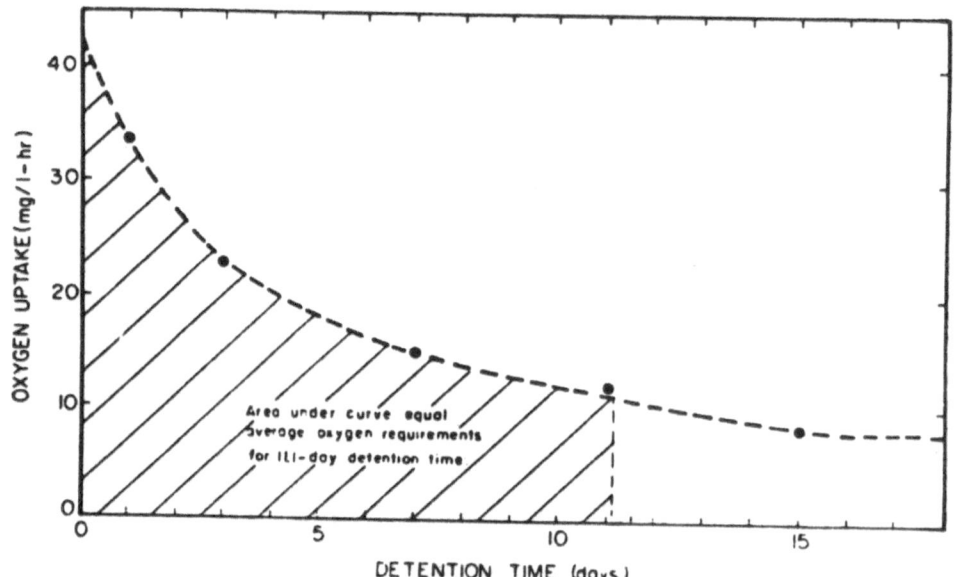

Fig. 5 : Chronological Data from
Batch Aerobic Digester

3. Calculate the effluent VSS
   a. Calculate the degradable VSS

   $$(X_d)_e = (X_d)_o - (X_d) \text{ removed}$$

   $$(X_d)_o = X_o - X_n$$

   $$(X_d) \text{ removed} = (\% \text{ removal desired}) (X_d)_o$$

   b. Calculate effluent VSS

   $$X_e = (X_d)_e + X_n$$

4. Calculate detention time

   $$t = \frac{X_o - X_e}{k_b (X_e - X_n)}$$

5. Calculate oxygen requirements
   a. Calculate area under O_2 uptake curve for design detention time (*Fig. 5*).
   b. Correct oxygen uptake rate for VSS concentration removed (refer to example problem for method).
   c. Calculate total oxygen required based on design detention time and calculated VSS concentrations (refer to example problem for method).

6. Calculate horsepower

   $$N = N_o \left(\frac{\beta C_{sw} - C_L}{C_s}\right) \alpha \theta^{T_w - 20}$$

   $$Kw = \frac{Kg \ O_2/day}{N(24)}$$

where:

N = actual oxygen transfer efficiency at field conditions, Kg O_2/Kw-hr

N_o = standard oxygen transfer efficiency for pure water, Kg O_2/Kw-hr

β = ratio of dissolved oxygen concentration at saturation in waste to that in pure water

C_{sw} = actual saturation dissolved oxygen concentration, mg/ℓ

C_L = design dissolved oxygen concentration, mg/ℓ

C_s = saturation dissolved oxygen concentration of pure water at 20°C, 1 atmosphere, mg/ℓ

α = ratio of oxygen transfer rate in waste to that in pure water

7. Check power level and determine if it is sufficient for mixing

$$PL = Kw/m^3$$

The mixing PL for aerobic digesters is dependent on solids concentration. At low solids levels, 14-20 Kw/1000 m^3 will suffice; however, at high solids levels (greater than 20,000 mg/ℓ), up to 40 Kw/1000 m^3 may be required.

8. Estimate SS leaving digester

    a. Calculate non-volatiles of influent

        Non-volatiles = Influent SS - Influent VSS

    b. Calculate SS of effluent

        Effluent SS = Effluent VSS + Non-volatiles

EXAMPLE PROBLEM

The excess biological sludge from the activated sludge system has been estimated to be approximately 3227 Kg/day at a concentration of 1.28 percent and a flow of 251 m^3/day. It is desired to reduce the degradable portion of the excess sludge by 85 percent prior to vacuum filtration and subsequent land disposal. The volatile content of the influent sludge is 85 percent.

Design an aerobic digester to obtain 85 percent destruction of the degradable suspended solids under summer conditions. Then calculate the percent destruction which would occur under winter situations at the design detention time for summer.

DESIGN INFORMATION

1. Influent flow   = 251 m^3/day
2. Influent suspended solids   = 12,800 mg/ℓ

3. Reaction rate coefficient, k_b = 0.141 (*Fig. 4*)
4. Temperature coefficient, θ = 1.05
5. Ambient Temperatures;

                        summer   = 29.5°C
                        Winter   = 4.5°C

6. Oxygen requirements (*Fig. 5*)

DESIGN APPROACH

1. Correct k_b for temperature

$$k_b = 0.141 \text{ (base 10) from } Fig.\ 4$$
$$= 2.305\ (0.141)$$
$$= 0.325 \text{ day}^{-1} \text{ (base e)}$$

Summer

$$k_b = 0.325\ (1.05)^{29.5-20}$$
$$= 0.325\ (1.58)$$
$$= 0.514 \text{ day}^{-1}$$

Winter

$$k_b = 0.325\ (1.05)^{4.5-20}$$
$$= 0.325\ (0.47)$$
$$= 0.153 \text{ day}^{-1}$$

2. Calculate non-degradable content from *Fig. 3*

$$X_o = 6,115 \text{ mg/}\ell$$
$$X_n = 1,760 \text{ mg/}\ell$$
$$X_n/X_o = 0.29$$

Therefore,

$$X_o = (12,800)(0.85) \text{ (from experimental data)}$$
$$= 10,900 \text{ mg/}\ell \quad \text{(influent VSS)}$$
$$X_n = 0.29\ (10,900)$$
$$= 3,160 \quad \text{(non-degradable VSS)}$$

3. Calculate desired effluent VSS. Desire 85 percent removal of degradable content

$$(X_d)_o = X_o - X_n$$
$$= 10,900 - 3,160$$
$$= 7,740 \text{ mg/}\ell$$

$$(X_d) = 0.85 \ (7,740)$$
$$= 6,590 \text{ mg/}\ell$$

$$(X_d)_e = 7,740 - 6,590$$
$$= 1,150 \text{ mg/}\ell$$

$$X_e = (X_d)_e + X_n$$
$$= 1,150 + 3,160$$
$$= 4,310 \text{ mg/}\ell$$

Percent vss destruction (at 85 percent degradable vss destruction)

$$= (10,900 - 4,310)(100)/10,900$$
$$= 60.5 \text{ percent}$$

4. Calculate detention time (Summer)

$$t = \frac{X_o - X_e}{k_b(X_e - X_n)}$$

$$= \frac{10,900 - 4,310}{0.514(4,310-3,160)}$$

$$= 11.1 \text{ days}$$

5. Calculate volume, v and surface area, A

$$V = Qt$$
$$= 251 \text{ m}^3/\text{day} \times 11.1 \text{ days}$$
$$= 2786 \text{ m}^3$$

Select depth at 3.65 m

$$A = V/d$$
$$= 2786 \text{ m}^3/3.65 \text{ m}$$
$$= 763.28 \text{ m}^2$$

6. Calculate vss destruction during winter

$$k_b = 0.153 \text{ day}^{-1}$$

$$X_e = \frac{X_o + k_b t X_n}{1 + k_b t}$$

$$= \frac{10,900 + 5,430}{1 + 1.72}$$

$$= 6,030 \text{ mg}/\ell$$
$$(X_d)_e = 6,020 - 3,160$$
$$= 2,870 \text{ mg}/\ell$$

Therefore,

Present VSS destruction $= (10,900-6,030)(100)/10,900$
$$= 44.7 \text{ percent}$$

Present degradable VSS
destruction $= (7,740-2,870)(100)/7,740$
$$= 62.9 \text{ percent}$$

Similarly, the percent VSS removal or destruction can be calculated for any temperature as shown in *Fig. 6*.

7. Calculate oxygen requirements based on biological utilization from *Fig. 5*, the area under the oxygen uptake curve is equal to 6,512 mg/ℓ. Therefore, the average daily oxygen use is:

$$\frac{6,512 \text{ mg}/\ell}{11.1 \text{ days}} = 587 \text{ mg}/\ell\text{- day}$$

This usage is only for the VSS concentration examined during the laboratory study. From *Fig. 3*, the VSS reduction is estimated at 4,010 mg/ℓ for 11.1 days. Consequently, the specific oxygen use rate, K_r, is:

$$K_r = \frac{587 \text{ mg utilized}/\ell\text{-day}}{4010 \text{ mg}/\ell \text{ VSS removed}}$$

$$= 0.146 \text{ mg } O_2/\text{day-mg VSS destroyed}$$

The oxygen requirements for the design conditions are calculated as:

$$X_o-X_e = 10,900 - 4,310$$
$$= 6,590 \text{ mg}/\ell \text{ of total VSS destroyed}$$

Therefore,

$$\text{Total Oxygen required} = 6,590 \text{ mg}/\ell \ (0.141 \ \frac{\text{mg } O_2/\text{day}}{\text{mg VSS}})$$

$$= 929 \text{ mg/\ell-day of } O_2$$

Therefore,

$$\text{Kg } O_2/\text{day} = 929 \text{ mg/\ell-day} \times 2786 \text{ m}^3 \times 1000\ell/1 \text{ m}^3 \times 1 \text{ Kg}/10^6\text{mg}$$
$$= 2588 \text{ Kg/day}$$

8. Calculate horsepower (Summer)

$$N = N_o \left(\frac{\beta C_{sw} - C_L}{C_s}\right) \alpha\theta^{Tw-20}$$

$$= 1.95 \left(\frac{0.98(7.7) - 2.0}{9.1}\right) (0.75)(1.028)^{29.5-20}$$

$$= 1.16 \text{ Kg } O_2/\text{Kw-hr}$$

$$= 2588 \text{ Kg } O_2/\text{day}/(1.16 \text{ Kg } O_2/\text{Kw-hr})(24 \text{ hr/day})$$

$$= 93 \text{ Kw}$$

Therefore, use one 93 Kw mechanical slow-speed aerator for biological oxygen requirements. For conservatism two 46 Kw units may be desired to ensure operation if one unit fails.

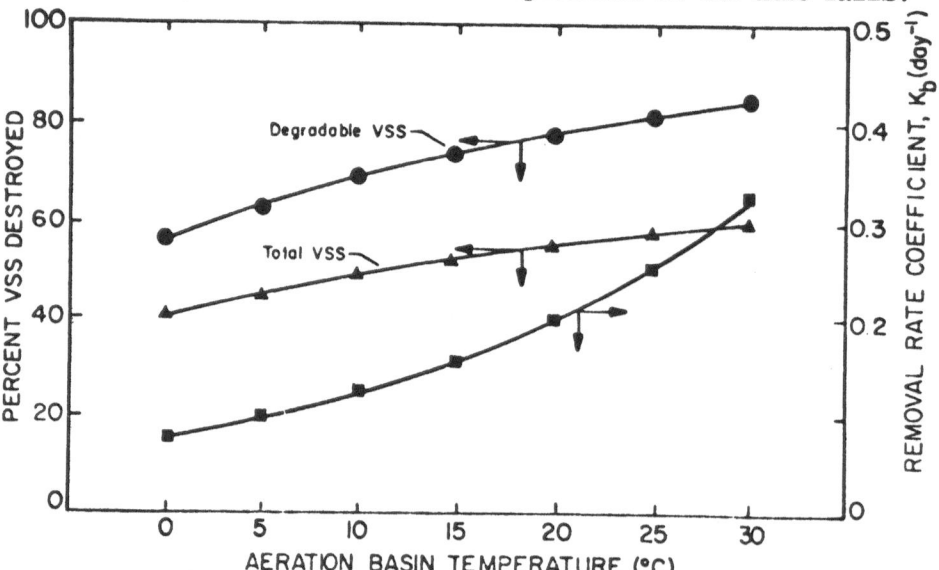

Fig. 6 : Completely Mixed Aerobic Digester Performance as a Function of Temperature

9. Check Power Level for Mixing

$$Kw/m^3 = 93 \; Kw/2786 \; m^3$$
$$= 33.38 \; Kw/1000 \; m^3$$

This power level is adequate for mixing (minimum level recommended is 14-20 $Kw/1000 \; m^3$)

10. Estimate the SS levels leaving the digester

Summer

$$
\begin{aligned}
\text{Influent SS} &= 12{,}800 \; mg/\ell \\
\text{Influent VSS} &= 10{,}900 \; mg/\ell \\
\text{Non-volatiles} &= 12{,}800 - 10{,}900 \\
&= 1{,}900 \; mg/\ell \\
\text{Effluent VSS} &= 4{,}310 \; mg/\ell \\
\text{Effluent SS} &= 4{,}310 + 1{,}900 \\
&= 6{,}210 \; mg/\ell \\
&= (6210 \; mg/\ell)(251/m^3/day)(1000\ell/m^3)(\frac{1 \; Kg}{10^6 mg}) \\
&= 1558 \; Kg/day
\end{aligned}
$$

Winter

$$
\begin{aligned}
\text{Non-volatile} &= 1{,}900 \; mg/\ell \; \text{(as shown above for summer)} \\
\text{Effluent VSS} &= 6{,}030 \; mg/\ell \\
\text{Effluent SS} &= 6{,}030 + 1{,}900 \\
&= 7{,}930 \; mg/\ell \\
&= (7930 \; mg/\ell)(251 \; m^3/day)(1000\ell/ \; 1 \; m^3)(\frac{1 \; kg}{10^6 \; g}) \\
&= 1990 \; Kg/day
\end{aligned}
$$

PROCESS DESIGN SUMMARY

Influent Flow, m^3/day	251
Influent Solids, Kg/day	3227
Influent Solids, %	1.28
Influent Solids Volatile Content, %	85
Reaction Rate Coefficient, k_b (*Fig. 4*)	0.141
Temperature Coefficient, θ	1.05
Ambient Temperatures	
Summer °C	29.5
Winter °C	4.5
Digester Detention Time, days	11.1
Digester Volume, m^3	2786 m^3
Digester Area, m^2	763.28 m^2

Effluent Solids

Summer, Kg/day		1558
Summer, mg/ℓ		6,120
Winter, Kg/day		1990
Winter, mg/ℓ		7,930

Percent Degradable Suspended Solids
  Destruction

Summer, %		85.0
Winter, %		62.9

Oxygen Requirement, Kg/day	2588
Power Requirement, Kw	93
Power Level, Kw/1000 m^3	33.38

REFERENCES

1. Adams, C.E., W.W. Eckenfelder, Jr. and R. M. Stein, Modifications to Aerobic Digester Design, *Water Research* 8:2, 1974.

2. Eckenfelder, W.W. and D.L. Ford, *Water Pollution Control,* The Pemberton Press, Jenkins Publishing Co., Austin, TX., 1970.

PROCESS AND ECONOMIC CONSIDERATIONS OF PONDS FOR THE TREATMENT OF INDUSTRIAL WASTEWATERS

Davis L. Ford, Ph.D., P.E.

Senior Vice-President, Engineering-Science, Inc.

Lial F. Tischler, Ph.D., P.E.

Manager, Austin Office, Engineering-Science, Inc.

The inherent attractiveness of waste stabilization pond treatment of selected industrial wastes has long been recognized and the performance reasonably well documented. The applicability of these systems has been predicated primarily on land availability, climatic conditions, and amenability to the industrial raw waste load. There are approximately one thousand waste stabilization pond systems treating industrial wastewaters in the United States, about half of which are used in the poultry and food processing industry. There are several such installations treating refinery, petrochemical, and chemical wastewaters - particularly in the Southwestern United States.

With the passage of the Federal Water Pollution Control Act Amendments in 1972 (P.L. 92-500), the rising cost of energy, the diminishing availability of land, and fluctuations in the national economy, a continuous reassessment of the pond advantages and disadvantages is merited. It is the purpose of this treatise to objectively evaluate waste stabilization ponds as an applicable mode of industrial wastewater treatment, both in terms of process capabilities and economic incentives.

INTRODUCTION

Basic process and design concepts for waste stabilization ponds are sufficiently described in the literature and will not be considered here [1, 2, 3, 4]. There are many "adaptive" factors which must be integrated into ponds treating industrial wastewater, however, such as algal toxicity allowances, nutrient availability, transient load constraints, and reaction times necessary to achieve required reduction of residual organic compounds [5, 6]. Process simulation techniques using bench - or pilot-scale pond systems are often required to establish ponds as a viable process alternative in treating industrial wastes. Although such studies are time-consuming, they generally can be justified in the establishment of a rational basis for designing the facility, minimizing capital costs, and predicting effluent quality. The latter item takes on added significance when considering that the various industrial point source categories have effluent limitations as a function of production capacity in accordance with P.L. 92-500. The parameters included in these Guidelines vary between categories, but basically include BOD_5, COD or TOC, oil and grease, total suspended solids, and pH. The ability of existing or proposed pond systems to meet these quality limitations for the appropriate industrial category or subcategory must be firmly established in order to predict the probability of compliance or noncompliance with Federal or state permits. This evaluation may lead to an algal removal requirement in order to fully comply, thus significantly impacting the cost-effectiveness of waste stabilization ponds as a method for treating industrial wastewaters.

PROCESS CONSIDERATIONS

Waste stabilization ponds are generally classified as anaerobic, facultative, or polishing. Anaerobic ponds receive a high organic loading and are therefore devoid of oxygen. Facultative ponds, which are most commonly used, provide an aquatic environment in which photosynthetic and surface-supplied oxygen exceeds the demand in the aerobic upper layers while the facultative zone, in which aerobic and anaerobic conditions alternate, prevails throughout most of the pond depth and the bottom layers are totally anaerobic. Polishing, or aerobic ponds, are shallow and are specifically designed to allow oxygen penetration throughout the depth.

Organic loading and hydraulic detention time are paramount in determining the classification of a waste stabilization pond. The average loading factors and performance data for a wide range of industrial categories are shown for anaerobic ponds in *Table 1*, for facultative/aerobic ponds in *Table 2*, and for anaerobic/aerobic ponds in *Table 3* [7, 8, 26]. These data underscore the process capability of ponds to successfully treat many industrial wastewaters, particularly using series combinations of anaerobic and facultative pond systems. This history of performance, positive treatability study results, availability of land, and proper climatological conditions all would tend to bias the selection of waste stabilization ponds as the most applicable treatment process.

Other process advantages are recognized in operation and maintenance considerations. Operation and maintenance costs for pond sys-

TABLE 1 :   Summary of Average Date from Anaerobic Ponds

INDUSTRY	Area (Hectares)	Depth (meters)	Detention Time (days)	Loading (Kg/ Ha-day)	BOD Removal (percent)
Canning	1.00	1.82	15	440	51
Meat & Poultry	0.40	2.22	16	1415	80
Chemical	0.60	1.06	65	60	89
Paper	28.60	1.82	18.4	389	50
Textile	0.89	1.76	3.5	1609	44
Sugar	14.13	2.13	50	269	61
Wine	1.49	1.22	8.8	-	-
Rendering	0.40	1.82	245	179	37
Leather	1.05	1.28	6.2	3370	68
Potato	4.04	1.22	3.9	-	-
Average Values				966	60

TABLE 2 :   Summary of Average Data from
            Facultative-Aerobic Ponds

INDUSTRY	Area (Hectares)	Depth (meters)	Detention Time (days)	Loading (Kg/ Ha-day)	BOD Removal (percent)
Meat & Poultry	0.52	0.91	70	81	80
Canning	2.78	1.76	37.5	156	98
Chemical	12.51	1.52	10	176	87
Paper	33.90	1.52	30	118	80
Petroleum	6.26	1.52	25	31	76
Petrochemical	-	-	-	112	95
Wine	2.82	0.46	25	248	-
Dairy	3.03	1.52	98	25	95
Textile	1.25	1.22	14	185	45
Sugar	8.07	0.46	2	97	67
Rendering	0.89	1.28	48	40	76
Hog Feeding	0.24	0.91	8	400	-
Laundry	0.80	0.91	94	58	-
Miscellaneous	6.05	1.22	88	63	95
Potato	10.21	1.52	105	125	-
Average Value				128	81

TABLE 3 : Summary of Average Data from Combined
Anaerobic-Aerobic Ponds

INDUSTRY	Area (Hectares)	Depth (meters)	Detention Time (days)	Loading (Kg/ Ha-day)	BOD Removal (percent)
Canning	2.22	1.52	22	693	91
Meat & Poultry	0.32	1.22	43	300	94
Paper	1020	1.67	136	31	94
Leather	1.86	1.22	152	56	92
Miscellaneous Industrial Wastes	57	1.25	66	144	-
Resins, Alcohols Amines, Esters, Styrene, Ethylene	308	-	-	46	99
Olefins, Glycols, Aldehydes, Polyolefins	190	-	240	46	99
Average Values				112	97

tems are low compared to other methods of biological treatment due
to the inherent simplicity of these systems. Of considerable importance is the energy savings associated with the use of a pond system in the warmer climates. Since the oxygen utilized in these
systems is not introduced by mechanical means but relies on plant
photosynthesis and surface reaeration, no energy is required to provide the oxygen necessary for the biological degradation of organic
compounds. Considering the high cost of energy and the depletion
of fossil fuels associated with energy use, a system such as the
waste stabilization pond, when it can provide high effluent quality
and protection of the receiving waters, must be considered to be
beneficial to the overall national energy program and the environment. In terms of maintenance, the fact that no artificial aeration
devices or mechanical equipment are required results in a considerable
operational savings to the using entity. The only mechanical equipment involved in a typical pond facility are pumps which may be required in some situations when gravity feed cannot be used through-

294

out the entire system or when recirculation of pond contents is desired.

Another important advantage accrues to the long residence time
which is typical of waste stabilization ponds as currently applied
to wastewater treatment in the organic chemicals industry. Many of
the raw materials, intermediates, and products entering a process
wastewater stream from an industrial complex are resistant to short-
term biodegradation. Thus, long periods of treatment are often necessary to reduce contaminant levels to acceptable concentrations.
This is especially true when considering COD or TOC reduction efficiency. The extended detention times provided by waste stabilization pond systems provide equalization and have a higher potential
for removal of refractory organic materials than do the high-rate,
low-detention-time biological treatment processes. The long retention times typical of waste stabilization pond systems enable the
development of bacterial species capable of breaking down some or
all of the refractory compounds which may be found in many industrial
wastewaters.

ALGAE REMOVAL CONCEPTS

In the discussion to this point, the merits of waste stabilization pond systems have been emphasized in treating industrial wastewaters qualified with favorable climate, treatability, design and
operational conditions. It should be recognized, however, that positive process control cannot always be maintained, ponds are subject to upset, and the effluent quality susceptible to both gradual
seasonal and abrupt weather changes. Nonetheless, properly designed
waste stabilization pond systems are capable of producing an effluent
quality consistent with the Environmental Protection Agency's definition of "best practicable control technology currently available"
(BPCTCA) for most industrial categories. The one major exception
to this is the ability to meet the TSS requirements established under the Guidelines. In order to meet these currently established
effluent limitations, a method for removing effluent solids, which
are primarily algal cells in the pond systems, would have to be selected. As this step significantly alters the incentive for selecting waste stabilization ponds in terms of cost-effectiveness, the
various algae removal concepts will be discussed in some detail.

There have been extensive investigations of algae removal methods
during the past several years. These studies have centered not only
on effluent quality improvement but also on algae harvesting as a
livestock food source. A review of the candidate unit processes for
algal TSS removal is presented as follows.

CENTRIFUGATION

A comprehensive evaluation of centrifugation using chemical additives indicated a 77 percent removal of algae, although the concentrate suspended solids would still not meet Guideline objectives [9]. The investigators concluded that centrifugation would not only produce an effluent incapable of meeting most TSS quality criteria, but also was highly energy-consumptive with intensive operation and maintenance requirements.

MICROSTRAINING

Microstrainers have been used primarily in water treatment plants to remove algae from lake and reservoir waters. In a pilot-plant study operated at flow rates of 3.15 to 6.3 liters/sec, an extremely low degree of algae removal was observed, even with the use of filter aids [9]. Microstraining operations in Canada have achieved up to 89 percent reduction of plankton [10]. None of the above applications, however, have been directed strictly to algae removal. It should be noted that these applications represent treatment of lake waters with much lower algae concentrations than typical waste stabilization ponds. Water reclamation treatability studies conducted in California included operation of a 30 gpm microstrainer pilot plant [11]. Typical influent algae counts were 50,000 to 700,000 organisms per milliliter. Even though the filtering fabric had a pore size of 23 microns, algae removal was uneconomical. Growths of algal and bacterial slimes on screens results in extremely low flow rates through the units [12]. Others have found microstraining to be too complex and too expensive for routine use [13].

COAGULATION-FLOCCULATION-SEDIMENTATION

Many investigations have been directed toward the chemical removal of algae using various coagulants and polyelectrolytes. An optimum dosage of alum for algae removal was determined to be 70 to 100 mg/ℓ according to one investigator [10]. Several researchers have found effective algal flocculation in the pH 2-4 range by adding 10 mg/ℓ of cationic polyelectrolyte [10]. One study evaluated the use of alum, lime, nonionic polyelectrolytes and cationic polyelectrolytes [14]. Alum was determined to be most effective at concentrations of 125-170 mg/ℓ. Excess lime was required to obtain equivalent treatment, and polyelectrolyte addition in this study seemed to have little effect on process performance. Although opinions vary, cationic polyelectrolytes appear to be more effective than nonionic or anionic ones when applied in addition to coagulants. Most have been deemed ineffective when used alone.

Treatability studies conducted prior to the design of a wastewater reclamation facility considered the use of alum, lime, ferric sulfate, cationic polyelectrolytes, and sulfuric acid. Alum was found to be the most effective coagulant, requiring dosages of 300 mg/ℓ [11]. Another investigation used both bench-and pilot-scale reactors to evaluate the chemical precipitation of algae. A combination of 110 mg/ℓ of ferrous sulfate and 150 mg/ℓ of lime resulted in a 90 percent removal of algal cells at an initial concentration of 1,000,000 cells per milliliter. The TSS reduction was from 163 mg/ℓ to 25 mg/ℓ. Alum, however, was determined to be the most desirable coagulant, removing 83 to 98 percent of the algae at an optimum dosage of 250 mg/ℓ [15].

It should be recognized that this approach toward algae removal creates a large mass of mixed organic and inorganic sludge which must be handled in an appropriate manner. Most studies have shown that these hybrid sludges are unfit for fertilizer or livestock feed.

DISSOLVED AIR FLOTATION

Dissolved air flotation (DAF) has some inherently attractive features as applied to algae removal. The DAF allows close operational control through varying air-solids ratios, and one would expect less chemical requirements than for the coagulation-flocculation-sedimentation process since the algal cells do not have to be weighted to assist in settling. One study indicated 95 percent algae removal with a DAF using a controlled pH of 3.0 and a dosage of 51 mg/ℓ ferric sulfate. In bench-scale tests, 90 percent TSS removals were obtained using 85 mg/ℓ ferric sulfate or 75 mg/ℓ alum. Pilot-plant tests indicated that 175 mg/ℓ of alum would be required for equivalent removal. Differences between the bench- and pilot-scale tests were attributed to changes in algal species and more realistic test conditions in the pilot plant. A maximum hydraulic loading rate to the DAF was determined to be 95 liter/min-m^2 [16].

The DAF unit is the key process in the recently developed Accelerated Photosynthetic System (APS). This process, designed to develop maximum algae growth in a controlled basin, utilizes DAF as the final algae separation step. Large pilot-scale studies have indicated that TSS levels in the APS basin range from 200 to 300 mg/ℓ, and this is reportedly reduced to 20 to 30 mg/ℓ following flotation using lime or alum coagulants [17].

FILTRATION

The possibility of using granular media filtration for algae removal has been thoroughly investigated with only marginal success. One pilot-plant program evaluated filtration of a municipal wastewater pond effluent using a hydraulic loading rate of 40.7 liters/min-m^2. The average results are presented in Table 4 [18].

TABLE 4 : Pilot Upflow Filter Results

Parameter	Influent	Effluent	Percent Removal
BOD$_5$	6	5	16
COD	57	47	18
TSS	19	8	58

Another study evaluated algae removal through an experimental rapid sand filter operated at 81.3 liters/min-m^2. It was noted that discrete particles such as silt, clay, organic colloids, and algae have displayed considerable filter-penetrating ability. During the study, the electrophoretic mobilities of algae and sand were varied through pH adjustment and sand coating with ferric iron and thallium. These ions were thought to adsorb on the sand surfaces and reverse sand particle charge. At low influent pH values, the coatings increased removal by as much as 10 percent. Algae removals of 90 percent or more were obtained when the influent pH was adjusted to a value between 2.0 and 3.0, using initial algae concentrations of 8,900 to 81,600 cells per milliliter (TSS of three to 16 mg/ℓ). The removal mechanisms was thought to be one of organism-sand interaction. It is important to note that algae in the filter backwash were dispersed and did not settle.

Algae removal through a diatomaceous earth reversible cake filter pilot plant was the subject of another investigation [19]. The influent, a domestic wastewater lagoon effluent, contained 75 mg/ℓ TSS and was filtered at a rate of 46.36 liters/min-m^2. The filter effluent indicated an average of 80 percent removal of TSS, but the filtration cycle lasted only 30 minutes, with a backwash cycle of 1.5 minutes which is not practical for a full-scale operation. Pressure drops through the filter had to be maintained at a level less than 0.28 kg/cm^2 or the algae would pack and seal the cake, releasing soluble organics in the process. Separation of algae from recovered diatomaceous earth was extremely difficult. The filter employed a monofilament polypropylene filter cloth and a diatomite body feed of 45 mg/ℓ. Only 76 percent of the body feed was recovered during backwashing. Similar results were reported in an earlier study which noted that the algal cells form an impenetrable mat over the surface of the filter [15]. This resulted in very short filter runs and large wash water requirements.

It is apparent from these and other investigations that although algae removal through granular filters can be accomplished, the inordinately low filtration rates combined with difficulty in handling backwash waters renders this approach of dubious practicality as an algae removal process for waste stabilization pond effluents.

CHEMICAL DESTRUCTION

The use of various chemicals for altering the form of algae and rendering them more susceptible for removal have been shown to be relatively successful. A dosage of 0.3 to 2.0 mg/ℓ of chlorine has been reported as effective for algae control [20]. Others have shown no apparent correlation between chlorine addition and algal concentrations [21]. At a water reclamation facility, chlorination was thought to aid in the flocculation of motile algae [22]. In other instances, the chlorination of a waste stabilization pond effluent resulted in increasing the BOD_5 values and forming potentially toxic chlorinated hydrocarbon compounds [23].

The algicidal properties of copper sulfate ($CuSO_4$) have been utilized since 1904 for controlling algae in lakes and reservoirs [20]. The compound takes several days to eliminate the algae and is usually applied from boats. Chlorine is generally added at the same time to destroy protozoa and oxidize odor-causing compounds. However, the destruction of algal blooms by this method is usually accompanied by an intensification of odors and an increase in the number of saprophytic bacteria feeding on the dead algal cells. In other words, the soluble organic material in the water increases significantly which is certainly detrimental to the efficiency of a wastewater treatment facility. Dosages have to be increased for high alkalinity waters and for those containing significant quantities of organic matter. Other methods of chemical destruction include the addition of powdered activated carbon to shut out sunlight and the addition of lime to deprive the algae of carbon dioxide. Since copper is in itself toxic to various aquatic organisms, none of these three methods is applicable to a flow-through system. Recently, two nontoxic algicides have received initial screening tests, but are suitable only for destruction of blue-green algae. The experiments were of limited value as far as practical applications are concerned [24].

OTHER METHODS

The use of ion exchange resins in the removal of algae from wastewater stabilization pond effluent was evaluated as a possible removal process [9]. Both strong and weak cationic resins were capable of removing algae through a change in surface charge. The process was not particularly effective due to fouling of the exchange resin and frequent regeneration requirements.

The use of disks, baffles, and raceways, supporting only attached growth, has been suggested as a possible means of algae removal;

however, further particulate removal would still be required to produce an effluent of reasonable quality. Complete containment of wastewater has been utilized in the case of slaughterhouse and meat-packing wastes, requiring large land areas and deep anaerobic ponds. Such a mode of removal is applicable only in the case of high-strength, low-volume wastewaters, and results in odors and filling of ponds.

Chemical precipitation in ponds has been offered as a candidate method for algae removal, but decay of settled matter produces soluble BOD and the process lacks positive control. Soil mantle disposal of spray irrigation requires large land areas and is limited by clogging of soil pores. Moreover, such a method of elimination may result in pollution of ground water and surface runoff. None of the methods described above appear to be particularly promising in terms of efficient application to algae removal from stabilization pond effluents.

DEMONSTRATION of ALGAE REMOVAL- CASE HISTORIES

The only prototype-scale technology demonstrated for algae removal from stabilization effluents to date involves integrated water reclamation facilities. Performance data for two of these plants are available and are discussed below:

Windhoek Plant

A plant was operated in Windhoek, South West Africa to investigate reclamation of polishing pond effluent [14]. Raw domestic wastewater first underwent primary clarification and then was routed through a trickling filter. Filter effluent was sent to a polishing pond with a detention time of seven days. Note that this is not a true facultative waste stabilization pond system. The pilot plant consisted of the following unit processes: coagulation-flocculation, sedimentation, sand filtration, activated carbon filtration, and breakpoint chlorination. The required alum dosage depended on the TSS and algae concentrations in the pond effluent, but ranged from 125 to 170 mg/ℓ. Optimum lime dosages ranged from 300 to 400 mg/ℓ.

Suspended solids were reduced from 85 to 17 mg/ℓ through the entire plant. Algae concentrations in the polishing pond were not given, but can be expected to be much lower than that typical of a facultative pond system due to the lower organic loading in the polishing pond. The effluent from the sand filter still contained some residual color, and chlorination removed the major portion of that residual. Performance data for the individual processes are not available. Problems were encountered with floating algae in the sedimentation basin, often amounting to as much as 50 percent of the influent algae loading.

Lancaster Tertiary Treatment Plant

This water reclamation facility was constructed to treat 1890 m³/day of municipal stabilization pond effluent [22, 11, 25]. The treatment train consists of coagulation-flocculation with 300 mg/ℓ of alum addition, sedimentation, dual-media filtration, and chlorination, as shown in *Fig. 1*.

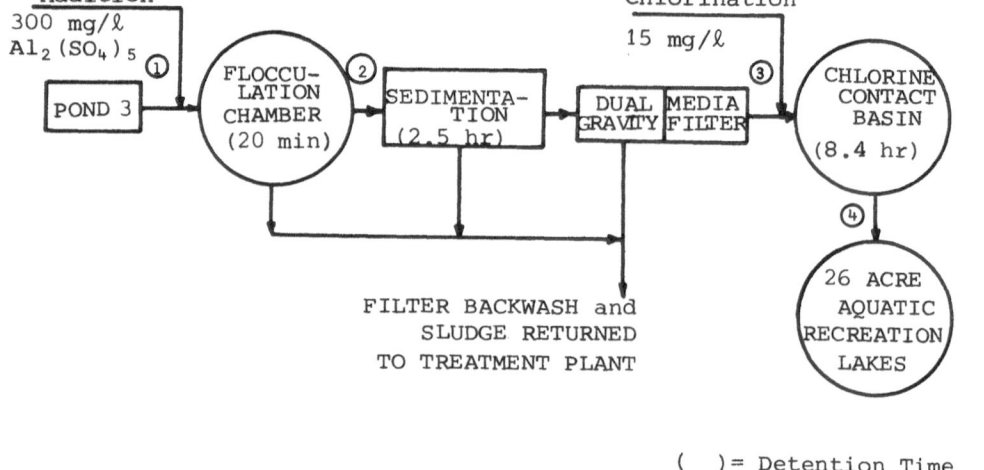

( ) = Detention Time
○ = Sampling Station

Fig. 1: Process Flow Diagram Lancaster
Tertiary Treatment Plant

Design criteria are tabulated in *Table 5*. As a result of treatment through this integrated plant, suspended solids are reduced from 75 mg/ℓ to 10 mg/ℓ, and algae are reduced from 200,000 organisms/milliliter to 1,000 organisms/milliliter. Typical treatment performance is summarized in *Table 6*. Blue-green algae do not flocculate readily and often penetrate the dual-media filter.

The two plans described above involve a number of unit operations and produce high-quality effluents. Indeed, they are basically water treatment plants and operations are extremely costly. They are included in this discussion as examples of reliable, efficient algae removal systems, but obviously exceed the treatment intensity conceived as BPCTCA treatment requirements under P.L. 92-500.

TABLE 5 : Lancaster Treatment Plant Design Criteria

1. Flocculation Tank	
Flow	83 m^3/hr
Length	2.45 m
Width	4.90 m
Depth	2.45 m
Detention Time	20 min
Alum dose	300 mg/ℓ (50% solution)

2. Sedimentation Tank	
Length	20.75 m
Width	4.90 m
Depth	2.45 m
Overflow rate	20.37 m^3/day/m^2
Detention Time	2 - 3 hours

3. Filter		
Design flow		82 m^3/hr
Area		16.7 m^2
Loading		4.9 m^3/hr/m^2
Final Headloss		17.7 m^3/hr (max)
Maximum backwash cycle		24 hrs
Maximum backwash rate		44.1 m^3/hr/m^2
Surface wash		0.45 m /hr at 3.4 atm
Maximum bed expansion	#1.5 Anthrafill	4.1 m^3/hr
Bed composition	#20 sand	2.05 m^3/hr
	graded gravel	4.1 m^3/hr

4. Chlorination Station	
Design Flow	79.5 m^3/hr
Contact pond volume	622 m^3
Detention time	8 - 10 hrs
Maximum chlorine dosage	15 mg/ℓ
Water depth	2.15 m

TABLE 6 :   Lancaster Tertiary Treatment
            Plant Performance Data

	Pond 3 Water	Sedimentation Effluent	Sand Filter Effluent	Final Effluent
Sampling Station	1	2	3	4
pH	8.3	6.8	6.8	6.8
Turbidity (JTU)	90	15	8	6
Total Alkalinity (mg/l as $CaCO_3$)	260	140	140	140
Suspended Solids (mg/l)	75	25	12	10
Dissolved Solids (mg/l)	600	600	600	600
COD (mg/l)	190	90	75	75
BOD (mg/l)	38	--	--	11
Hardness (mg/l as $CaCO_3$)	80	85	90	90
Ammonia-Nitrogen (mg/l as N)	0.1-20	0.1-20	0.1-20	0.1-20
Organic Nitrogen (mg/l as N)	7	3	2	2
Nitrate (mg/l as N)	1.0	1.0	1.0	1.0
Nitrite (mg/l as N)	0.1	0.1	0.1	0.1
Total Phosphate (mg/l as PO_4)	40	0.5-2	0.5-2	0.5-2
Carbon Dioxide (mg/l)	1	1	1	1
Dissolved Oxygen (mg/l)	0.5-35	7-15	7-15	7-15
ABS (mg/l)	5	5	5	5
Algae (counts/mil)	200,000	20,000	10,000	1,000
Coliform (MPN/100 ml)	150,000	10,000	10,000	0

ECONOMIC CONSIDERATIONS

Waste stabilization ponds, as previously stated, have always been attractive economically when compared to alternative secondary treatment systems, both in terms of capital cost and annualized cost. This attractiveness diminishes as land costs become higher but decreases very significantly if an algae removal step is required. This is illustrated by taking an example treatment system designed to receive 3780 m^3/day of industrial wastewater with a BOD_5 loading of 1136 Kg/day. Using a range of organic surface loading values similar to that shown in *Tables 2* and *3* of 128 Kg/ Ha-day and 56.1 Kg/Ha-day for facultative aerobic and anaerobic-aerobic ponds, respectively, surface areas and capital costs can be estimated. This loading range is assumed to be suitable for treating most industrial wastes to soluble BOD concentrations near Guideline levels. A capital cost of $ 61744/Ha which includes ordinary earthwork, dike construction, conveyance facilities, and appurtenances but is exclusive of land costs. Land costs are then added in at various prices per acre as shown in *Table 7*.

TABLE 7 : Estimated Capital Costs

	Facultative-Aerobic Pond	Anaerobic-Aerobic Pond
Design Surface Loading	128 Kg BOD_5/Ha-day	56.1 Kg/Ha-day
Area Required	8.9 Ha	20.24 Ha
Base Cost	$ 550,000	$ 990,000
Land at:		
$ 2470/Ha	$ 572,000	$ 1,029,000
$ 24700/Ha	$ 792,000	$ 1,425,000
$ 74100/Ha	$ 1,210,000	$ 2,180,000

Using a capital recovery factor of 0.12 and assuming that the operational and maintenance costs are equal to the amortized capital, the cost per lb of BOD_5 applied can be calculated, which is an appropriate method for presenting unit costs for industrial wastewater treatment facilities. The cost of algae removal is additive, and a unit cost of $ 2.2 per kg of stabilization pond TSS removed is assumed [26]. Using a TSS concentration of 100 mg/ℓ in the pond effluent, the equivalent cost is $ 0.75 per kg BOD_5 applied for the example plant. This has a pronounced pond approach as illustrated in *Fig. 2*. For example, there possibly would be no capital cost in-

centive to use ponds if algae removal is required over ,activated
sludge/post-filtration systems, as implied in the aforementioned
Figure.

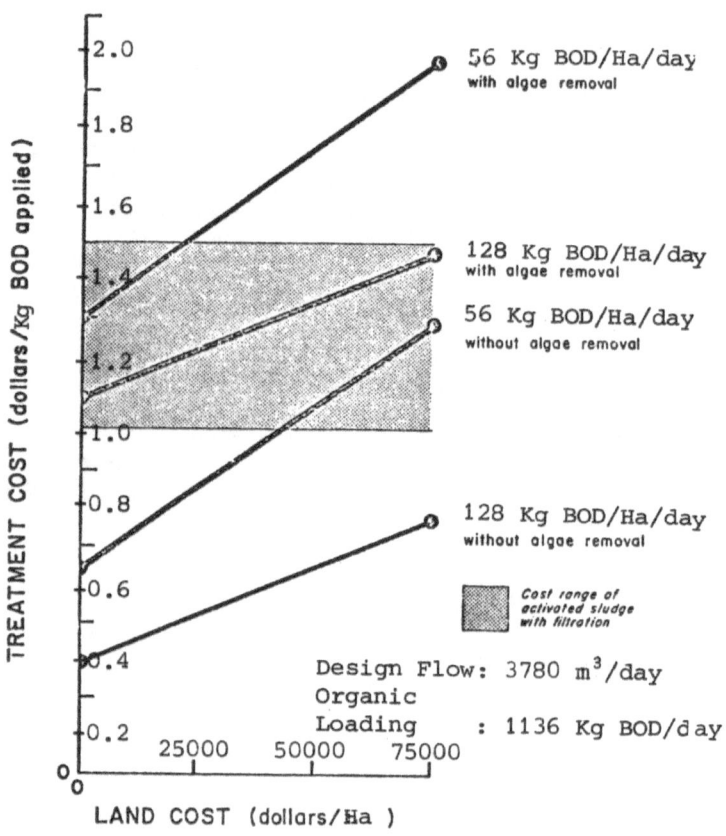

Fig. 2 :   Cost Curves Waste
Stabilization Ponds

Probably the most attractive features of waste stabilization
ponds are the limited energy demands in a period of diminishing
energy resources and increasing energy costs. For most secondary
biological treatment systems, the energy consumption rate is between 0.4 and 0.5 kilowatt-hours per pound of BODs removed [6, 26].
Using the surface organic loading rates and removal efficiencies
for ponds as indicated in *Tables 2* and *3*, a surface area-power equi-

valent relationship can be developed. This is illustrated in *Figure 3*, indicating an incentive to design toward a higher loaded pond concept if efficiency can be maintained in the strict sense of a surface area-power trade-off. An equivalent daily savings in power costs is also shown using a baseline cost of $100/year/HP (1.5¢ per kw-hr). This savings of energy and cost, of course, is significantly reduced if algae removal facilities are required to polish the pond effluent. This is attributable to additional pumping requirements, unit process power demands, chemical costs, and sludge handling facilities. A cost-benefit analysis should therefore be thoroughly developed before requiring this step to meet TSS limits based on high-rate system biological solids rather than algal cells.

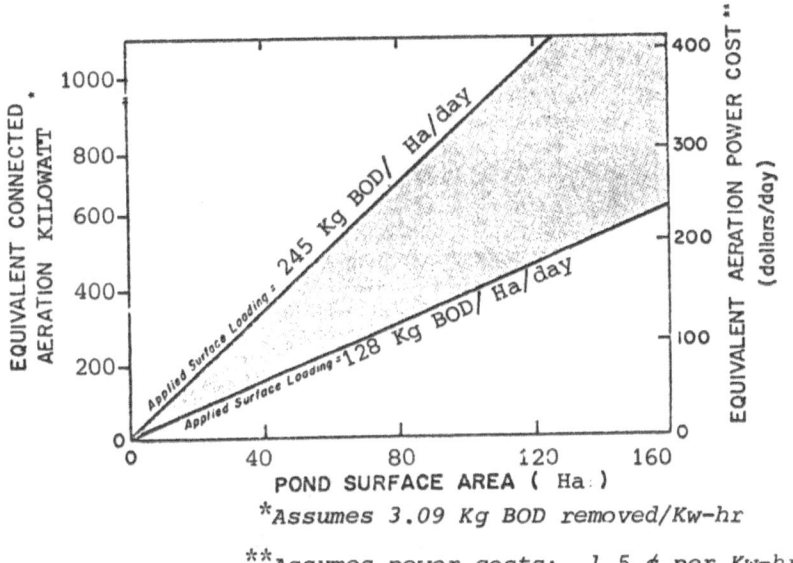

*Assumes 3.09 Kg BOD removed/Kw-hr

**Assumes power costs: 1.5 ¢ per Kw-hr

Fig. 3 :   Surface Area-
Power Equivalent Relationship

SUMMARY

Waste stabilization ponds, given sufficient area and proper climatological conditions, are capable of treating many industrial wastewaters to quality levels acceptable for discharge. These quality levels are commensurate with BPCTCA Guideline levels for most industrial categories with the exception of suspended solids, attributable primarily to algae cells. This leads to an evaluation of algae removal processes. A brief summary of treatment process applicability is presented in Table 8 (26). Most of the applications to date have been experimental in nature with mixed results. There is a noticeable scarcity of algae removal data from full-scale operating plants, and what data are available indicate inordinately low hydraulic application rates to obtain reasonable removal efficiencies. This problem is compounded by the fact that most algae removal systems become less efficient as the algae TSS feed concentration decreases. The highest removal efficiencies have been observed at 300-400 mg/ℓ TSS, whereas stabilization ponds normally operate at 75 to 100 mg/ℓ. These process questions, combined with energy and operating debits, underscore the need to carefully evaluate the efficiency of polishing waste stabilization pond effluents by removing algal suspended solids as related to the potential impacts of algae on the receiving waters.

TABLE 8 : Summary of Process Applicability for Algae Removal from Waste Stabilization Pond Effluents

Process(es)	Relevant Design Criteria Overflow Rate/ Surface Loading	Chemical Additions (mg/1)				Attainable Minimum Effluent TSS Concentration (mg/1)
		$Al_2(SO_4)$	Poly	$Ca(OH)_2$	$FeSO_4$	
1. Flocculation/ Precipitation	20.33 m^3/m^2-day	100-300	5-10	-- 150-200	-- 100-150	>25 >25
2. Dissolved Air Flotation	0.081 m^3/m^2-day	150-200	--	--	--	>10
3. Granular Media* Filtration (Upflow)	0.040 m^3/m^2-day	--	--	--	--	>10
4. Diatomaceous Earth Filtration	0.020 m^3/m^2- day	Diatomaceous earth body feed = 45 mg/1				>15

* Must be preceded by air flotation or sedimentation to reduce solids loadings.

DESIGN PROCEDURE

The BOD in the effluent from a single completely mixed pond can be expressed as

$$Y_1 = \frac{Y}{1 + k_1 t} \qquad \qquad \ldots \text{(1)}$$

and the BOD in the effluent from a second pond, connected in series, as

$$Y_2 = \frac{Y_1}{1 + k_2 t_2} = \frac{Y}{(1+k_1 t_1)(1+k_2 t_2)} \qquad \dots \quad (2)$$

in which

Y = the influent BOD_s (mg/ℓ)

Y_1 = the effluent BOD_s from the first pond (mg/ℓ)

Y_2 = the effluent BOD_s from the second pond (mg/ℓ)

t_1, t_2 = the retention times for ponds 1 and 2 (days)

k_1, k_2 = the reaction rate constant (base e) for a given temperature, for ponds 1 & 2 (day^{-1}) (5)

Unfortunately, open and usually uncontrolled pond performance cannot be described as precisely as implied in Equations 1 and 2 and, consequently, it is necessary to use a proven empirical approach, Equations 3 and 4.

For a fixed percentage reduction in BOD, it can be shown that for single ponds, the ratios of the reaction constants are equal to the ratios of the retention times,

$$\frac{k_{35}}{k_T} = \frac{t_T}{t_{35}} = \theta^{(35-T)} \qquad \dots \quad (3)$$

where

k = the reaction rate constants for various temperature,

t = reaction times,

T = temperature

Laboratory data taken at 9°, 20°, 24°, and 35°c show that $\theta = 1.085$ and $k_{35} = 1.2$ day^{-1} for a synthetic non-settleable sewage.

A pond can function very well when the entire contents are not oxygenated photosynthetically and, more significantly, the biological degradation rate in ponds is temperature dependent. Practical design criteria necessitate careful selection of reaction rates and minimum temperatures, as well as the common considerations of light intensities, food, etc. For many domestic wastes, the following empirical relationship has considerable merit:

$$V = CQL_aff' \left[\theta^{(35-T)}\right] \qquad \dots \quad (4)$$

where

V = pond volume (m^3)

Q = influent flow (m^3/day)

L_a = ultimate influent BOD (mg/ℓ)

θ = temperature coefficient

T = average temperature of the coldest month (^{0}C)

C = 3.55 x 10^{-2} (used where temperature fluctuations are large and designs are based on a depth of 1.52 m and one extra foot for solids storage)

f = algal toxicity factor=1 for domestic wastes (2)

f' = sulfide correction = 1 for $SO_4^=$ concentrations of less than 500 mg/ℓ (3)

The recommended depth in the middle U.S.A. is 1.83 m and a surface area based on a 1.22 to 1.52 meter aerobic depth. However, additional depth, particularly where settleable solids are present, provides added flexibility. Shallow ponds may produce poor results, particularly during hot and dry summer periods when the wastes contain large amounts of settleable solids and sulfate ion concentrations in excess of 500 mg/ℓ. By keeping the ponds sufficiently deep, some periodically occurring nuisance conditions can be prevented. However, when the pond is placed in a tropical area where the temperature is uniform, evaporation is negligible, sunshine is abundant, and the wastewater has undergone primary treatment, it is possible to use the original Hermann-Gloyna formulation, with C = 1.78 x 10^{-2} (4) and a total depth of 0.91 to 1.22 m.

In a facultative pond, only the upper portion will be more-or-less continuously aerobic. This aerobic layer is usually one meter deep. For this reason, a depth of about 0.91 m should be used in calculating surface areas required if the volume and detention time are calculated by the Marais Formula, Equations 1 and 2 (5). The total constructed depth, which includes the anaerobic pond volume and the allowance for sludge buildup on the bottom, should still be at least 1.52 to 1.83 m. Since Equation 4 and the common areal loading rate standards are empirical, they take into account the fact that only part of the pond will be aerobic.

The above equations consider the important parameters as related to the biodegradation of a typical wastewater. Where wastes contain relatively high concentrations of BOD, adjustments must be made to keep the surface area, depth, and volume in proper proportions. Similarly, adjustments must be made if toxic wastes are added to a system.

Notably the above equations are based on a single pond. For large systems it is best to have parallel ponds. Also, if a high degree of bacterial reduction or if a minimal algal content in the effluent is desired, it is necessary to follow the facultative pond with one or two shallow (0.91-1.22 m deep) and short detention (5 to 10 days) aerobic ponds, called maturation ponds. These ponds will help reduce short-circuiting and promote a more uniform effluent.

EXAMPLE DESIGN PROBLEM

Given the following information, design a pond system to produce a 90 percent reduction in BOD.

TOPIC	Value
BOD, Influent 5-day, 20^0C	250 mg/ℓ
BOD, Influent, Ultimate, 20^0C	305 mg/ℓ
BOD, Effluent 5-day 20^0C (90% removal)	25 mg/ℓ
BOD, Effluent, Ultimate, 15^0C	27 mg/ℓ
Photosynthetic Efficiency (f)	4%
Quantity	7560 m^3/day
Reaction Coefficient (Base e) k_1 (5°C)	0.102/day
Reaction Coefficient k_1 (15°C)	0.24/day
Reaction Coefficient k_1 (20°C)	0.35/day
Reaction Coefficient k_1 (30°C)	0.80/day
Reaction Coefficient k_1 (35°C)	1.2/day
Temperature (mean air, summer)	30^0C
Temperature (mean air, winter)	15^0C
Temperature (mean air, coldest month)	5^0C
Temperature Coefficient θ	1.072 to 1.085
Visible solar energy (Summer)	250
Visible solar energy (Winter)	150

DESIGN BASED on WIDE SPREAD USAGE

In the southwestern U.S.A., many ponds have been built on the basis of 56 Kg of BOD_5, 20°C per Ha per day. For the northern areas in the U.S.A., these loadings have been reduced to as little as 11.2 Kg per Ha per day. The organic load to the example pond is

$$250 \text{ mg}/\ell \cdot 7560 \text{ m}^3/\text{day} \cdot 1000\ell/1 \text{ m}^3 \cdot 1 \text{ Kg}/10^6\text{mg} = 1890 \text{ Kg/day}$$

Using a loading of 56 Kg/Ha per day and a depth of 1.83 m the surface area and volume are

$$A = \frac{1890 \text{ Kg/day}}{56 \text{ Kg/Ha-day}} \qquad = \quad 33.75 \text{ Ha} = 337\ 500 \text{ m}^2$$

$$V = \quad 337\ 500 \text{ m}^2 \cdot 1.83 \text{ m} \qquad = \quad 617\ 625 \text{ m}^3$$

The retention time t is

$$t = \frac{617\ 625 \text{ m}^3}{7560 \text{ m}^3/\text{day}} \quad = \quad 82 \text{ days}$$

The pond layout should be one rectangular pond or two equally-sized rectangular ponds, with the flexibility of introducing wastes into either pond. Ponds should always be as close to rectangular as possible to reduce dead spaces.

Design Based on Equation 4

The use of this equation is particularly applicable for rapid estimates involving the effects of temperature. Note in this problem it is suggested that the design temperature be based on the coldest month and on the 5-day BOD, 20°C.

It is necessary to correct for temperature and convert to ultimate FOD. Also, to compensate for a possible sludge effect, estimates should be based ultimate BOD values. However, it should be recalled that retention will be increased slightly as a result of evaporation in some areas of the country where evaporation exceeds rainfall.

$$V = \quad 3.55 \cdot 10^{-2} (7560 \text{ m}^3/\text{day}) (250 \text{ mg/\textit{l}}) \left[1.072^{(35-5)} \right]$$

$$V = 539\ 487 \text{ m}^3$$

With a depth of 1.83 m, the surface area, as based on a facultative zone of 1.52 m, is 539483 m³/1.52 m = 354923 m² = 35.4 Ha, and the retention time is 72 days. The surface loading is

$$\frac{1890 \text{ Kg/day}}{35.4 \text{ Ha}} = 53.38 \text{ Kg BOD/Ha-day}$$

Design Based on Equations 1 and 2

For comparative purposes, assume that winter conditions control and that the design is based on the coldest month, with $k_T = 0.102$/day.

$$25 \text{ mg/}\ell = \frac{250 \text{ mg/}\ell}{1 + (0.102/\text{day})t}$$

$$t = 88 \text{ days}$$

Using the relationships from the first example and an assumed aerobic layer depth of 1.0 m

$$\text{Volume} = 88 \text{ day} \quad 7560 \text{ m}^3/\text{day} = 665\ 280 \text{ m}^3$$

$$\text{Surface Area} = 665\ 280 \text{ m}\ell^3/1.0 \text{ mt} = 665\ 280 \text{ m}^2 = 66.52 \text{ Ha}$$

$$\text{Surface Loading} = \frac{1900 \text{ Kg/day}}{66.52 \text{ Ha}} = 29 \text{ Kg BOD/Ha-day}$$

Several ponds in series may produce better results. Assume a series of three ponds. As a first approximation, assume that the first pond will remove 70 percent of the BOD_5. However, such an assumption cannot be made with the risk of producing odors. Therefore, the detention time in the first pond is calculated as

$$(0.30)(250 \text{ mg/}\ell) = \frac{250 \text{ mg/}\ell}{1 + (0.102/\text{day})t}$$

$$t = 23 \text{ days}$$

The volume is then calculated as 173 670 m^3, and, assuming a depth of 1.0 m, the surface area is 173 670 m^2 = 17.36 Ha.

The second pond should have a retention period of at least 7 days (2). The influent BOD to the second pond is 0.30(250)=75 mg/ℓ, so the effluent BOD_5 will be

$$Y = \frac{75 \text{ mg/}\ell}{1 + (0.102/\text{day})(7 \text{ days})} = 44 \text{ mg/}\ell$$

The pond efficiency is

$$(\frac{75 - 44}{75}) 100 = 42 \text{ percent}$$

and the overall efficiency is

$$(\frac{250 - 44}{250}) 100 = 84 \text{ percent}$$

The volume of the second pond is

$$7 \text{ day} \cdot 7560 \text{ m}^3/\text{day} = 52920 \text{ m}^3$$

and the area, assuming an aerobic depth of 1.0 m is

$$\frac{52920 \text{ m}^3}{1.0 \text{ m}} = 52920 \text{ m}^2 = 5.29 \text{ Ha}$$

Assuming the third pond also to have a retention period of 7 days, the final effluent BOD_5 will be

$$Y = \frac{44 \text{ mg}/\ell}{1 + (0.102/\text{day}) (7 \text{ days})} = 25 \text{ mg}/\ell$$

The volume will be 52920 m^3 and the area 5.29 Ha, as in the second pond. The efficiency will be

$$(\frac{44 - 25}{44}) 100 = 42 \text{ percent}$$

and the overall efficiency will be

$$(\frac{250 - 25}{250}) 100 = 90 \text{ percent}$$

The efficiency of 90 percent is equal to that required. If it had been less, the designer would have had to go back and increase the retention time of one or more of the ponds. If the retention time of the first pond is increased to 38 days to achieve 80 percent BOD_5 removal, the overall efficiency will increase to 93 percent.

REFERENCES

1. Gloyna, E.F., *Waste Stabilization Ponds*, Monograph No. 60, WHO, Geneva, Switzerland, 1971.

2. Oswald, W.J., "Fundamental Factors in Waste Stabilization Pond Design", *Proceedings of the 3rd Conference on Biological Waste Treatment*, Manhattan College, New York, 1960.

3. Thirumurthi, D.,"Design Criteria for Waste Stabilization Ponds", *WPCF Journal*, 46-9, 1974.

4. Marais, G.V.R., New Factors in the Design, Operation and Performance of Waste Stabilization Ponds with Special Reference to Health, Expert Committee Meeting on Environmental Change and Resulting Impacts on Health, WHO, 1964.

5. Gloyna, E.F., and Espino, E., Sulfide Production in Waste Stabilization Ponds, *ASCE Sanitary Engineering Division*, 95, 1969.

6. Gloyna, E.F., Facultative Waste Stabilization Pond Design, Seminar on Ponds as a Waste Treatment Alternative, The Univ. of Texas at Austin, July, 1975.

7. Eckenfelder, W.W., *Industrial Pollution Control*, McGraw Hill Book Company, New York, 1966.

8. Federal Water Pollution Control Association, *Petrochemical Effluents Treatment Practices*, Report No. 12020. 1970.

9. Golueke, C.F., and Oswald, W.J., Harvesting and Processing Sewage-Grown Planktonic Algae, *WPCF Journal*, 37, 1965.

10. Middlebrooks, E.J., et. al., Techniques for Algae Removal from Wastewater Stabilization Ponds, *WPCF Journal*, 46, 1974.

11. Dryden, F.D., and Stern, G., Renovated Wastewater Creates Recreational Lake, *Environmental Science and Technology*, 2, 1968.

12. Young, F.L., Vacuum Filtration of Algae, M.S. Thesis, The University of Texas at Austin, 1965.

13. Missouri Basin Engineering Health Council, Waste Treatment Lagoons - estate of the Art, EPA Report No. 17090, Washington, D.C., 1971.

14. Van Vurren, L., and Van Duuren, F., Removal of Algae from Wastewater Maturation Pond Effluent, *WPCF Journal*, 37, 1965.

15. Drynen, W.R., Methods of Concentrating Algae, M.S. Thesis, The University of Texas at Austin, 1956.

314

16. Bare, W.F., Jones, W.B., and Middlebrooks, E.J., Algae Removal Using Dissolved Air Flotation, *WPCF Journal*, 47, 1975.

17. Shelef, G., and Schwarz, M., Prediction of Photosynthetic Biomass Production in Accelerated Algal-Bacterial Wastewater Systems, Dept. of Medical Ecology, The Hebrew University, Jerusalum, 1972.

18. McGhee, T.J., and Patterson, R.K., Upflow Filtration Improves Oxidation Pond Effluent, *Water and Sewage Works*, 121, 1974.

19. Brown, J.C., Algae Removal Using a Diatomaceous Earth Reversible Cake Filter, 64th National AIChE Meeting, New Orleans, Louisiana, 1969.

20. Fair, G.M., Geyer, J.C., and Okum, D.A., Water Purification and Wastewater Treatment and Disposal, *Water and Wastewater Engineering*, 2, John Wiley and Sons, 1968.

21. Echelberger, W.R., et. al., Disinfection of Algal Laden Waters, *ASCE Sanitary Engineering Division*, 97, 1971.

22. SCS Engineers, Survey of Treated Municipal Reuse, Long Beach, California.

23. Missouri Basin Engineering Health Council, Waste Treatment Lagoons - State of the Art, EPA Report No. 17090 EHX 07/71, Washington, D.C. 1971.

24. Prows, B.L., and McIlhenny, W.F., Research and Development of a Selective Algaecide to Control Nuisance Algale Growth, EPA Report No. 660/3-74-019, 1974.

25. Department of Country Engineers, Los Angeles County, Final Report - Wastewater Reclamation Project for Antelope Valley Area, 1968.

26. Engineering - Science /Texas, Confidential Report for Texas Eastman Company, 1975.

GENERAL CONCEPTS OF ANAEROBIC TREATMENT

A. L. Downing

Partner and Senior Engineer

A. D. K. Kell

Binnie and Partners, Westminster, London, U.K.

INTRODUCTION

In some natural habitats, such as can be found in the bottom sediments in ponds and lakes, organic matter in the absence of air and in the presence of water decays anaerobically to form methane and carbon dioxide. Anaerobic digestion represents the controlled application of this naturally occurring process. This type of decay is due to the action of a complex flora of anaerobic bacteria which break down carbonaceous matter in a number of stages to the final end products of methane and carbon dioxide.

Anaerobic digestion is widely used for the stabilization of municipal wastewater sludges and is finding increasing acceptance as an alternative to aerobic processes for the treatment of industrial wastes - particularly those characterized by high organic strength. At present the methane fermentation process is the only anaerobic biological process in common use at sewage works. However, in recent years, concern has been growing over rising nitrate levels in receiving waters used as a source of drinking water, and in lakes vulnerable to eutrophication. This has resulted in a considerable research effort into the denitrification of sewage effluent. It is likely that many works whose effluent is reused will adopt this stage of treatment in the future. The topic of denitrification, however, is discussed on a separate occasion and so this paper will be concerned solely with anaerobic digestion processes designed for destruction of carbonaceous matter.

ADVANTAGES and DISADVANTAGES of
THE ANAEROBIC PROCESS in WASTEWATER TREATMENT

Although only in common use for the treatment of sewage sludges
and limited use for certain industrial wastewaters the anaerobic
digestion process could find wider application in the field of
waste water treatment. It has a number of advantages over aerobic
treatment processes, such as those employed in activated sludge or
trickling filter treatment, and in certain conditions these can
outweight the disadvantages. The differences between the two processes can be highlighted by comparing their principal features.

Aerobic processes involve mixing waste water with large quantities of microorganisms and air. The microorganisms use the organic matter for cell growth and also for energy (by converting the
organic carbon to carbon dioxide using the oxygen in the air). The
process involves the use of large quantities of energy resulting in
rapid cell growth. The conversion of organic matter to cells represents a change in its form, enabling its subsequent removal from
the waste stream by physical means. The anaerobic process also involves the mixture of waste water with large quantities of microorganisms but in this case air is absent. The microorganisms convert the organic material to methane and carbon dioxide without external input of oxygen. In doing so the microorganisms take up relatively little energy and thus their growth yield is small. Only
a small proportion of the degradable waste is converted to new cell
material the majority being converted to methane gas. Thus a higher
proportion of the waste is stabilised than in the aerobic process.

The fact that anaerobic treatment has a distinct advantage over
aerobic treatment in its energy requirements may become of increasing importance as energy resources become more limited. Aerobic
treatment, particularly of strong wastes, involves the expenditure
of considerable amounts of power. Further about half of the COD is
converted into biomass which still requires treatment. This biomass appears as an unstable and smelly sludge. Anaerobic systems,
in contrast can remove a much higher proportion of the COD, often
over 80 percent, with the expenditure of relatively little power.
Furthermore the production of methane presents the opportunity for
the recovery of energy. The synthesis of biomass is usually equivalent to little more than 15 percent of the COD removed and the
waste sludge is stable and free from offensive smell. It has been
estimated by Cillie et al.,[1] that waste water with a COD of 20,000
mg/ℓ can be treated by anaerobic processes for a quarter of the cost
of treatment by aerobic processes. Anaerobic treatment was considered cheaper for CODs down to 4000 mg/ℓ at which level the costs
are similar. The development of the anaerobic filtration process
may provide a cost advantage at even lower COD levels.

As mentioned above the anaerobic process gives rise to a much smaller sludge disposal problem. Not only is the mass of sludge produced per unit mass of degradable organic matter smaller but the resulting sludge is more tractable. Also the reduced cell production results in a proportional reduction in the requirement for nutrients such as nitrogen and phosphorus. While this factor is generally not of significance in the digestion of municipal sludges it can be of importance in the treatment of industrial wastes, lacking these materials to which the appropriate nutrients must be fed.

It has been noted that the waste sludge produced from anaerobic digestion is stable and inoffensive. It is often suitable for use directly on agricultural land. Organic nitrogen is brought into solution and the carbon to nitrogen ratio reduced, effects which improve the properties of the sludge as a fertilizer. Most pathogenic organisms are destroyed if digestion is mesophilic or thermophilic and most lipids and other constituents of waste which might attract flies and vermin when applied to agricultural land are degraded.

However, the anaerobic process does have some disadvantages which affect its potential for treating certain types of waste. One major disadvantage is that the process takes place very slowly at ambient temperatures. However the reaction rate roughly doubles for each ten degree centigrade rise in temperature over a considerable range and the difference in rates of destruction of organic matter at the preferred operating temperatures and those in aerobic plants at ambient temperatures is not nearly so great.

The successful operation of the anaerobic process depends on the efficient functioning of methane producing bacteria. These bacteria suffer from the disadvantage that they have a very slow growth rate and are susceptible to inhibition due to changes in pH, temperature, and the presence of toxic materials. These characteristics have the effect that the process takes a long time to start up and is unable to adapt quickly to change in waste load, temperature and other environmental conditions.

A further disadvantage is that the effluent from the digestion process often has high levels of suspended solids and BOD. The supernatant from treatment of sewage sludge contains a much higher concentration of oxidisable matter than that for aerobic digestion of sludge and has been found to be more difficult to treat than settled sewage.

To generalize the advantages of anaerobic treatment over aerobic treatment, for wastes amenable to such treatment, are likely to outweigh the disadvantages when the BOD is greater than 5,000 mg/ℓ. For less concentrated wastes the disadvantages may predominate and may limit the use of the process. A noted exception is the successful anaerobic treatment of meat packing wastes with BOD concentrations as low as 1000 mg/ℓ [2].

CHEMISTRY and MICROBIOLOGY

GENERAL DESCRIPTION of DIGESTION PROCESS

The anaerobic digestion process can be divided into two main stages of which the first may involve processes of both liquefaction (when suspended impurities are present) and acidification; the second consists primarily of gasification. The sequence of the phases of the process is shown in *Fig. 1.*

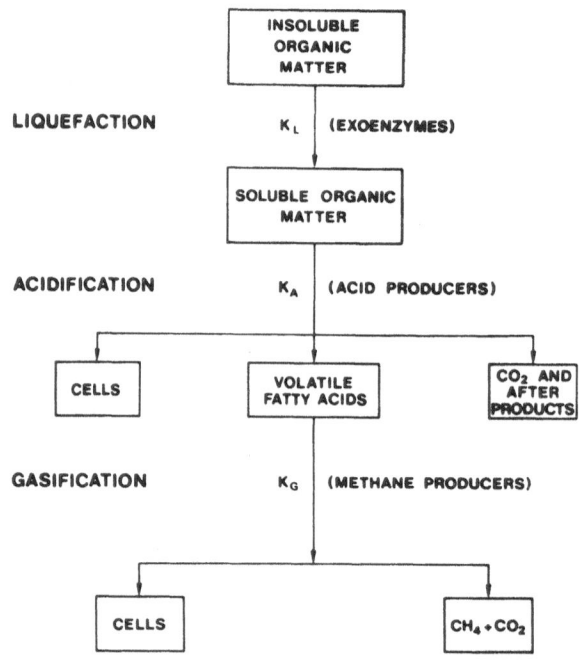

Fig. 1 :   Sequence of Anaerobic
Digestion Process
(Edeline (18)).

In the first stage predominantly insoluble complex organic components such as fats, proteins and carbohydrates are first hydrolysed into smaller soluble compounds and then broken down further to produce mostly short chain fatty acids. The liquefaction process is carried out by the agency of extra cellular enzymes. Proteins are degraded into their component amino acids ⁻ ⁻ saccharides are broken down to monosaccharides while fats are hydrolysed to a mixture of long chain fatty acids and glycerol. Following hydrolysis a diverse and variable assortment of both anaerobic and facultative bacteria known collectively as the "acid-forming bacteria" break down the products of hydrolysis to produce mostly short-chain fatty acids. This phase is known as acidification. During this rearrangement both carbon dioxide and hydrogen gases are formed and energy is released for cell growth. A small proportion of the organic waste is converted to cell material. In addition a substantial proportion of the cell growth. A small proportion of the organic waste is converted to cell material. In addition a substantial proportion of the organic nitrogen is converted to ammonium ions and organic sulphur appears as sulphide.

Anaerobic digestion is distinguished from putrefaction by the action of a group of bacteria known as "methane forming bacteria". In the second stage of the process these bacteria break down the products of acidification yielding methane and carbon dioxide. It is during this stage that stabilization of the waste occurs. Methane bacteria are species of fastidiously anaerobic bacteria which can produce methane and carbon dioxide from substances of acetic acid, formic acid, formaldehyde, propionic acid, butyric acid, some alcohols and carbon dioxide and hydrogen. These substrates are the major and perhaps the only yielders of methane.

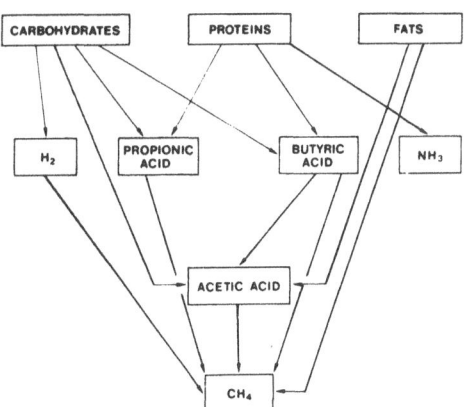

Fig. 2 : Pathways for Methane
Production
[McCarthy [5]]
(Courtesy of Public
Works Magazine)

Fig. 2 indicates, in simplified form, the major pathways for the production of methane from complex organic compounds. The principal features of these pathways are described in the following sections.

Degradation of Fats

The first stage in the degradation of fats is their hydrolysis to fatty acids and glycerol. It has been shown by Jeris and McCarty [3] that the degradation of long chain fatty acids follows the usual pathway of β oxidation. In the production of each molecule of acetic acid some hydrogen - carrier molecules (flavin adenine dinucleotide (FAD) and nicotinamide adenine dinucleotide(NAD)) are transferred. These carrier molecules are subsequently returned to their oxidized state by liberation of extra cellular hydrogen gas, which in its turn reacts with carbon dioxide to form methane and water.

Degradation of Proteins

The degradation of proteins is generally considered to follow the following sequence. First they are hydrolysed into peptides under the agency of extra cellular enzymes and then broken down into their component amino acids. These amino acids are then absorbed into the appropriate "acid-forming" bacteria and are degraded intracellularly forming a variety of short chain saturated and hydroxy acids with the liberation of ammonia. The principal products of the process are acetic acid and ammonia.

Degradation of Carbohydrates

Complex carbohydrates are first hydrolysed to form monosaccharides and then degraded to a variety of intermediate and end products depending on the bacterial species involved and on the environmental conditions. The principal pathways included the "Emden-Meyerhof" pathway and the "Hexose Monophosphate shunt" of which the former appears to be the more important.

Formation of Methane

It has proved very difficult to isolate and study the methane producing bacteria and consequently the pathways for methane formation are only partially understood. Häusler [4] has described the succession of microbial processes in anaerobic decomposition. He found that methane fermentation consisted of several metabiotic processes which could be observed on a gas production curve as three characteristic peaks. In a mixed substrate containing the products of acidification the bacteria *Methanobacterium Suboxydans* acts on higher organic acids, principally butyric acid, to produce acetic acid. As acetic acid is produced this is immediately further decomposed by *Methanosarcina Methanica* to methane and carbon dioxide. Lastly propionic acid, which is present principally from the decomposition of amino acids, is decomposed by *Methanobacterium Propionicum* first to acetic acid and then to methane and carbon dioxide. Any alcohols present in the substrate are first oxidised to the corresponding acids by *Methanobacterium Omelianskii* and then decomposed as described previously.

According to McCarty [5] the most important methane formers which are the agents for the degradation of acetic and propionic acids grow very slowly and sludge retention times of four days or longer are required for their growth. These bacteria carry out the major portion of waste stabilization. Their slow growth and their rate of acid utilization normally represent the limiting step around which the anaerobic treatment process must be designed.

McCarty [5] has estimated that about 70 percent of the methane produced by sewage sludge digesters derives from acetic acid. Stadtman and Barker [6] have reported how isotopic labelling has shown that the methyl group of acetic acid is transferred to the methane direct.

$$C^{14}H_3COOH \longrightarrow C^{14}H_4 + CO_2$$

$$CH_3C^{14}OOH \longrightarrow CH_4 + C^{14}O_2$$

Most of the remaining methane produced in the process is formed by the reduction of carbon dioxide. Here, hydrogen, which is removed from organic compounds by the action of enzymes, reduces carbon dioxide to methane gas. In this instance the carbon dioxide functions as a hydrogen or electron acceptor in a similar manner to oxygen in the aerobic process. Free carbon dioxide is always present during anaerobic treatment and very little gaseous hydrogen appears in the digester gas.

$$CO_2 + 8H \longrightarrow CH_4 + 2H_2O$$

Current knowledge on the pathways for methane formation has been summarized by Barker [7]. Although much remains to be discovered it is known that both tetrahydrofolate and vitamin B_{12} coenzyme are involved in the process.

GAS PRODUCTION and PROCESS EFFICIENCY

The quantity and composition of gas produced in the digestion process depends on the composition of the feed. Buswell [8] produced an equation from which the quantity of the gas may be determined if the waste composition is known:

$$C_nH_aO_b + (n-a/4 - b/2) \ H_2O = (n/2 - a/8 + b/4) \ CO_2$$
$$+ \ (n/2 + a/8 - b/4) \ CH_4$$

It follows from this equation that for each gram of chemical oxygen demand (COD) removed, assuming complete oxidation to carbon dioxide and water, approximately 350 ml of methane would be produced at STP.

Thus if the loading of a plant in terms of the COD of the waste is known and the quantity of methane production is measured then the efficiency of waste stabilization can be calculated.

A more elaborate equation has been given by McCarty [5] for the case when the organic compund degraded contains nitrogen.

ENVIRONMENTAL FACTORS

The digestion process described in the previous section operates most efficiently under specific conditions of temperature, pH and nutrient supply. Some materials are toxic when present at certain concentrations and inhibit the process. These environmental factors are discussed in the following sections.

ANAEROBIC CONDITIONS

It is essential that the anaerobic digestion process should be carried out in the complete absence of oxygen as the methane producing bacteria essential to the second stage of the process are fastidious anaerobes and even low concentrations of dissolved oxygen are toxic to them. This is seldom a practical problem for even if the influent waste contains oxygen at low levels the facultative organisms which take part in the acidification stage of the process rapidly remove any traces. The practice of carrying out the digestion process in a closed tank in an atmosphere of methane and carbon dioxide is also helpful in maintaining anaerobic conditions.

Studies by Mosey and Hughes [9] indicate that the redox potential of healthy "balanced" digestion is −265 ± 25 mV at pH values between 7.0 and 7.2. Each unit increase in pH value produces a change of -60 mV. The measurement of redox potential can be used to monitor the performance of a digester.

Oxidising agents such a nitrate, hexavalent chromium and ferric iron can prove toxic in large "shock" doses. When introduced in small concentrations they are rapidly reduced to form nitrogen and ammonia, trivalent chromium and ferrous iron respectively.

TEMPERATURE

The anaerobic digestion process can take place over a wide range of temperature (5°C to 60°C). Up to about 37°C the rate of reaction increases with increase of temperature. Beyond this temperature opinion is divided whether a continuous increase in reaction rate takes place. However, it is generally accepted that a second optimum operational temperature is reached at 55°C. Operation of the process can be placed within three separate temperature ranges having different characteristics. These are described in the following sections.

Psychrophilic Digestion (5°C to 25°C)

This range applies to anaerobic waste treatment plants which operate unheated. Septic tanks, Imhoff tanks, sludge lagoons and some sludge digesters fall into this category. The characteristics of operation within the range are that reaction rates are low and the residence times of the microorganisms can vary between 100 and 300 days. There is no problem with acclimatisation of bacteria when operating within this range of temperatures.

Mesophilic Digestion (25°C to 38°C)

This is the common range of temperature for operation of sludge digesters. It is necessary to supply heat to maintain the operating temperature at a rate which varies with the ambient temperature. In many cases use is made of the methane produced during the process to provide this heat input. The reaction rate is much quicker than for psychrophilic digestion and residence times are in the order of 20 to 40 days. Bacterial species used in the process occur naturally at mesophilic temperatures in such environments as the rumen of herbivores. However, some species are sensitive to s u d d e n changes in temperature and in general these should be limited to about 2°C a day if inhibition is to be avoided.

Thermophilic Digestion (50°C to 60°C)

In theory operation of sludge digesters in this range should give faster reaction times than for mesophilic digestion and should thus provide an even more efficient process. However, attempts to operate digesters at these temperatures have largely proved unsuccessful and thermophilic digestion is now rarely practiced. Anaerobic environments in this temperature range are rarely met in nature and it appears that there is considerable difficulty in acclimatising methane bacteria. This difficulty in acclimatisation may be the cause of the apparent discontinuity of increase in reaction rate reported by some observers.

NUTRIENTS

Biological activity cannot take place unless inorganic nutrients required by the bacteria for growth are present. The most important nutrients are nitrogen and phosphorus but trace amounts of other materials are also required. Municipal waste sludges normally contain sufficient concentrations of all the necessary nutrients for unhindered growth to take place but industrial wastes are frequently more specific in composition and may require nutrients to be added to the substrate for optimum operation of the digestion process.

The requirements for nitrogen may be determined from the rate of cell growth and the proportion of nitrogen in the cell material. Biological cells such as constitute the normal flora of a sludge digester are estimated to have an average chemical formula approximating to $C_5H_9O_3N$. This indicates that nitrogen makes up about 11 percent of the total weight of the cells. McCarty [5] reports that the requirement for phosphorus is about one-fifth of that for nitrogen (i.e. about 2 percent of the biological solids weight). The requirement for sulphur is not known although it is likely to be less than that for phosphorus. In addition traces of other elements such as iron and cobalt are known to be necessary for cell metabolism.

pH and BUFFER CAPACITY

One of the most important factors affecting the functioning of the anaerobic digestion process is the pH under which it takes place. The process can be operated successfully within the pH range 6.0 to 8.0 and the optimum is about pH 7.0. At pH values below 6.0 the efficiency drops off rapidly and as the acidity increases conditions become more and more toxic to methane bacteria.

Steady state conditions are maintained when the methane forming organisms break down the acids formed by the acid-formers at the same rate as they are produced. If anything happens to inhibit the action of the methane formers the acid concentration will increase, the pH will tend to fall and may become low enough to limit further the effectiveness of the methane formers, causing the process to fail.

Under normal operating conditions maintenance of a balance between the two stages of the process usually presents no difficulties

325

but during start up of the process, when excessive sudden shock loads are introduced, when rapid temperature variations occur or when inhibitory materials enter the system the balance between the methane and the acid producing organisms may be upset.

The overall pH of the wastewater undergoing anaerobic treatment is dependent on the large number of acid-base equilibria of its constituents. However, at the pH values close to neutrality which are of importance in the digestion process the carbon dioxide-bicarbonate system is predominant in determining the pH. The interdependence of the main parameters of the system is illustrated in *Fig. 3*. The concentration of carbonic acid in the waste water is dependent on the partial pressure of carbon dioxide in the reactor.

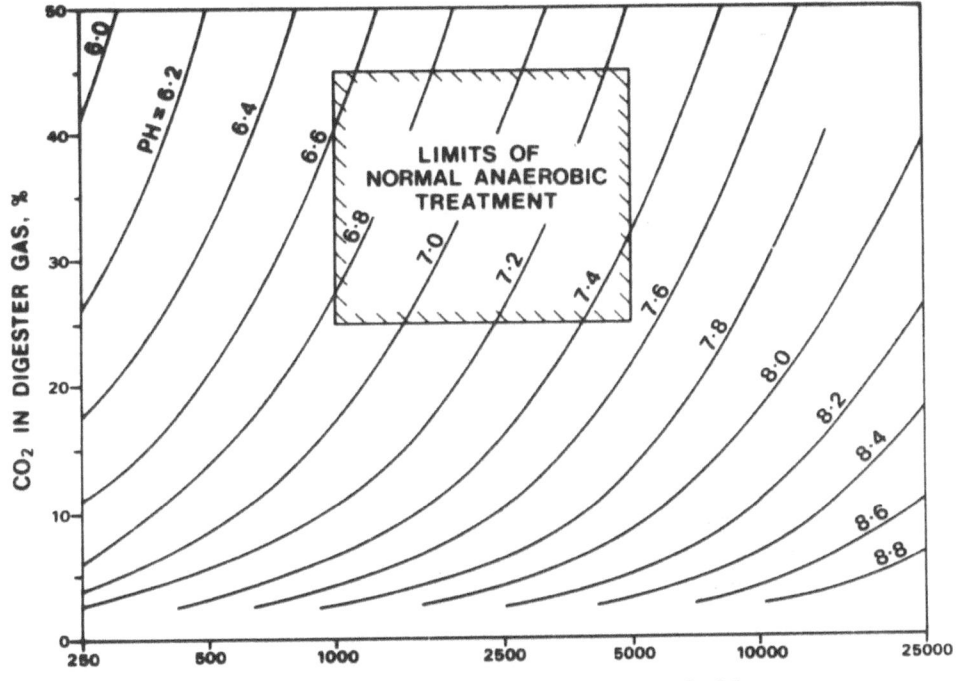

BICARBONATE ALKALINITY – mg/l as CaCO₃

Fig. 3 : Relationship between Bicarbonate Alkalinity, pH, and Carbon Dioxide percentage in Anaerobic Treatment (Courtesy of Public Works Magazine) (McCarty [5])

The hydrogen ion concentration is then determined by the following relationship:

$$(H^+) = \frac{K_c(H_2CO_3)}{(HCO_3^-)}$$

Under normal conditions of gas production and bicarbonate concentration in a digester it can be seen from *Fig. 3* that the pH will fall in the range 6.6 to 7.4. If the digester becomes unbalanced the volatile acid concentration will increase destroying the bicarbonate alkalinity according to the following relationship:

$$HCO_3^- + H\ Ac = H_2O + CO_2 + Ac^-$$

and resulting in a reduction in pH.

The larger the initial concentration of bicarbonate the greater the "buffer capacity" in the waste against an unacceptable lowering of the pH at times of digestion imbalance.

When municipal waste water sludge is digested one of the products of the breakdown is ammonium bicarbonate - the ammonia deriving mainly from proteins in the raw waste. This alkalinity helps to maintain the pH near neutrality. However, many industrial wastes are lacking in alkalinity or contain little organic nitrogen and it may be necessary to add alkalinity to maintain a satisfactory pH level.

Lime is commonly used to increase the alkalinity of trade waste or to remedy imbalance in sewage sludges while the cause of the imbalance is identified and corrected. Although lime is most suitable for neutralising short-chain fatty acids it is only effective at pH values of up to about 7. Above this pH calcium carbonate is precipitated and further addition of lime is ineffective in acid neutralisation. As an alternative to lime the hydroxide or carbonate salts of sodium or potassium may be used and these materials do not suffer from the same disadvantage of forming insoluble products. However, care must be taken not to introduce inhibitory concentrations of sodium or potassium ions.

INHIBITION

There are many materials both organic and inorganic which may be toxic or inhibitory to the anaerobic waste treatment process. The term "toxic" is a relative one and depends on the environmental conditions and the concentrations of the inhibitory material. At low concentrations a "toxic" material often has a stimulatory effect and as its concentration increases a maximum rate of biological activity is reached beyond which the activity reduces to its original level and then is inhibited until eventually it approaches zero.

Another factor is the ability of microorganisms to adapt to some extent to inhibitory concentrations of many materials. The extent

of such adaptation is variable - in some cases after acclimatisation activity may almost regain its original level, in other cases
the improvement may be relatively small.

The content of inhibitory material in sewage sludge will vary a
great deal with the nature and proportion of industrial waste in
the sewage; that in industrial wastes will depend on the process
for which these wastes are derived.

INHIBITION by ALKALI and ALKALINE-EARTH METALS

Inhibition by salts of alkali and alkaline earth-metals is generally not a problem in the treatment of municipal sewage sludge unless high concentrations of such salts are added to the waste water
for pH control. Industrial wastes, however, may contain appreciable
concentrations of such salts and in particular those of sodium, potassium, calcium or magnesium. Table 1 shows the concentrations
of the cations of these salts which are generally regarded as "stimulatory", "moderately inhibitory" and "strongly inhibitory".

TABLE 1 : Stimulatory and Inhibitory Concentrations of
Alkali and Alkaline-Earth Cations
(McCarty [5])

CATION	Concentrations in mg/ℓ		
	Stimulatory	Moderately Inhibitory	Inhibitory
Sodium	100-200	3500-5500	8,000
Potassium	200-400	2500-4500	12,000
Calcium	100-200	2500-4500	8,000
Magnesium	75-150	1000-1500	3,000

When combinations of these cations are present together the nature of the effect on the system becomes more complex. Some combinations act antagonistically reducing the total affect while
others act synergistically to increase the total toxicity. If an
ion is present in a waste at a concentration that produces inhibition the effect can be reduced if an antagonistic ion is added to
the waste. Sodium and potassium have been found effective antagonists especially when present at the stimulatory concentrations
listed in Table 1. Calcium and magnesium are poor antagonists and
will normally increase the toxicity caused by other cations although

in certain circumstances they have been found to have a stimulatory effect [5].

INHIBITION by HEAVY METALS

Digesters are vulnerable to inhibition by the ions of heavy metals although the insoluble precipitates of heavy metal salts are not toxic. This inhibition is due to the action of the heavy metals in deactivating enzymes essential to the process. The soluble salts of copper, zinc and nickel are those which give rise to the majority of problems in anaerobic treatment processes. The low solubility of iron and aluminium salts in the normal pH range minimises their effect on the waste stream.

In practical cases a combination of heavy metals is commonly present in a waste stream. Mosey [10] has proposed that the total toxicity of all heavy metals with the exception of chromium can be related to their cumulative concentration in the waste stream where this is expressed in milliequivalents of metal per kilogram of solids (meq/kg). Within the normal pH range of digester operation the effect of iron can be ignored as it is virtually insoluble and, taking into account the partial reduction of copper to the cuprous state, the total concentration, K_h, in meq/kg can be expressed as:

$$K_h = \frac{Zn/32.7 + Ni/29.4 + Pb/103.6 + Cd/56.2 + Cu/47.4}{\text{solids concentration}}$$

where "solids concentration" is the dry solids content of digested sludge expressed as kg/ℓ. From examination of available data Mosey [10] concluded that digestion failure is probable when K_h exceeds 400 meq/kg and almost certain when K_h is more than 800 meq/kg.

The effect of chromium is best considered separately as it has special characteristics. Hexavalent chromium in the waste stream is rapidly reduced to the trivalent state within the digester. Mosey [10] has suggested that the toxicity of chromium should be related to its total concentration expressed as a percentage of the sludge solids. From the limited data available he concluded that chromium is inhibitory when the concentration exceeds about 2.5 percent of that of the dry solids in the digester sludge.

There are several methods by which the content of heavy metal ions can be reduced in digesters. One of the most effective procedures is to precipitate the metals as insoluble sulphide salts. This is effective for iron, zinc, nickel, lead, cadmium and copper but not for chromium. Sulphide may be present in the waste stream e.g., from the degradation of certain organics, or if not available in sufficient quantities sodium sulphide or a sulphate salt (which will be reduced to sulphide in the anaerobic conditions) may be added. It is interesting to note that sulphides in themselves are quite toxic but when combined with heavy metals to form insoluble salts both the sulphide and the metal ions become inactive. Another method used in the control of heavy metal toxicity is the precipitation of the metals as sparingly soluble carbonate salts.

INHIBITION by AMMONIA

Ammonia, both in the form of ammonium ions (NH_4^+) or as dissolved ammonia gas (NH_3) has been found to be toxic to the digestion process. Inhibitory concentrations have been reported in some farm slurries and in highly concentrated municipal waste sludges. Dissolved ammonia is far more toxic than ammonium ions (available information indicates that inhibition is likely when the concentration of ammonium ions exceeds about 3000 mg N/ℓ or the concentration of dissolved ammonia exceeds about 150 mg N/ℓ) and so the degree of inhibition is strongly dependent on the pH of the system. To some extent the inhibition is self-regulating in that it tends to cause an increase of volatile acids which lower the pH thus converting dissolved ammonia to the less toxic ammonium ions.

INHIBITION by SULPHIDES

Sulphides may be present in wastes undergoing digestion e.g., due to the reduction of sulphades present in the original waste and as a result of biological processes within the digester. Some food trade wastes contain appreciable concentrations of sulphade. When present in dissolved form sulphides can be quite toxic. It has been found that concentrations above about 200 mg/ℓ can produce an inhibitory effect. As mentioned earlier sulphides react with heavy metals to form insoluble and non-toxic precipitates and thus one common method of control is the addition of a suitable quantity of an iron salt to the waste.

INHIBITION by ORGANIC COMPOUNDS

The effects of detergents on wastewater treatment have long been a concern of those involved and much has been done to study and control these effects [11]. Both "hard" and "soft" alkyl benzene sulphonate (ABS) detergents inhibit anaerobic digestion of sewage sludge when their concentration exceeds about 2.5 percent of the digested sludge solids (corresponding to 15-23 mg/ℓ in the sewage). In the U.K. average concentrations of detergents in settled sewage are approaching the level at which inhibition can be expected and problems could become widespread if the use of ABS type detergents continued to increase. Control can be effected by addition of long chain amines (such as stearine amine, a commercial mixture of stearyl-

330

amine and palmitylamine) which form stable complexed with ABS detergents. Introduction of about half the quantity equivalent to the detergent content of inhibited digesters appears to correct their condition.

A number of complete or partial failures of anaerobic digestion have been attributed to the presence of chlorinated hydrocarbons in the sludge. The toxicity of these compounds has been investigated by Swanwick and Foulkes [12].

They found chloroform to be extremely toxic and 1, 1, 1 - trichloroethane (Genklene) was found to be equally toxic at low concentrations but less toxic than chloroform at higher concentrations. Four chlorobenzenes in common commercial use were found to be about as toxic as anionic detergents.

The question as to whether the toxicity of the organic acids produced during digestion derives from their anionic acid radicals or from the lowering of pH value that the acids produce has long been a source of controversy. Andrews [13] has put forward the proposition that undissociated volatile acids act as both substrates and inhibitors while their salts are biologically inert. Thus the toxicity of the acids is pH dependent. Other investigators maintain that it is solely the acidity that affects the microorganisms.

INDICATORS of TOXICITY

There are a number of ways in which imbalance in the system, and hence the possible pressure of toxic substances, are indicated. If conditions become out of balance the volatile acids concentration and the carbon dioxide percentage in the digester gas will rise and the pH, the total gas production, and the proportion of waste stabilized will decrease. Any or all of the parameters may be used to monitor the performance of a reactor. A sudden increase in volatile acid concentration is frequently one of the first indicators of digester imbalance and the monitoring of this parameter often gives the best indication of the state of the process.

TYPES of TREATMENT PLANT

The three principal types of treatment plant for the anaerobic treatment of sewage sludge and industrial wastes are shown in *Fig. 4* and are described broadly in the following sections.

THE CONVENTIONAL PROCESS

Treatment by the conventional process takes place in a completely mixed reactor where the solids retention time, (T), equals the hydraulic retention time. Solids retention time can be defined as

$$T \text{ (days)} = M_t/M_e \qquad \text{.... (1)}$$

where;

M_t = the total weight of suspended solids in the treatment system

M_e = the total weight of suspended solids leaving the system per day in the plant effluent

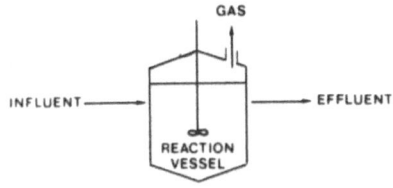

1. CONVENTIONAL PLANT

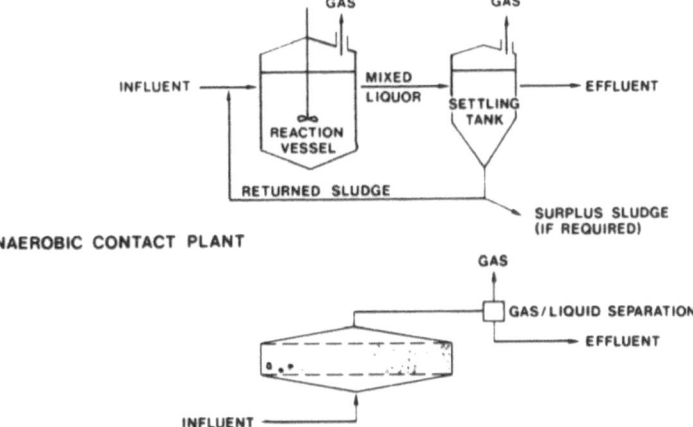

2. ANAEROBIC CONTACT PLANT

3. ANAEROBIC FILTER

Fig. 4 : Basic Types of Plant for
Anaerobic Treatment
(Courtesy of Public Works
Magazine)
(McCarty [5])

The process is most suited to the digestion of concentrated wastes, particularly those in which the organic matter comprises principally settleable solids which can be more economically concentrated before anaerobic treatment than afterwards. Municipal

sewage sludges are a good example of a suitable waste for anaerobic treatment. The digesting sludge is usually heated to about
37°C and the necessary heat is often provided by combustion of the
digester gas.

The principal advantage of this type of plant is its simplicity
in design and operation. However, a long hydraulic detention time
is necessary and the effluent has the same high suspended solids
concentration that obtains within the reactor.

The digestion of trade wastes using this process is very similar in operation to the treatment of sewage sludge. Most trade
wastes require an initial inoculum of about 10 percent of digester
capacity, of actively digesting sewage sludge to initiate the process. Trade wastes free from fats and greases may be treatable in
shorter retention times than sewage sludges as these substances are
slow to break down. Also,the efficient mixing of some trade wastes
which have a large proportion of dissolved material is particularly
important as they tend to produce "thin" organic sludges which are
vulnerable to preferential wasting.

Mosey [14] has summarized performance data for laboratory-scale
digesters treating wastewaters from the processing of apples, toma-
toes and peaches; penicillin broth; wine pot still liquor; yeast
waste liquor;and diluted molasses slop. Percentage removal achieved
either of BOD or organic carbon (C) with retention times in days in
brackets were respectively 77 BOD, 86 C (7); 73 BOD, 88 C (7); 51
BOD, 87 C (7); 75 BOD (20); 95 C (2); 82 C (2); and 66.5 C (3.75).

THE ANAEROBIC CONTACT PROCESS

This process differs from the conventional process in that a
settlement tank is used to concentrate and return the biologically
active anaerobic sludge to the digester. This permits higher hydraulic loadings for a given reactor size and process efficiency. It
is used in preference to the conventional process in cases when the
raw waste contains a substantial proportion of dissolved or colloidal organic matter which cannot be concentrated by settlement of
the raw waste before digestion.

For the treatment of dilute wastes hydraulic detention times
should be very short if the process is to be economical. With the
contact process wastes having a BOD of about 1000 mg/ℓ have been
successfully treated using detention times of between six and twelve
hours.

An operational drawback of plants of this type is that gas continues to form in the settling tank and the gas bubbles tend to prevent efficient settlement of solids. Some plants have adopted va-

cuum degassing which has produced an improvement but has by no means solved the problem.

Typical plant performance data collated by Mosey [14] are reproduced in *Table 2*.

THE ANAEROBIC FILTER

This type of plant is still being developed and the published information available relates exclusively to laboratory-scale plants. However, the form of plant shows much promise for the treatment of dilute industrial wastes containing mostly dissolved or colloidal matter. The principal feature of the plant is a flooded bed of inert filter medium (for example stones of 20 to 40 mm diameter) through which the wastewater is passed upwards. The microorganisms either adhere to or are entrapped between the particles of inert medium enabling a long solids retention time to be maintained at high hydraulic loadings. High rate treatment can be carried out even at ambient temperatures which is a useful feature for dilute wastes which do not produce sufficient gas for heating.

A particular facet of the process which is advantageous is that the majority of gas production is in the lower part of the filter. As the gas rises it produces turbulence which helps to maintain clear passageways for the wastewater and to promote "rinsing" of the filter.

However, the process is not suitable for the treatment of wastes containing high concentrations of suspended solids as such wastes would rapidly clog the filter.

Some performance data for laboratory scale anaerobic filters collated by Mosey [14] are reproduced in *Table 3*.

BASIC KINETICS

Although the kinetics of the reactions that take place in any actual anaerobic treatment process are very complicated, the use of simplified relationships which take account of the most important variables can be most useful in gaining an understanding of the system and in process design.

In settling up a series of basic equations to describe the system it is considered reasonable to assume that microorganisms use substrate for two purposes only – that is for growth and for maintenance. Substrate refers to the concentration of biodegradable organics in the waste stream. Thus within any given system under steady state conditions the rate of change of total substrate con-

TABLE 2 . Performance Data for the Anaerobic Contact Process   Mosey (14)

Type of waste	Scale of plant	Digestion temperature (°C)	Hydraulic retention time (d)	Parameter	Influent (mg/l)	Effluent (mg/l)	Percentage removal
Slaughterhouse ..	Laboratory	33	0.69	BOD	1 500	100	93
" ..	"	33	2.94	"	1 500	84	94
" ..	Pilot	33	1.25	BOD	2 100	90	96
" ..	"	33	1.25	Org.C	940	92	90
Meat packing ..	Pilot	35	0.5	BOD	1 600	80	95
" ..	Full	32	0.54	"	1 380	130	91
" ..	"	32	0.34	SS	990	200	80
Maize-starch ..	Full	Ambient	3.3	BOD	6 280	755	88
" ..	"	Ambient	3.3	Org.C	3 250	317	90
" ..	"	"	3.3	VS	6 556	623	90
Brewery ..	Pilot	not stated	2.23	BOD	3 280	130	96
Distillery ..	Full	30	7.2	COD	22 400	540	98
" ..	"	30	5.3	"	12 600	400	97
" ..	Laboratory	33	6.2	BOD	25 000	986	96
" ..	"	33	6.2	Org.C	12 000	1 812	85
" ..	Laboratory	35	0.92	BOD	845	60	93
" ..	"	35	0.92	TS	1 820	850	53
Citrus ..	Laboratory	34	1.38	BOD	2 670	130	95
" ..	Pilot	34	2.32	"	3 440	1 100	68
Yeast ..	Laboratory	30	2	BOD	3 042	391	87
" ..	Pilot	30	1.7	"	5 076	761	85
Chewing gum ..	Full	not stated	11.7	BOD	1 840	740	60
Milk ..	Laboratory	31	6	BOD	3 300	10-20	99.5
" ..	"	31	6	"	380	20-50	90
" ..	"	31	6	VS	3 750	260	93
" ..	"	31	6	"	310	140	55

Table 3:   Performance Data for Laboratory Scale Anaerobic Filters   Mosey (14)

Type of waste	Digestion temperature (°C)	Hydraulic retention time (d)	Parameter	Influent (mg/l)	Effluent (mg/l)	Percentage removal
Raw domestic sewage ..	4	1.5	BOD	180	40-40	67
"	25	1.5	"	180	10-35	82
"	25	1.5	SS		9	>95
Protein/carbohydrate	25	0.75	COD	1 500	122	91.5
"	25	0.375	"	1 500	312	79.3
"	25	0.187	"	1 500	950	36.7
"	25	3.00	"	3 000	204	93.4
"	25	1.50	"	3 000	247	88.4
"	25	0.375	"	3 000	1 100	63.0
Volatile acids	25	1.5	COD	1 500	24	99.4
"	25	0.75	"	1 500	139	90.5
"	25	0.375	"	1 500	314	79.0
"	25	3.0	"	3 000	42	98.6
"	25	1.5	"	3 000	240	92.0
"	25	1.5	"	6 000	139	97.7
"	25	0.75	"	6 000	794	86.9
Food processing	35	3.56	COD	8 475	546	93.5
"	35	3.56	BOD	5 200	975	81.4
"	35	3.56	Org.C	2 400	115	95.2
"	35	3.56	SS	1 508	455	70.0
"	35	0.54	COD	8 475	5 000	41.0
"	35	0.54	BOD	5 200	3 890	25.2
"	35	0.54	Org.C	2 400	1 720	28.3
"	35	0.54	SS	1 508	1 855	-

336

centration is equal to the sum of the rate of change of concentration of substrate used for growth of microorganisms plus the rate of change of concentration of substrate used for maintenance of microorganisms or

$$\frac{dS}{dt} = \frac{dS_g}{dt} + \frac{dS_m}{dt} \qquad \cdots \quad (2)$$

where;

S = the total concentration of substrate
S_g = the concentration of substrate used for growth
S_m = the concentration of substrate used for maintenance

The rate of change of total concentration of substrate can be expressed in terms of the concentration of microorganisms present if it is assumed that certain simple relationships hold true. Firstly, it is assumed that the growth requirement of the microorganisms is directly proportional to the increase in microbial mass and secondly that the maintenance requirement is directly proportional to the microbial mass itself.

These relationships can be expressed mathematically as follows:

$$-\frac{dS_g}{dt} = \frac{1}{a}\frac{dX}{dt} \qquad \cdots \quad (3)$$

and

$$-\frac{dS_m}{dt} = k_2 X \qquad \cdots \quad (4)$$

where;

X = the concentration of microorganisms present

a , k_2 = proportionally constants

- = indicates a rate of decrease in substrate concentration.

Substituting equations (3) and (4) back into equation (2) gives:

$$-\frac{dS}{dt} = \frac{1}{a}\frac{dX}{dt} + k_2 X \qquad \cdots \quad (5)$$

Another kinetic expression that has been frequently used to describe the biological waste treatment process was first used by Michaelis and Menton [15] to describe enzyme reactions. It can be shown that the rate of change of substrate concentration is related to the concentration of microorganisms and the concentration of substrate in the following way.

$$-\frac{dS}{dt} = \frac{k \, X \, S}{K_S + S} \qquad \cdots \cdots (6)$$

where k and K_S are constants. *Fig. 5* shows graphically the relationship between the rate of utilisation of substrate per unit mass of microorganism (1/X.dS/dt) and the substrate concentration, S. From the graph it can be seen that the constant k (which has units of 1/T) gives the maximum rate of substrate removal when the substrate concentration is very much greater than the value of K_S. The constant K_S (which has units of M/L and is known as the Michaelis constant) is equal to the substrate concentration when the reaction rate is at half its maximum value.

If Equations (5) and (6) are combined then

$$\frac{1}{a} \frac{dX}{dt} + k_2 X = \frac{k \, X \, S}{(K_S + S)}$$

which can be rearranged to give

$$\frac{dX}{dt} \frac{1}{X} = \frac{akS}{(K_S + S)} - \frac{k_2}{a}$$

This expression can be simplified by combining the constants a and k_2 to give

$$\frac{dX}{dt} \cdot \frac{1}{X} = \frac{akS}{(K_S + S)} - b \qquad \cdots \cdots (7)$$

However, if the process is being carried out under steady state conditions the fractional growth rate of the bacteria dX/dt·1/X must equal their fractional removal rate in the effluent. But for a completely mixed conventional digester the fractional removal rate is the reciprocal of the nominal mean retention time of the

338

biological solids within the system, T, (see equation (1)). Thus,

$$\frac{1}{T} = \frac{akS}{(K_s+S)} - b \qquad \dots \quad (8)$$

It can be seen that equation (8) gives a relationship between digestion tank capacity (for a given waste flow), and the concentration of substrate in the digester. In the case of the completely mixed system S also represents the concentration of substrate in the effluent. From the foregoing it can be seen that if the characteristics of a particular waste are known (i.e. the values of a, b, and K_s are known) then the equation can be used as the basis of conventional digester design. The equation can also be applied to the other types of treatment plant - the contact process and the anaerobic filter process - where T becomes the nominal retention time of the mixed liquor suspended solids or other parameter reflecting the fractional growth of the microorganisms.

Table 4 gives some experimental values quoted by Mosey[14] that have been determined for constants a, b, and K_s for particular organic substances. The associated digester performance curves obtained by substituting these experimental values into equation (8) are shown in *Fig. 6*.

TABLE 4 : Experimental Values of Kinetic Constants

Constant	Propionate	Acetate	Stearate
Temp (°C)	35	35	37
T_{min} (d)	2.6	3.28	9.4
k (d^{-1})	9.6	8.1	0.77
K_s (mg/ℓ)	32	154	417
	(as acetic)		
a	0.042	0.04	0.11
b (d^{-1})	0.01	0.01	0.01

See also refs (16) and (17).

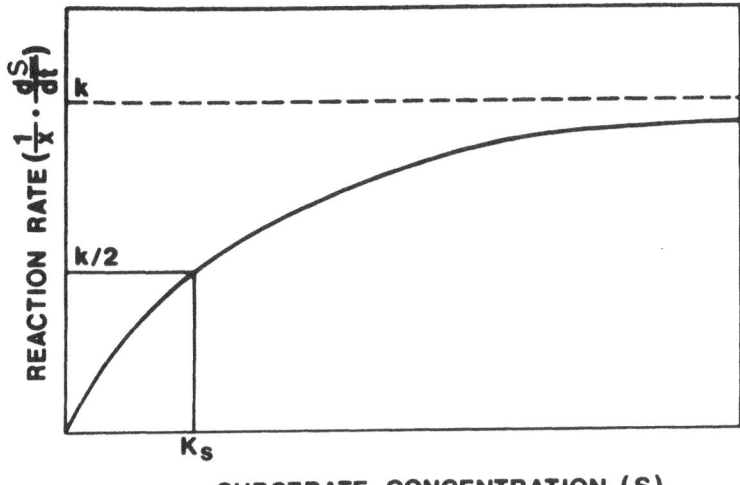

Fig. 5 : Relationship between Reaction
Rate and Substrate Concentration

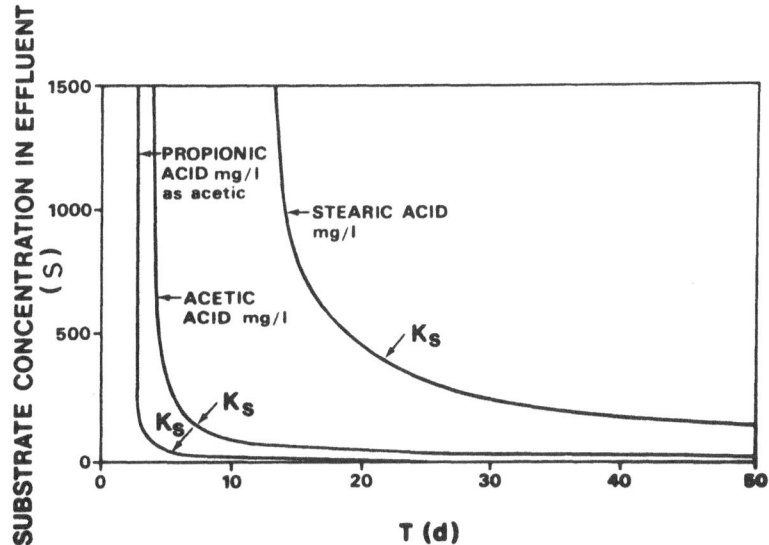

Fig. 6 : Steady-State Relationship for
Methane Fermentation of Fatty
Acids (Mosey [14])

It will be noted from the curves shown in *Fig. 6* that the theoretical minimum retention time in the reactor (T_{min}) - the "washout" time - occurs when the concentration of substrate in the reactor is very large compared with K_s. In this case $S/(K_s+S)$ tends to unity and equation (8) can be modified to give

$$T_{min} = \frac{1}{a\,k - b} \qquad \qquad \dots \quad (9)$$

In practice, however, washout of the microorganisms occurs when the concentration of substrate in the reactor approaches the concentration in the influent. The difference is usually small for concentrated washes but becomes significant for dilute wastes where the concentration of the influent substrate can be of the same order as the Michaelis constant (K_s).

Although operation of the process near the minimum solids retention time is possible efficiencies are low and reliability of the process is poor. McCarty [5] recommends that design solids retention times should be at least two and a half times the minimum. Efficiency at these design times is some 90 to 95 percent of the maximum obtainable.

It is possible to take the effects of inhibitors into account in the use of the formula given above. Their effects can be included by modification of the substrate constants for different conditions of inhibition. This method is simple but necessitates large amounts of data to describe fully systems which are affected by more than one inhibitor.

Alternatively new kinetic relationships can be derived which incorporate the functions of inhibiting materials. The disadvantage of this method is that the kinetic relationships become more unwidely and their theoretical basis more uncertain as the systems become more complex. It is evident that this topic is one in which further research effort would be beneficial in obtaining a more accurate picture of the processes at work and their interrelationships.

NOMENCLATURE

a	=	microorganisms growth yield coefficient	(mg/mg)
b	=	microorganisms decay coefficient	(d^{-1})
k	=	maximum rate of substrate utilisation	(mg/mg-d)
K_h	=	total heavy metal concentration per kilogram solids	(m.eq/kg)

K_s	=	Michaelis constant	(mg/ℓ)
M_e	=	total weight suspended solids leaving system in day	(kg/d)
M_t	=	total weight of suspended solids in system	(kg)
S	=	total concentration in substrate	(mg/ℓ)
S_g	=	concentration of substrate used for growth	(mg/ℓ)
S_m	=	concentration of substrate used for maintenance	(mg/ℓ)
T	=	solids retention time	(d)
T_{min}	=	theoretical minimum solids retention time	(d)
X	=	concentration of microorganisms	(mg/ℓ)
S	=	Total concentration of substrate	(mg/ℓ)
S_g	=	Concentration of substrate used for growth	(mg/ℓ)
S_m	=	Concentration of substrate used for maintenance	(mg/ℓ)
T	=	Solids retention time	(d)
T_{min}	=	Theoretical minimum solids retention time	(d)
X	=	Concentration of microorganisms	(mg/ℓ)

REFERENCES

1. G.C. Cillie, M.R. Henzen, G.J. Stander and R.D. Baillie, Anaerobic Digestion-IV, The Application of the Process in Waste Purification, Water Research,3, 623, 1969
2. G.J. Schroepfer, W.J. Fuller, A.S. Johnson, N.R. Zienke and J.J. Anderson, The Anaerobic Contact Process as Applied to Packinghouse Wastes, Sewage ind. Wastes, 27, 460, 1955
3. J.S. Jeris and P.L. McCarty, The Biochemistry of Methane Formation Using C14 Tracers, J. Wat. Pollut. Control Fed., 37, 2, 178, 1965
4. J. Häusler, The Succession of Microbial Processes in the Anaerobic Decomposition of Organic Compounds, Proc. 4th Conf. of Int.Ass. Wat. Pollut. Research, 1969
5. P.L. McCarty, Anaerobic Waste Treatment Fundamentals, Parts 1 to 4, Public Works,95, 9, 107 and 95, 10, 123 and 95, 11, 91 and 95, 12,95, 1964
6. T.C. Stadtman and H.A. Barker, Tracer Experiments on the Mechanism of Methane Formation, Arch. Biochem, 21, 256, 1949

7. H.A. Barker, Biochemical Functions of Corrinoid Compounds, Biochem J., 105, 1, 1967

8. A.M. Buswell and H.F. Mueller, Mechanisms of Methane Fermentation, Ind. Eng. Chem., 44, 550, 1952

9. F.E. Mosey and D.A. Hughes, The Toxicity of Heavy Metal Ions to Anaerobic Digestion, J. Inst. Wat. Pollut. Control, 74, 1, 1975

10. F.E. Mosey, Assessment of the Maximum Concentration of Heavy Metals in Crude Sewage Which Will Not Inhibit the Anaerobic Digestion of Sludge, J. Inst. Wat. Pollut. Control, 75, 1, 1976

11. Water Pollution Research Laboratory, Synthetic Detergents and Sludge Digestion, W.P.R.L. Notes on Water Pollution, 29, 1965

12. J.D. Swanwick and M. Foulkes, Inhibition of Anaerobic Digestion of Sewage Sludge by Chlorinated Hydrocarbons, J. Inst. Wat. Pollut. Control, 70, 1, 1971

13. J.F. Andrews, A Dynamic Model of the Anaerobic Digestion Process, Proc. 23rd Purdue Waste Conference, 285, 1968

14. F.E. Mosey, Anaerobic Biological Treatment, Inst. Wat. Pollut. Control Symposium on the Treatment of Wastes from the Food and Drink Industry, Newcastle, 1974

15. Michaelis, and Menton, Biochem Zeit, 49, 333, 1913

16. A.E. Lawrence and P.L. McCarty, Kinetics of Methane Fermentation in Anaerobic Treatment, J. Wat. Pollut. Control Fed.

17. J.T. Novak and D.A. Carlson, The Kinetics of Anaerobic Long Chain Fatty Acid Degradation, J. Wat. Pollut. Control Fed., 42, 11, 1932, 1970

18. F. Ediline, La Digestion Anaerobie dans l'Epuration des Eaux, 17 e Cycle de Perfectionnement , Geni Chimique 1976 Branch Belge de la Societe de Chimie Industrielle, Bruxelles, 1976

MECHANISMS OF SLUDGE DIGESTION

W. Wesley Eckenfelder, Jr.
Distinguished Professor
Vanderbilt University, Nashville, Tennessee

The anaerobic conversion of organic solids to inoffensive end products is very complex, and is the result of many reactions, but can be illustrated in a simplified manner as shown in *Figure 1*. In the conventional high rate digestion system all of the reactions shown in Figure 1 are occurring simultaneously in the same tank. Under equilibrium operating conditions (steady state) all of the reactions must be occurring at the same rate, since there is no build up of intermediate products. While many factors such as sludge composition and concentration, pH, temperature, mixing, etc., influence in the reaction rates, it is generally assumed that the

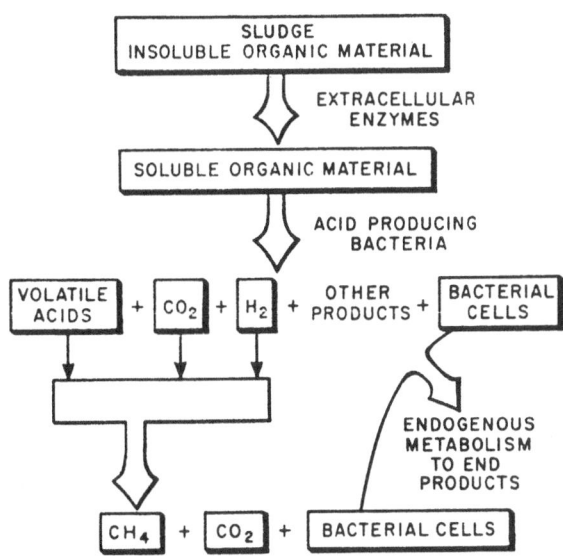

Fig. 1 : Mechanism of Anaerobic
Sludge Digestion

overall rate is controlled by the rate of conversion of volatile
acids to methane and carbon dioxide. Digester upset and failure
occurs when there is an imbalance in the rate of conversion of volatile acids to methane and carbon dioxide.resulting in a build up
of intermediate volatile acids. Since digester performance depends
on all three rates occurring simultaneously, a discussion of the
factors influencing each is pertinent.

LIQUEFACTION

Relatively little data is available on the rate of solubiliza-
tion of organic solids in a sludge digester. It can be assumed
that the rate is related to the particle size of the sludge solids
and the degree of agitation in the tank. In most cases, it is generally assumed that liquefaction is more rapid than the conversion
of volatile acids to methane so that this is not a rate limiting
step. Available data would indicate that 70% of the volatile solids destroyed in the digestion process occurs in less than five
days detention. The one exception is grease which in many cases
will accumulate in the digester particularly at short residence
times.

ACID FERMENTATION

The solubilized organic matter is rapidly converted to organic
acids under anaerobic conditions. The primary acids produced are
acetic, propanic, and butyric with trace amounts of formic, valeric, isovaleric and caproic.

Acid fermentation is characterized by a drop in pH from near
neutral to about pH 5.0. (The subsequent conversion of the acids
to methane and carbon dioxide results in a rise in pH to 6.8 - 7.4).
It is very significant to note that there is no appreciable reduction in COD or BOD in the waste mixture through the acid fermentation stage, since there is merely a conversion in the type of organic compounds. There is, of course, a reduction in volatile suspended solids content, through liquefaction as previously noted.

METHANE FERMENTATION

Very little detailed information is available on the methane organisms. As a group these organisms ferment only a few compounds,
most of these being the products of other bacterial fermentations
such as alcohols, volatile acids and a few gases. Several species
may be required for the conversion of higher carbon acids such as
valeric. The postulated mechanism of methane fermentation is shown
in *Figure 2*.

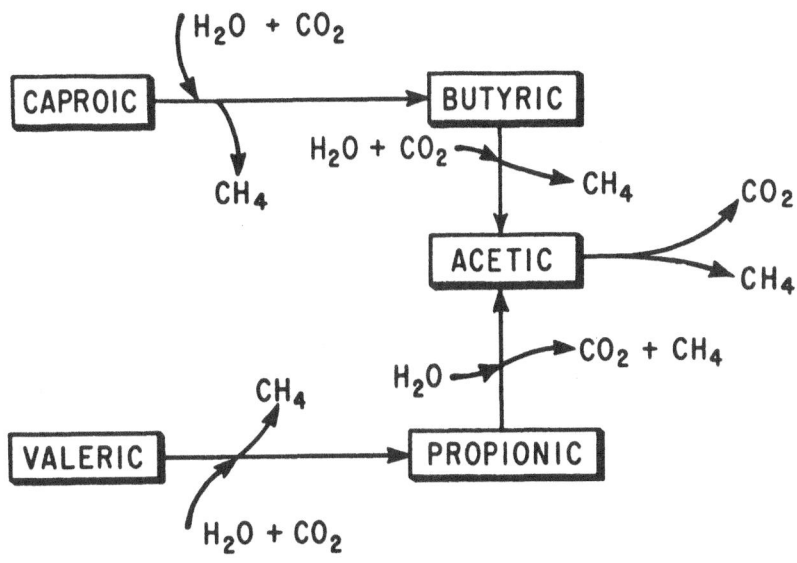

Fig. 2 :   Methane Fermentation From
             Volatile Acids

Although methane is produced from all of the volatile acids, it
is ultimately derived from only two major sources, carbon dioxide
reduction and acetic acid fermentation (acetic acid is an eventual
product in the fermentation of all volatile acids having three or
more carbons). The primary reactions involved can be summarized:

(1)   Acetic Acid Fermentation

$$CH_3COOH \rightarrow CH_4 + CO_2$$

(2)   Carbon Dioxide Reduction

$$CO_2 + 8H \rightarrow CH_4 + 2H_2O$$

Since the methane fermentation is the rate controlling step in the
digestion process, environmental control for this fermentation is
of primary importance.

All methane bacteria are strictly anaerobic and therefore function in the absence of oxygen and at a low Oxidation Reduction Potential (OPR). The result of Dirasian, Molof and Borchardt [2],
indicated that optimum digestion occurred at ORP values between
-520 and -530 mv. The functional range was -490 to -550 with activity decreasing rapidly at the extreme ends of these ranges.

pH is also critical to optimum methane fermentation. The optimum range of pH is 6.8 to 7.4 with an extreme range of 6.4 to 7.8.

High concentrations of inorganic salts may result in temporary or permanent inhibition of the fermentation process. Heavy metals such as copper, zinc, nickel and chromium may cause inhibition depending on the state of the material, its solubility and possible precipitation in the process by combination with sulfide.

The optimum conditions to maintain maximum rates of methane fermentation are summarized in Table 1.

TABLE 1 : Environmental Conditions for Methane Fermentation

Variable	Optimum	Extreme
pH	6.8 to 7.4	6.4 to 7.8
Oxidation Reduction Potential (mv)	-520 to -530	-490 to -550
Volatile Acids (mg/ℓ as acetic)	50 - 500	> 2,000
Total Alkalinity (mg/ℓ as $CaCO_3$)	1500 - 5000	1000 - 3000
Salts NH$_4$ (mg/ℓ as N)		300
Na (mg/ℓ)		3500-5500
K (mg/ℓ)		2500-4500
Ca (mg/ℓ)		2500-4500
Mg (mg/ℓ)		1000-1500
Gas Production Destroyed (mg^3/Kg VS)	1.06 - 1.37	----
Gas Compositions (%CH$_4$)	65 - 70	-
Temperature (°C)	32 - 38	-

RATES of REACTION

As previously indicated, successful digestion depends upon maintaining a balance between the various rates of reaction occurring in the digester. These rates are sequentially shown in Fig. 3 for a batch operation. Since the rate of methane fermentation must control the overall rate to avoid process failure, further consideration of the rate of this fermentation is important.

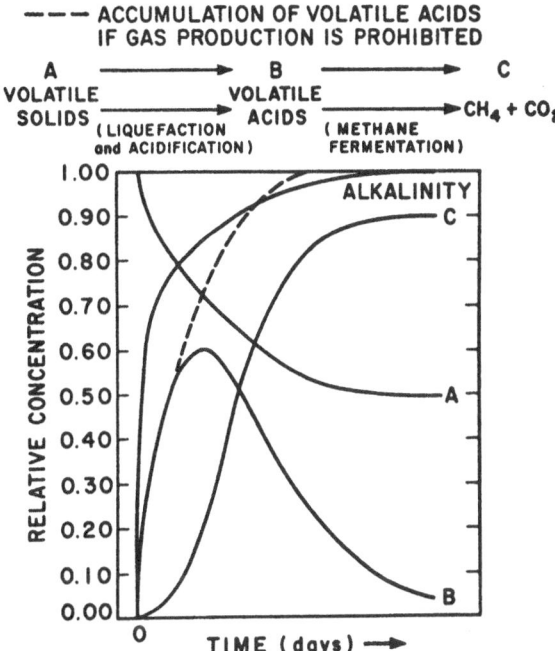

Fig. 3 : Sequential Mechanism of Anaerobic Sludge Digestion (after Malina [4]).

In order to effect methane fermentation, sufficient time must be available in the reactor to permit growth of the organisms or else they will be washed out of the system. In a completely mixed flow through digestion tank this means that the detention time in the unit must be greater than the growth rate of the organisms.It is significant to note that there are several species of methane organisms active in a digestion system, all having different growth rates. Andrews has shown that some organisms with a high growth rate (< 2 days) can produce methane, probably from the fermentation of formate, methanol, CO_2 and H_2 and possibly some volatile acid fermentation. Other organisms require residence times of up to 20 days. This is illustrated by the data of Andrews from continuous digestion units employing a soluble waste feed as shown in Fig. 4. While data is limited, some results have been reported relative to the growth rate of methane organisms. These data are summarized in Table 2.

In a continuously operating, mixed digester, substantially complete methane fermentation should occur providing the residence time

TABLE 2 : Growth Rate of Methane Organisms

Substrate	Temp, °C	Residence Time, days	Reference
Methanol	35	2	Speece & McCarty
Formate	35	3	Speece & McCarty
Acetate	35	5	Speece & McCarty
Propionate	35	7.5	Speece & McCarty
Primary & Activated Sludge	37	3.2	Torpey

exceeds the growth rate of the essential methane producing organisms and optimum environmental conditions are maintained. The digester reactions under these conditions are shown in *Fig. 5*.

As shown in *Fig. 5*, at low residence times there will be volatile solids reduction due to the liquefaction of the solids and the subsequent conversion to volatile acids by acidification. During this period a small amount of methane fermentation may occur (depending on environmental conditions such as pH) primarily due to reduction of formate, methanol, CO_2 and H_2 as shown in *Fig. 5*. There will be a decrease in pH and a corresponding increase in the volatile acid concentration. There will be very little COD reduction however, since the organisms have merely been converted from a solid form to a soluble form in the supernatant liquor. When the detention time in the digester exceeds the growth rate of the principal methane organisms there will be a rapid increase in methane production with a corresponding decrease in volatile acid concentration and an increase in pH. There are probably several methane organisms responsible for the volatile acid conversion, each of which will have a different generation time or growth rate; the methane production curve is relatively flat as shown in *Fig. 5*.

The acetic acids, at higher residence times are obtained from two sources, direct fermentation and the breakdown of higher carbon acids to acetics. The major part of the methane production comes from acetic acid fermentation although some is generated from the breakdown of the higher acids as shown in *Fig. 2*. At high residence periods, substantially all of the volatile acids are converted to methane and carbon dioxide. It should be noted that since all of the volatile solids present are not degradable in the digestion unit, a portion will remain, even after long periods of retention. For sewage sludge this fraction is approximately 40%.

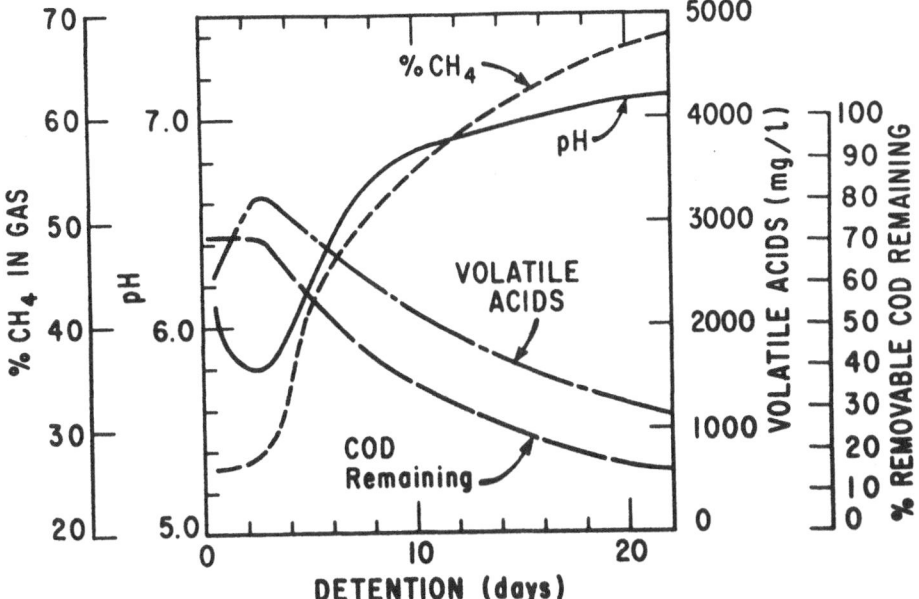

Fig. 4 : Anaerobic Degradation of
Organic Waste

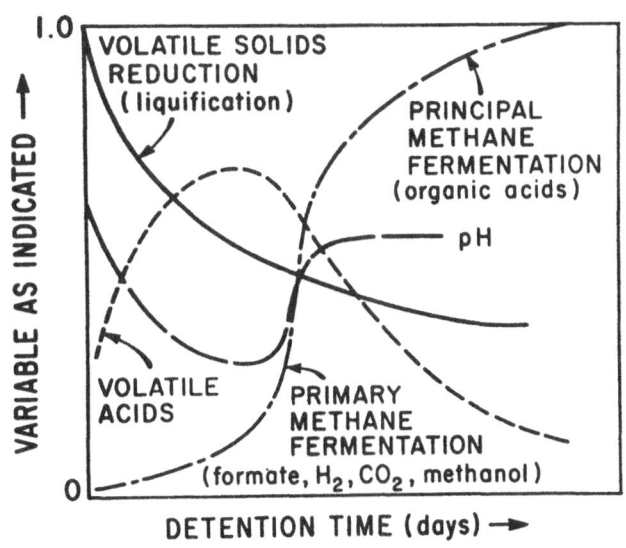

Fig. 5 : Mechanism of Continuous
Mixed Anaerobic Digestion

GAS PRODUCTION

The major part of the gas produced in a sludge digester comes from the breakdown of volatile acids as shown in *Fig. 2*. Some gas is produced by the early stages of methane fermentation of CO_2 and H_2, methanol, etc., but this contribution is probably very small in a sludge digester. The gas will be composed of CH_4, CO_2 with small quantities of H_2S and H_2. The percentage of CH_4 in the gas will depend in large measure on the residence time, the percentage of CO_2 being higher at the lower residence times with corresponding lesser numbers of methane bacteria. McCarty has shown from theoretical considerations supported by experimental evidence that 0.35 m³ of methane gas will be produced per kg COD reduced. The reported gas production for volatile solids reduction in a well operating anaerobic digestion tank is 1.06-1.25 m³/kg, vs. destroyed with a methane content of about 65%. This is about equivalent to 0.31-0.44 m³ CH_4/Kg COD destroyed which is close to the value reported by McCarty. It is significant to note at this point that these values are a maximum, assuming complete conversion of the solids to methane. Volatile solids reduction can, of course, occur by liquefaction and conversion to volatile acids without any COD reduction. Under these conditions the methane yield per unit of volatile solids reduction may be very low.

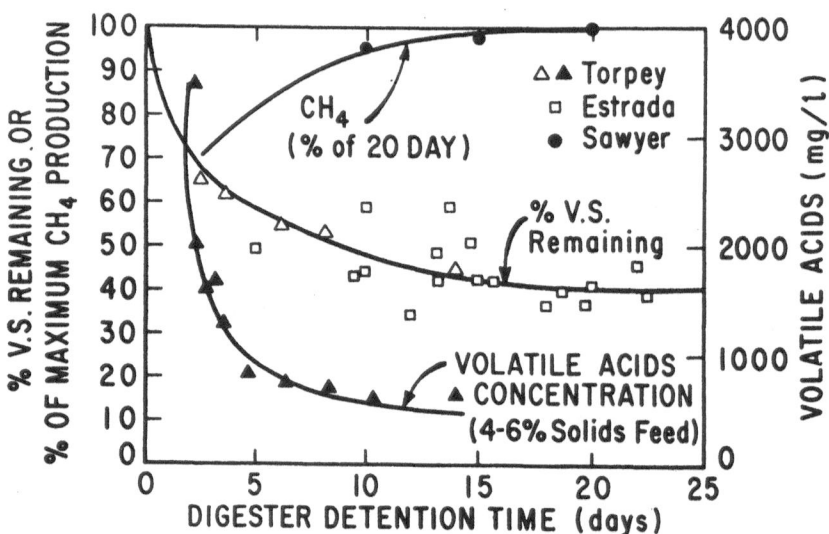

Fig. 6 : High Rate Anaerobic
Digester Performance

DIGESTER LOADING and PERFORMANCE

The available data would indicate that effective digestion should be possible with detention periods as low as five days providing other environmental conditions are maintained. Under these conditions, about 70 percent of the degradable organic solids will be liquified with the major portion being converted to gas. Increasing the detention time to 10 days should result in 90 percent of degradable organics being liquefied with over 90 percent of the degradation by-products being converted to gas. These data are graphically shown in *Fig. 6*. While methane fermentation should be related only to sludge age or digester detention time (as related to methane organisms growth rate), Sawyer has shown that at solids feed levels in excess of 8 percent methane fermentation decreases. This is probably due to the mixing problems encountered with the high solid levels. Grease reduction, however, improves with a higher solid concentration in the feed sludge.

REFERENCES

1. Andrews, J.F., Cole, R. D. and Pearson, E.A., Kinetics and Characteristics of Multistage Methane Fermentations, Engr. Res. Lab., SERL Report No. 64-11, University of California, Berkeley.

2. Dirasian, H.A., Molof, H.A., and Borchardt, J.A., Electrode Potentials Developed During Sludge Digestion, *Jr. Water Pollution Control Fed.*, 35, 424, 1963.

3. Estrada, A., Cost and Performance of Sludge Digestion Systems, *Proc. Am. Soc. of Civil Engrs.*, 86, SA 3, 111, 1960.

4. Malina, J., *Advances in Water Quality Management,* University of Texas Press, Austin, Texas, 1967.

5. Sawyer, C. N., An Evaluation of High Rate Digestion, *Biological Treatment of Sewage and Industrial Wastes,* Vol. II, (J. McCabe and W.W. Eckenfelder, Editors), Reinhold Pub. Co., New York, N.Y., 1958.

6. Speece, R. E. and McCarty, P.L., Nutrient Requirements and Biological Solids Accumulation in Anaerobic Digestion, *Advances in Water Pollution Research,* Vol. II, Pergamon Press, London, England, 1964.

7. Torpey, W.N., Loading to Failure of a Pilot High Rate Digester, *Sewage and Industrial Wastes,* 27, 121, 1955.

ADVANCED BIOLOGICAL TREATMENT PROCESSES

A. L. Downing
Partner and Senior Engineer

A. D. K. Kell
Binnie and Partners, London, U.K.

INTRODUCTION

In recent years there have been considerable developments in the field of biological treatment of liquid wastes both to meet new or increasing needs and to improve the efficiency of existing processes. There has been much concern with the quality of effluents from waste treatment plants due to problems of eutrophication arising from increasing levels of nutrients in receiving waters and to the dangers of discharge of potentially toxic substances which have not been removed during treatment or have been created in the course of the process. This concern has led to the development of 'polishing' processes and the use of algae for the removal of nutrients such as nitrates and phosphates. Modifications to the activated sludge process have been developed that encourage the hydrolysis of organic nitrogen to ammonia and the denitrification of nitrate to nitrogen gas. Other developments have been concerned with the efficiency of biological oxidation processes. These have included modifications to the conventional activated sludge process by the use of pure oxygen in place of air and by carrying out the process in a deep shaft rather than in horizontal tanks. A third area of development has been in the conversion of wastes into more useful products. An example of this type of development is the microbial conversion of waste rich in carbohydrates into single cell protein by submerged aerobic fermentation.

This paper does not attempt to cover the whole field of advanced biological treatment processes but concerns itself with nitrogen removal by nitrification and denitrification, with the use of pure oxygen in the biological oxidation process, and with the Deep Shaft process.

NITROGEN REMOVAL

General

Nitrogen occurs in a number of different forms in municipal waste waters at considerably higher concentrations than are found in natural waters. This additional nitrogen derives principally from constituents of human wastes such as unassimilated proteinaceous matter, breakdown products of protein and urea. Nitrogenous compounds are also common constituents of industrial wastes.

The most common forms in which nitrogen occurs in municipal waste waters are dissolved elementary nitrogen, ammonia, nitrite, nitrate, and nitrogenous organic compounds especially urea. Nitrite is not often present in raw waste waters in significant concentrations nor is nitrate unless there is an appreciable concentration in local natural waters or in the public supply; both may be present, however, in much larger concentrations in biologically treated effluents.

Effects of Nitrogen
on Reuse of Water

Waste water effluents are commonly discharged into a receiving body of water which has a subsequent municipal, commercial or recreational use. The presence of some forms of nitrogen may have adverse effects on these uses. The most important effects of the forms of nitrogen usually found in municipal waste waters are summarized below.

The presence of dissolved nitrogen is normally unobjectionable and the question of its removal does not usually arise. Ammonia can cause problems particularly because of:

a) the depletion of oxygen resources that can result from its oxidation by nitrifying bacteria to nitrite and nitrate;

b) its toxicity to fish;

c) its role as a nutrient promoting the growth of unwanted algae in natural waters and water supply reservoirs; and

d) the more direct interference which its presence can cause in the treatment of water for public supply and especially in disinfection.

Nitrite can cause adverse effects of the type mentioned under all the above headings and, in relation to (d), an additional objection arises because if its presence were to give rise to excessive concentrations in water put into public supply it could cause methaemoglobinemia in young babies drinking the water; and because of the possibility that it could form carcinogenic nitrosamines with amines present in the human diet. Nitrate is objectionable for the same reasons as nitrite except that being fully oxidised it does not constitute a potential source of demand on oxygen resources nor does it give rise to formation of nitrosamines. Nitrogenous organic compounds may be objectionable specifically because of their content of nitrogen (as distinct from their general properties) on account of the possibility of their susceptibility to chemical or biochemical processes giving rise to the release of ammonia, nitrite or nitrate. The World Health Organization[1] has recommended that the concentration of nitrate in drinking water should preferably not exceed 11.3 mg N/ℓ and has suggested a maximum acceptable level of 22.6 mg N/ℓ.

The lowland waters in many countries are experiencing rising concentrations of nitrogenous compounds, particularly nitrate, due principally to increasing use of fertilisers and increasing discharges of more nitrified effluent. In a number of places, including some parts of the Thames basin, the existing and projected nitrate levels are causing concern and in recent years much research and development effort has been devoted to the study of nitrogen removal from waste waters.

Removal of nitrogen can be achieved by many means including such general processes as distillation, reverse osmosis, electrodialysis, and ion-exchange and particular processes such as desorption of ammonia by air-stripping or removal of various forms by algal culture. However, we shall concern ourselves here solely with the biological process of nitrification followed by denitrification.

CHEMISTRY and MICROBIOLOGY

The removal of nitrogenous compounds from waste waters by biological methods is effected by the conversion of these compounds into gaseous nitrogen or less importantly to nitrous oxide. Organic compounds containing nitrogen usually breakdown to produce ammonium ions. There is no direct pathway for the biological oxidation of ammonia to produce nitrogen. This can only be effected by the oxidation of ammonia to nitrite or nitrate (nitrification) followed by the reduction of nitrite or nitrate to nitrogen (denitrification). The steps of the process are shown schematically in *Fig. 1*.

356

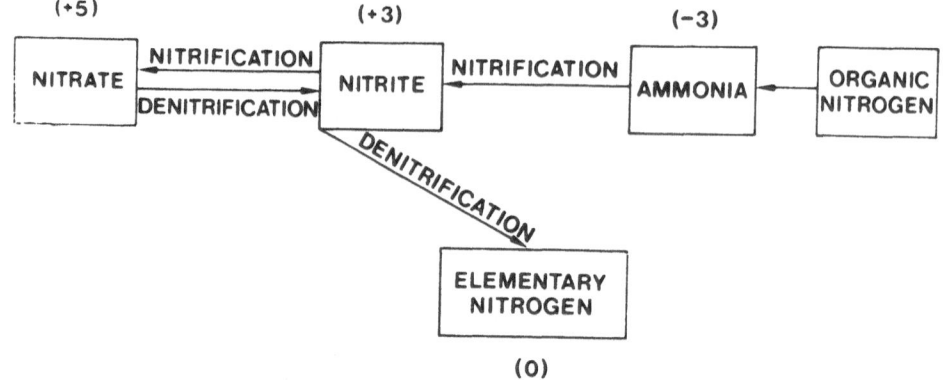

NUMBERS IN BRACKETS REFER TO REDOX STATE.

Fig. 1 : Process Involved in Biological
Conversion of Organic Nitrogen
to Nitrogen Gas

NITRIFICATION

The biological oxidation of ammonia takes place in two principal stages. Firstly oxidation of ammonia to nitrite is brought a-
bout by the action of *Nitrosomonas* which is a strict autotroph requiring dissolved carbon dioxide or its ionised forms as a source
of carbon. This is the only important organism involved in the
process although *Nitrosococcus* and perhaps a few others have the
same ability to produce nitrite. Secondly oxidation of nitrite to
nitrate is brought about mainly by the action of the autotroph *Nitrobacter* although *Nitrocystis* can also produce the same action.

The nutritional requirements of these bacteria are not known in
detail. In addition to carbon dioxide and ammonia or nitrate a
minimum concentration of dissolved oxygen is an absolute requirement for growth. Also *Nitrosomonas* is known to require sodium,
phosphate, magnesium, iron, calcium and copper while *Nitrobacter*
requires sodium, phosphate, magnesium, iron, molybdenum, zinc and
copper. The optimum growth temperature for *Nitrosomonas* has been
reported to be in the range 30 - 36°C and little or no growth occurs below 5°C. The effects of temperature on the growth of *Nitrobacter* are similar.

A number of substances inhibit the growth of both organisms including thiourea, cyanide, phenols, certain alkaloids, streptomy-
cin, and salts of heavy metals (such as mercury, chromium and copper) in more than three amounts. *Nitrobacter* is also inhibited by
chlorate ions. Other metal ions (such as sodium, calcium, and potassium) can cause inhibition if present at high concentrations.

These growth factors and others not discussed here all have an effect on the nitrification process in waste water purification. It has been found that the effects of various factors on nitrifying organisms in a mixed medium can be markedly different from the effects in pure culture.

The nutritional requirements of *Nitrosomonas* are normally present in adequate quantities in domestic sewage. The energy source for *Nitrobacter* is usually absent but is formed by *Nitrosomonas*. The buffering capacity of alkalis in sewage and sludge is an important factor in nitrification as the reaction rate is lowered at pH levels less than about 7.2. It has proved difficult to establish the importance of oxygen concentration on the degree of nitrification which can be achieved in biological oxidation processes. Nitrification can take place even under dissolved oxygen concentrations of less than 0.5 mg/ℓ [2] but tests reported by the U.K. Ministry of Technology [3] indicate a significant reduction in the degree of nitrification at oxygen concentrations below about the 2 mg/ℓ.

DENITRIFICATION

Nitrite and nitrate ions are used by many facultative aerobic microorganisms as final electron acceptors in the respiration chain when the oxygen tension in the medium is sufficiently low. The normal end product of this "nitrite or nitrate respiration" is elementary nitrogen, N_2, or nitrous oxide, N_2O. Both gases are relatively inert and will be lost from the medium into the atmosphere. Common facultative heterotrophic bacteria which have the ability to use nitrate or nitrite ions include species of *Micrococcus*, *Pseudomonas*, *Denitrobacillus*, *Spirillum*, *Bacillus*, and *Achromobacter*.

The metabolic pathway followed by a nitrogen atom from its 5+ oxidation state in the nitrate ion, NO_3^-, to the zero oxidation state in molecular nitrogen, N_2, or the 1+ state in nitrous oxide, N_2O, is not known in detail. However, there is biochemical uniformity in so far as the first step is always the reduction of nitrate to nitrite, NO_2^-. The subsequent conversion of nitrite to nitrogen seems also to be fairly similar with all denitrifying microorganisms, although there is no general agreement about its details.

In addition to a source of carbon the various nutrients necessary to the process which have been identified include the "usual" mineral nutrients – sulphate, phosphate, chloride, sodium, potassium, magnesium and calcium – and various trace elements required for the successive enzymic steps – molybdenum, iron, copper and manganese.

The ecological conditions necessary for denitrification have been investigated by Wuhrmann and Mechsner [4] for some strains of bacteria isolated from activated sludge. A strong interdependence between the oxygen tension and the pH value in the medium on the one hand and the rate of denitrification on the other was consistently found. Within the range of pH values between 7 and 8 at which sewage treatment plants normally operate, nitrite and nitrate are reduced only under strictly anaerobic conditions. In an acidic environment the presence of dissolved oxygen has a negligible inhibitory effect on the process.

The rate of reaction of the denitrification process varies considerably within the range of temperatures that could be experienced in waste water treatment. The reaction rate at 5°C is only one-fifth of that at 20°C. However, even at 5°C the rate is higher than the rate of nitrification which obtains at normal operational temperatures in the U.K.

FORMS of TREATMENT

There is usually no difficulty in achieving nitrification of the great majority of ammonia during treatment of domestic sewage either by the activated-sludge process or by biological filtration, providing that the loading and other operating conditions of the plant are appropriately chosen. With sewage containing certain types of inhibitory materials derived from industrial processes it may prove difficult or even impossible to achieve nitrification, though the mere presence of powerful inhibitors does not of itself preclude the possibility of obtaining nitrification quite readily.

Since significant numbers of organisms normally occurring in biological treatment units (activated sludge or trickling filters) have the ability to use nitrite or nitrate ions as final electron acceptors the biological removal of nitrogen in waste treatment is feasible. In fact, appreciable losses of nitrogen attributable to these processes are often observed with both conventionally designed filters and activated-sludge plants, sometimes giving rise, especially in the case of the latter, to flotation of sludge in final sedimentation tanks.

Various forms of treatment have been investigated and are under development. There are three principal methods of which two are based on the activated-sludge system and one involves the use of flooded biological filters. The two activated sludge treatment processes have been termed the 'separate' and 'integrated' systems. A separate system is one in which denitrification is achieved using a sludge developed specifically for this purpose and usually includes the addition of an exogenous source of carbon. An integrated system is one in which the same sludge is used both for aerobic oxidation of the original carbonaceous and nitrogenous matter and for denitrification.

A number of variations of the integrated system have been investigated by various workers. The most simple form investigated by Wuhrmann [5], and by Hunnerberg and Saffert [6] comprises primary sedimentation followed by biological oxidation, then denitrification relying on sludge cellular material to provide the necessary carbon, and finally secondary sedimentation.

The investigation showed that efficient denitrification can easily be achieved, provided that the detention time in the denitrification basin is sufficient for the mixed liquor to ;

a)   reach completely anaerobic conditions, and

b)   fully utilize NO_2^- and NO_3^-

The time for (a) can be calculated from the mixed liquor flow and suspended-solids concentration, oxygen content in the inflowing liquor, and specific endogenous respiration rate of the activated sludge. The time for (b) is obtained from the suspended-solids concentration, nitrite plus nitrate concentration and specific endogenous denitrification rate of the activated sludge.

The area of air-water interface in the denitrification basin should be kept as small as possible and continuous stirring of its contents is an essential feature to prevent settling of the solids and to secure a rapid mass transfer from the soluble phase to the sludge organisms. The nitrogen gas released tends to cause flotation of the sludge solids as a foamy layer on the water surface. This solids entrapment may considerably affect the solids concentration in the outflow from this basin. It is possible however, to use a stirring device which sweeps up the surface layer without too much aeration.

The required retention time in the denitrification tank is very sensitive to the operating temperature. For treatment of a normal U.K. sludge it might range from about 10 hours at 25°C to nearly 40 hours at 5°C.

Clearly the provision of a tank with a detention time of the order needed to cater for cold conditions would involve exceptionally heavy capital expenditure.

A variation of the system, first proposed by Wuhrmann [7], with the objective of introducing a non-cellular carbon source thus increasing the rate of denitrification, is to divert a fraction of the incoming waste water to the denitrification stage. Results from the earliest pilot plants operated according to this principle

were rather unsatisfactory. It is noted furthermore, that the total nitrogen removed in such a system will be partly offset by the
nitrogen compounds especially the ammonia, added to the denitrification basin with the raw waste water because the majority of nitrogen will short circuit to the final effluent.

It is well known that in some conventional activated-sludge
plants significant losses of nitrate can occur at the inlet end of
the aeration tank. In these cases, the conditions of microbial
growth, low oxygen tension and substrate availability favour denitrification. This phenomenon led to the idea that it should be possible to provide deliberately an anoxic zone (i.e., a zone with
mixing but not aeration) at the inlet end of the aeration tank to
remove the nitrate in the returned sludge.

Bailey and Thomas [8] have described how this process is being
studied both in small scale tests at the Stevenage Laboratory of
the Water Research Centre and in full scale trials at Rye Meads se-
wage-treatment works U.K. At the Rye Meads works the inlet end of
the activated sludge unit was modified to provide the necessary a-
noxic zone. The zone is 50 m in length and has a capacity of $950\,m^3$
out of the total activated-sludge tank capacity of $3800\ m^3$. It is
equipped with sixteen stirrers which ensure intimate contact between the incoming sewage, nitrified returned liquor and sludge solids and which prevent deposition of solids within the zone. The
plant has been operated under steady flow conditions with a retention period (based on the settled sewage flow) of 10 h of which
2.5 h is within the anoxic zone. The sludge return rate was equal
to the flow of settled sewage. Under these conditions substantial
complete   denitrification of the returned sludge was achieved,
removal of organic matter was satisfactory and nitrification did
not appear to be impaired. Analytical data obtained from 15 composite samples are summarized in Table 1.

The alternative form of treatment based on the activated-sludge
process is the 'separate' system. In this system a separate sludge
is used and generally the rate of denitrification is increased by
the addition of an external substrate of readily oxidizable carbonaceous matter deficient in nitrogen.

Several carbon sources have been investigated by the various researchers in the field. These have included acetate, lactate, ethanol, methanol, acetone and molasses. These all proved to have
similar efficiency and consumptive ratios but methanol was found
to be the most convenient and economical in the economic conditions
then prevailing. The reduction of nitrate to nitrogen by methanol
can be expressed as follows:

$$5CH_3OH + 6NO_3^- \rightarrow 5CO_2 + 3N_2 + 7H_2O + 6OH^-$$

TABLE 1 : Treatment of Settled Sewage by Conventional
and Modified Activated-Sludge Units at Rye Meads
Source: Bailey and Thomas [8].

	Settled Sewage	Conventional Plant Effluent	Modified Plant Effluent
BOD	205	7.3	8.4
COD	375	38	49
Organic Carbon	118	17.7	23
Amm. N	34.7	<0.4	<0.4
Oxidized N	-	38.0	18.2
Nitrous N	-	0.11	0.18
Kjeldahl N	48.8	2.3	2.8
Total N	48.8	40.4	21.2
SS	103	7.6	16.5
pH Value	7.75	7.6	7.8

All results (except pH value) expressed in mg/ℓ.

i.e. five moles of methanol are required to reduce six moles of
nitrate - nitrogen to elementary nitrogen or 1.9 mg of methanol is
required for each mg of nitrate. The methanol demand of nitrate
nitrogen can be calculated in a similar way. Studies by McCarty
et al [9] indicate that in addition to the stoichiometric quantity
an additional 30 per cent is needed to satisfy bacterial growth
requirements. Also, any oxygen present in the feed will exert a
demand on the methanol. McCarty [9] has derived a formula to express the total methanol requirement including that required for
bacterial growth:

$$M = 2.47 \ N_o + 1.53 \ N_i + 0.87 \ D_o \qquad \cdots \quad (1)$$

where;

M = required methanol concentration (mg/ℓ)

N_o = initial nitrate concentration (mg N/ℓ)

N_i = initial nitrite concentration (mg N/ℓ)

D_o = initial dissolved oxygen concentration (mg/ℓ)

The quantity of biomass produced, X_n (mg/ℓ), can be determined
from the following expression

$$X_n = 0.53 \; N_o + 0.32 \; N_i + 0.19 \; D_o$$

Barth, Brenner and Lewis [10] demonstrated that the addition of
methanol plus sodium aluminate to the anaerobic stage of a pilot
plant could result in the almost complete removal of inorganic nitrogen at temperatures in the range 12 to 22°C. The basic flow sheet
is shown in *Fig. 2*. Inevitably the suspended matter in the settled
effluent contained organic nitrogen but this was reduced to concentrations below 0.5 mg/ℓ by sand filtration. It was found best to
split the anaerobic chamber into three compartments in series and
to add methanol to the first two, the optimum concentration being
judged to be four times the concentration of nitrate-nitrogen. No
methanol could be detected in the effluent.

Experiments carried out at the Water Research Center Laboratory
in Stevenage have been reported by Bailey and Thomas [8]. In these
experiments a fully nitrified effluent was mixed with an equal quantity of activated sludge. The system comprised three stages; an
anoxic tank in which the nitrified effluent and returned sludge
were mixed with methanol, a reaeration stage and final clarification. It was found necessary to include the reaeration stage to
aid sedimentation by removing fine bubbles to arrest further denitrification, to displace nitrogen gas from solution, and to oxidize
any surplus methanol. It was found that at retention times of 25
minutes (based on influent flow) a 90 per cent reduction of nitrate
was obtained for liquor having 8000 mg/ℓ suspended solids at temperatures between 15 and 18°C. Studies are continuing to determine
the minimum retention period achievable and the effect of variations in temperature.

Considerable development work has also been concerned with the
use of flooded biological filter systems for the denitrification
stage of nitrogen removal. In this case the filter replaces the
anaerobic stage of the activated-sludge process and can be operated with either upward or downward flows provided it is kept anaerobic. A number of workers have reported satisfactory results.
For example Amant and McCarty [11] obtained a 90 percent removal
of nitrogen in an upward flow filter with a detention time of 1
hour (based on the void fraction of the bed) using media of about
25 mm diameter while English et al. [12] achieved a similar percentage removal in a downward flow filter with a retention time of
less than 10 min using media of only 1 mm diameter. The fine media filter required far more frequent washing than the coarse media. Experimental work on flooded upward flow filters is also being carried out by the Stevenage Laboratory of the Water Research
Centre [8]. The effects of using media of different sizes, shapes,
and materials and also of operation at various temperatures have
been studied. A possible solution to clogging problems in filters
using sand media was found during the course of these studies. During the operation of a filter biological growth causes the par-

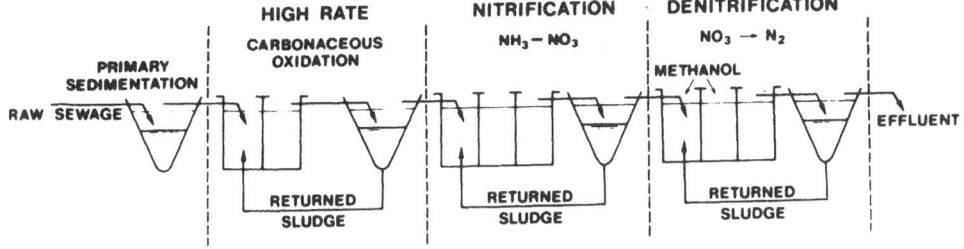

Fig. 2 :   Three-Stage Process for
Biological Removal of Nitrogen
from Municipal Waste Waters
(Barth [10]).

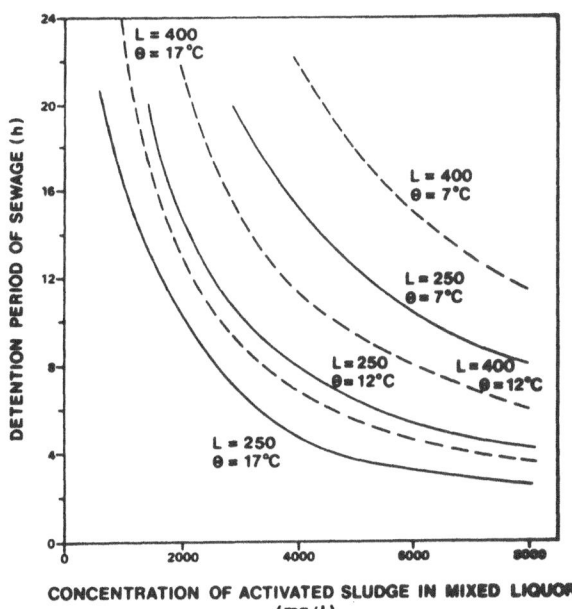

Fig. 3 :   Approximate Conditions for
Nitrification of U.K. Sewages
of Predominantly Domestic
Character.

ticle size to increase. In tests using a sand medium of 1 mm diameter the original bed volume was allowed to expand to accommodate an increase in particle diameter to 7 mm and to give a 20 minute detention time. The bed was maintained at this volume by a daily flushing action. Under these conditions it consistently gave a nitrate removal of 90 percent.

It has been suggested that the filter system has an advantage over the activated-sludge method because of low initial and operating costs, simplicity of operation, and the absence of a need for settlement and sludge recycling. However, in the absence of full-scale trials of the two types of process under the same conditions it is not possible to reach a firm conclusion.

It will be evident that one of the problems of this approach with the fluctuating conditions that are often experienced at municipal works might be in maintaining the correct addition of methanol. Too little would result in release of excessive quantities of nitrogen, and too much could lead to discharge of methanol. Clearly the consequences of the latter could possibly be serious for a variety of reasons but particularly if the final effluent were discharged to a river used as a source of water for drinking.

KINETICS of NITROGEN REMOVAL

Many attempts have been made to provide mathematical models of waste treatment processes. They vary from those which seek to relate operating conditions empirically to the overall rate of removal of polluting matter to those which attempt to describe the process in complex mixtures of substrates by relating them to the kinetics of growth and metabolism of pure cultures. However, when one is concerned with the state of a particular constituent of waste, and especially if it is one which is known to be metabolized only by a single or limited number of species of organism than the approach based on bacterial growth kinetics can have more to offer than the empirical one.

In the following sections it is shown how basic relationships describing bacterial growth can be used to describe the nitrification and denitrification processes especially as related to activated-sludge systems.

Consideration of the situation in activated-sludge plants leads to the conclusion that, since nitrifying bacteria in such plants are virtually wholly confined to the activated sludge, a fraction of them will be removed in any sludge withdrawn from the system or lost in the effluent, and under steady conditions their concentration will increase above that of the small inoculum which enters in the incoming waste only if the percentage increase in their concentration as a result of growth during aeration exceeds the percentage increase in the concentration of activated sludge as a whole.

If we consider the situation in a uniformly mixed activated-sludge plant, in which for simplicity it is assumed that the sludge is wholly confined to the aeration units, this condition can be expressed algebraically as

$$\frac{1}{C} \frac{dC}{dt} > \frac{1}{X} \frac{dX}{dt} \qquad \dots \quad (2)$$

where x is the concentration of activated sludge at any time t and c is that of *Nitrosomonas*. If the rate of growth of *Nitrosomonas* varies with concentration of cells and substrate according to the type of relationship postulated by Monod [13] then

$$\frac{dC}{dt} = k_n C \cdot \frac{A}{A + K_s} \qquad \dots \quad (3)$$

where k_n is the maximum growth-rate constant of the organism with ammonia in excess, A is the concentration of ammonia in the liquid phase, and K_s is the concentration of ammonia at which the growth-rate constant is half the maximum value.

If x is the increase in concentration of sludge that would result, if no sludge were removed, in a time, T_h equal to the period of retention of the mixed liquor in the aeration unit, then substituting into the inequality [2] the condition for development of a nitrifying population becomes

$$\frac{k_n A}{(A+k_s)} > \frac{\Delta X}{X T_h} \qquad \dots \quad (4)$$

Another condition that must be satisfied is that the steady-state concentration of ammonia be less than that in the incoming sewage, A_o.

Combining this condition with the inequality leads to the conclusion that for nitrifying organisms to accumulate in the plant

$$k_n T_h > \frac{\Delta X}{X} (1+K_s/A_o) \qquad \dots \quad (5)$$

It can be shown that the same relation is applicable to the situation in a plant through which the flow is piston-like and it may be noted that since A_o is commonly large by comparison with K_s the numerical value of terms within the bracket is close to unity. Taking this into account, together with the fact that $\Delta x/x T_h$ is effectively the reciprocal of the 'sludge-age' the condition for nitrification to be maintained can be stated as that the reciprocal of the sludge-age must be lower than the growth rate constant for *Nitrosomonas*. If this condition is satisfied the concentration of *Nitrosomonas* will begin to increase but it will reach a limit(roughly around 1 to 2 percent of the concentration of sludge) such that the fractional rate of increase resulting from complete or almost complete oxidation of ammonia entering the plant is equal to the fractional rate of increase in the sludge mass as a whole. If we now regard A and x as the steady-state concentrations at this limit then the equilibrium condition becomes

$$ \frac{k_n A}{(A+K_s)} = \frac{\Delta x}{x \, T_h} \qquad \qquad \dots \dots \quad (6) $$

In practice the growth-rate constant will be a complex function of the operating conditions and of the nature of the sewage treated. For a typical British domestic sewage it has been found sufficiently accurate for design purposes to assume that k_n is a function of temperature and pH value and that Δx is a function of the BOD of the sewage, the period of retention in the aeration unit, the temperature, and the concentration of sludge maintained under aeration; empirical equations for these relations have been determined [14].

It has been found that in mixtures of typical British domestic sewages the growth-constant of *Nitrosomonas* (in reciprocal days) varies with temperature Θ in the range of 5 to 25°C and pH value P, in the range 6.0 to 7.2 approximately according to the relation

$$ k_n = 0.18 \; e^{0.12 \, (\Theta - 15)} (1 - 0.83 \, (7.2 - P)) \qquad \dots \dots \quad (7) $$

The magnitude of k_n appears to be unaffected by pH value in the range 7.2 to 8.3 but begins to decrease with increasing pH above this range. Values of pH greater than 8.3 are, however, rarely encountered and are outside the scope of this discussion. It must be borne in mind that k_n can be greatly affected by the presence of inhibitory matter in sewage, especially certain types of substances which may be released in inudstrial effluents. In practice, therefore, it is desirable to measure k_n in the particular sewage for which the plant is being designed, rather than rely on equation 7.

The relation between Δx and operating conditions can be defined by the equations:

$$\Delta X = f_1 \, f_2 L (0.2 + 0.25 \, T_s^{-\frac{1}{2}}) / (1+p) \qquad \cdots \cdot (8)$$

$$f_1 = 1 - 0.00004 \, (X - 3000) \qquad \cdots \cdot (9)$$

$$f_2 = 1 - 0.02 \, (\theta - 16) \qquad \cdots \cdot (10)$$

where L is the 5-day carbonaceous BOD of the sewage, T_s the period of retention of the sewage in days; and p is the ratio of the returned sludge flow to that of sewage.

Combining equations (7) to (10) with the general condition (6) finally gives the means for calculating the conditions necessary to establish nitrification in continuous-flow plants.

The implications of these equations, are shown in *Fig. 3* which gives examples of the predicted effects of concentration of sludge temperature, and strength of sewage (expressed as 5-day BOD) on the minimum period of retention of sewage in the aeration units for consistent nitrification. It will be seen that this minimum period decreases with increasing concentration of sludge, with increasing temperature, and with decreasing BOD of the sewage.

In their study of the kinetics of denitrification Stensel et. al, [15] have applied the kinetic model developed by Lawrence and Mc-Carty [16] for the general description of biological treatment systems. The unifying parameter used by Lawrence and McCarty is the biological solids retention time, T, and this is related to microbial growth, substrate assimilation and process efficiency. The relationships they developed for complete mix systems with recycle of solids for solids and hydraulic retention times were:

$$\frac{1}{T} = \frac{a \, k \, S}{K_S + S} - b \qquad \cdots \cdot (11)$$

$$T_{min} = (a \, k - b)^{-1} \qquad \cdots \cdot (12)$$

$$T_s = T \, (1 + p - p \, X_r / X) \qquad \cdots \cdot (13)$$

These equations are applicable to plants having an anaerobic denitrification chamber with addition of methanol substrate and recycling of sludge.

In anaerobic systems organic carbon is the substrate that controls the microbial growth rate provided sufficient oxygen and inorganic nutrients are available. In the denitrification process assuming that sufficient inorganic nutrients are available, the substrate controlling growth rate will either be methanol or nitrite. Stensel et al [15] carried out a series of experiments to determine the growth-limiting substrate described by the kinetic model. They found that under normal conditions growth was limited by the methanol and that nitrate-nitrogen did not become growth-limiting until its concentration approached zero.

On the basis of a series of experiments they derived an empirical equation for the requirement of methanol.

$$M_C = 3.46 \ (N_O - N) + 1.5 \ D_O \qquad \qquad .... \ (14)$$

where;

M_C = methanol requirement measured as COD, mg/ℓ

N_O = influent nitrate-nitrogen concentration, mg/ℓ

N = effluent nitrate-nitrogen concentration, mg/ℓ

and D_O = initial dissolved oxygen concentration, mg/ℓ

This equation is comparable to that developed by McCarty (see equation (1)) which refers to required methanol concentrations, M, in mg/ℓ.

The methanol requirement, M_C, can be equated to the change in COD of the substrate $S_{CO} - S_C$, i.e.

$$S_{CO} - S_C = 3.46 \ (N_O - N) + 1.5 \ D_O \qquad \qquad .... \ (15)$$

where S_{CO} is the influent COD and S_C the effluent COD.

By combining equations [11] and [15] the proportion of nitrate removed, N', can be related to the influent COD of the substrate and the retention time.

$$N' = \{ S_{CO} - \frac{K_s (1 + b T)}{T (ak - b) - 1} - 1.5 D_o \} / (3.46 N_o) \quad \dots \quad (16)$$

where the process coefficients are related to the COD of the substrate in mg/ℓ.

This equation indicates that the nitrate removal efficiency can be controlled by varying the retention time and the quantity of organic carbon added to the process.

Stensel carried out a series of experiments to determine the values of K_s, k, b and a at various temperatures. He concluded that temperature has no significant effect on the process in the range 20° to 30°C but that biological activity decreases significantly below 10°C. Also bacteria will be washed out of the system at retention times of less than 0.5 days in the range 20 to 30°C and at about 2 days at 10°C.

BIOLOGICAL OXIDATION USING NEW OXYGEN

GENERAL

The purpose of biological oxidation systems is to enhance the natural process of self-purification of degradable organic wastes by bringing them into contact with a high concentration of microorganisms in the presence of excess dissolved oxygen. The economics of such systems are affected by the efficiency of oxygen transfer and use. In conventional aerobic treatment systems oxygen requirements are provided by bringing air into contact with the mixed liquor. However, it has long been noted that there could be certain advantages in using pure oxygen in place of air and during the past few years a number of manufacturers have developed systems based on the use of commercially prepared oxygen.

OXYGEN TRANSFER CHARACTERISTICS

Within a liquid the rate of increase of concentration of dissolved oxygen c, at any time, t, can be expressed by the equation

$$\frac{dC}{dt} = K_C \frac{A}{V} (C_s - c) \quad \dots \quad (17)$$

where K_c is a mass transfer coefficient, A is the area of interface between gas and liquid, v is the volume of liquid and c_s is the saturation concentration of the gas within the liquid. Also from Henry's Law

$$c_s = h \, P_p \qquad \qquad \dots (18)$$

where h is Henry's Law constant for the gas at a given temperature and P_p is the partial pressure of the gas above the liquid. By combining the two equations we have

$$\frac{dC}{dt} = K_c \frac{A}{V} (h \, P_p - C) \qquad \qquad \dots (19)$$

For operation at a given temperature in seeking to maximize of solution of oxygen, the ratio of the gas-liquid interface area to the volume of liquid should clearly be as large as possible other factors being the same. If a gas injection system is used this can be achieved by keeping the bubble size as small as possible (the effective K_c will be increased by encouraging turbulence). As a lower flow of gas is required for pure oxygen systems than for air systems bubble size from a given diffuser system tends at some airflow loadings to be smaller. Rates of solution will also increase with increasing partial pressure of oxygen. At a given total pressure the partial pressure of commercially produced oxygen is about five times the partial pressure of oxygen in air. Thus at low oxygen concentrations in the water the rate of oxygen transfer in a pure oxygen system is about five times that in an air system. Activated sludge systems operating under normal conditions give satisfactory results when the dissolved oxygen concentration DO is maintained at about $2 \, mg/\ell$. An increase in DO concentration has a proportionally smaller effect on the transfer rate for a pure oxygen system than for an air system.

APPLICATIONS to the ACTIVATED SLUDGE SYSTEM

Various techniques have been developed to provide an efficient means of pure oxygen transfer to activated-sludge mixed liquor.

These include:

(a) agitation of mixed liquor and oxygen in an enclosed tank for a period of 1 to 4 h before separation of the sludge and effluent in a conventional type of sedimentation tank;

(b) release of oxygen as fine bubbles at the bottom of a
     tank in order to achieve adequate mixing without exces-
     sive loss of oxygen at the surface;

(c) injection of oxygen under pressure into a side stream
     of liquor which is returned to the main flow of mixed
     liquor; and

(d) injection of oxygen into a down-flowing stream of mixed
     liquor so that the rapid upward movement of bubbles of
     oxygen is prevented.

The Union Carbide Corporation of the U.S.A. have developed a
system known as the "Unox" system which is based on the first of
these techniques. The stream has undergone extensive field testing both in the U.S.A. and in the U.K. over the past few years and
has been described by Lewandowski [17]. It comprises a covered
aeration tank and an on-site oxygen generator. The staged aeration
of the mixed liquor is achieved by dividing the aeration tank into
sections which are operated in series – the mixed liquor and the
process gas flowing concurrently from the first stage to the last.
Sewage and returned gas are fed into the first stage and oxygen is
introduced into the gas space above the liquor. Gas transfer is
effected by the use of mechanical surface aerators. The liquor
passes to the second stage where a second aerator effects gas transfer. The atmosphere above the liquor in the second stage is less
pure than the original oxygen stream as it contains carbon dioxide
resulting from the biochemical oxidation process. In successive
stages the proportion of carbon dioxide increases reducing the efficiency of transfer. Most Unox systems are designed to give about
90 percent utilization of oxygen. In the final stage waste gas
(usually about half carbon dioxide plus nitrogen and half oxygen)
is discharged to the air. The mixed liquor in the final stage of
the aeration tank is delivered to a conventional final settlement
tank.

Lewandowski [17] contends that the increase in mass transfer
rate that can be achieved by the Unox system can support a higher
concentration of MLSS than can be maintained in conventional plants.
If DO concentrations in the range 6-10 mg/ℓ are maintained it is
claimed that no sludge settlement problems arise from operation at
these MLSS levels. On the contrary improved settlement and dewatering characteristics are reported. In addition it is claimed
that sludge production is only some 75-80 percent of that obtained
in conventional air systems because (a) the higher concentration
of MLSS gives rise to an increase in the effect of endogenous respiration and (b) the higher DO concentration ensuring high activity
is maintained which results in more complete oxidation. Also a
high quality effluent can be produced.

The power necessary to dissolve one ton of oxygen in the mixed liquor is about 300 kWh [18] assuming that zero concentration of dissolved oxygen is maintained in the mixed liquor. This compares with corresponding figures of 500-700 kWh/ton oxygen dissolved for conventional aeration equipment using fine bubble, diffused air, or mechanical surface aeration. But the apparent power advantage is reversed if the power requirements of oxygen production are taken into account. The total power requirements for production and transfer of one ton of oxygen lie in the range 650 to 1000 kWh. However, the use of commercial oxygen does enable higher dissolved oxygen concentrations to be maintained in the mixed liquor. To raise the dissolved oxygen concentration in the mixed liquor from zero to 5 mg/ℓ increases the total energy requirements of a commercial oxygen system by about 5 percent but would double those of a conventionally aerated plant. In practice it is found that best use can be made of this advantage by operating at far higher MLSS concentrations (6000-8000 mg/ℓ) than are normal with conventional aeration thereby reducing the volume of tank (and hence land area) required for a given waste flow.

Other advantages claimed for the system are:

(a)   a reduction in nuisance caused by odours resulting from the use of closed tanks from which only a small volume of gas is discharged;

(b)   a reduction in the risk of sludge bulking as it is unlikely that deoxygenated pockets will occur within the reactor which could encourage the growth of filamentous organisms; and

(c)   a lower overall cost [19] than conventional aerator systems for treatment of flows exceeding 4.5 millions liters a day (dry wheather flow) [Mℓ (DWF)].

These claimed advantages may be offset by some disadvantages which will have a bearing on the choice of system to be adopted in particular cases. The results of pilot studies reported by Banks [20] indicate that the Unox system produces a less nitrified effluent than that given from a conventional plant. It has been suggested that the reason for this characteristic is that carbon dioxide produced during the process is less efficiently stripped from the mixed liquor due to the lower flow of gas within the liquor. This gives rise to a depression of the pH causing an inhibition of nitrification. Inhibition can be overcome by introduction of a second tank designed for conventional aeration in which nitrification occurs.

Other possible disadvantages are that if oxygen is made on site specially trained staff are needed to operate and maintain the equipment; also the use and storage of pure oxygen involves fire hazards not experienced with the use of air for aeration purposes, and there might be problems of corrosion and maintenance in sealed oxygenation tanks.

THE DEEP SHAFT PROCESS

GENERAL

The Deep Shaft process is a high-intensity biological treatment process for the purification of sewage or biodegradable industrial effluent. The process has been developed and patented by Imperial Chemical Industries Limited, and is now being marketed by them. At the time of writing the only operational data known to the authors are from a pilot plant at Billingham, U.K. However, commercial plants are now being put into operation which should soon provide additional information.

DESCRIPTION of the PROCESS

The principal features of the process are shown in *Fig. 4*. The main element is a Deep Shaft which is divided into two sections.

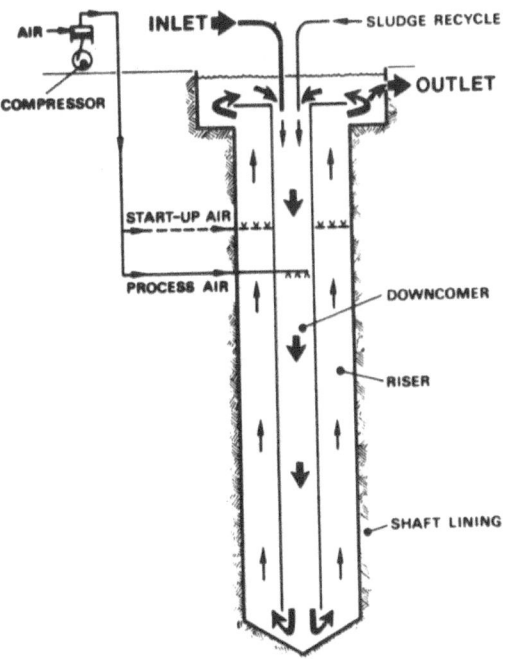

Fig. 4 : ICI Deep Shaft Unit

One of its main purposes is to provide the means for rapid and efficient transfer of oxygen to the liquor. Waste water and returned sludge are fed into the top of the shaft and are passed down one section of the shaft (the 'downcomer') to its base and then up the other section (the 'riser') to a disengagement chamber at the surface. Circulation is achieved by the injection of air into the downcomer at a sufficient depth to cause the mixed liquor to circulate at a velocity generally between one and two meters a second. Under these conditions the downward velocity of the mixed liquor is greater than the rate of rise of the air bubbles and the bubbles are circulated with the liquor. Depending on the particular process design the bubbles may be substantially or completely dissolved when they reach the base of the shaft. In the riser bubbles of residual oxygen, nitrogen, and carbon dioxide are released and grow in size as they rise to the surface. At the top of the riser the shaft discharges into the disengagement basin. In this tank the waste gases are released to the atmosphere and the circulating liquor is mixed with the incoming waste water and returned sludge before returning to the downcomer. A portion of the circulating liquor is withdrawn from the basin and is passed to a flotation chamber for solids separation. To achieve a good quality effluent it is normally necessary to provide vacuum degasification before sedimentation.

The innovators claim that an economically designed plant will have an oxygen uptake intensity in the order of 1 kg O_2/m^3hr and that levels of about 3 kg O_2/m^3hr can be achieved (see Table 2).

TABLE 2 : Comparative Oxygen Transfer Performance

	Oxygen Uptake Intensity kg O_2/m^3h	Oxygen Uptake per Unit Power Absorbed kg O_2/kWh	Oxygen Utilisation %
Conventional Activated Sludge Process (Normal Range)	0.05-0.1	1 - 1.5	5 - 15
Deep Shaft Process	3 (max)	6 (max)	90 (max)

The highest oxygen utilisation, uptake intensity, and uptake per kWh are not achievable simultaneously.

Source: ICI Technical Brochure

These figures and those given for oxygen uptake per unit power absorbed are considerably superior to those normally obtained by the conventional activated sludge process or by those claimed for the pure oxygen process. It would appear that there are two principal ways in which the deep shaft process achieves higher transfer rates than conventional processes. Firstly, some parts of the liquor are under considerable hydrostatic pressure and thus exhibit a correspondingly high oxygen solubility. Secondly, the bubble contact time is much greater than in a conventional diffused air process being typically in the order of three minutes, compared with say fifteen seconds.

The method of inducing circulation using air lift is as follows. At first compressed air is injected into the riser comparatively close to the surface (see *Fig. 4*). The action of the rising bubbles initiates circulation. Once circulation has been established the air supply is gradually transferred to the downcomer where the liquid flow round the system carries the air bubbles with it. By positioning the air injection point at an appropriate level the total gas voids in the riser are greater than in the downcomer. Thus circulation can be maintained once the flow regime has been established.

As an alternative to 'air lift' circulation flow through the shaft can be effected by mechanical means. This method however requires a greater energy input to maintain a given flow.

Study of the microorganisms present in the process has shown that the flora are not fundamentally different from those found in conventional plants although component species show significant differences in relative densities. The microbiological environment, however, is special in that oxygen is available at higher concentrations than occur in conventional processes and the microorganisms are subject to cyclic pressure variations. Initial fears that the pressure variations could cause cell damage appear unfounded and, on the contrary, pilot plant results indicate that an effluent of reasonable quality can be obtained from domestic sewage with retention times of less than two hours and with a carbon to cell conversion ratio of as little as 35% compared with 50% commonly experienced in other processes.

In common with other biological oxidation processes effluent from the shaft requires solids separation to complete the purification process and to obtain an active population of microorganisms at high concentration for recycling through the system. Vacuum degassing followed by conventional clarification by sedimentation is currently preferred but separation by flotation appears to be an alternative possibility.

PROCESS DESIGN

In their paper to the Institution of Chemical Engineers, Hines et.al.[21] showed that certain simplifying assumptions and empirical rules could be used in the explanation of the treatment process such that the basic design parameters could be roughly determined from a series of elementary equations.

For the Deep Shaft process it is not necessary to set up detailed equations explaining the mass transfer of oxygen as all oxygen in the air supply to the downcomer is available to the process and oxygen use is related to the biological capacity rather than the physico-chemical characteristics of the system.

Hines and his co-workers concluded that to avoid bubble coalescence the voidage fraction within the shaft should be kept below 0.2. Also, over the normal range of operation of the shaft the rising velocity of air bubbles in the water is almost independent of bubble size and the 'slip' velocity between the liquid and gas may be taken as 0.3 m/s for design purposes.

The required rate of oxygen supply to the shaft can be expressed in terms of the air supply to the downcomer (defined in terms of air velocity related to the total shaft cross section) and the depth of the shaft in meters, i.e.

$$R = \frac{3600 \ (0.25 \ G)}{H} \qquad \qquad \cdots \cdots (20)$$

where R is the rate of oxygen supply in kg O_2/h m^3, 0.25 is a constant relating oxygen supply in kg O_2 to air volume assuming STP and 90% absorption efficiency, G is the superficial air velocity related to the total shaft cross section in m/s, and H is the depth of shaft in meters.

It has been found that the optimum G is about 0.15 which gives:

$$R = 135/H \qquad \qquad \cdots \cdots (21)$$

This indicates that for a G value of 0.15 a shaft 135 m deep would give efficient transfer of 1 kg O_2/h m^3. Higher transfer rates can be achieved by increasing G and reducing H at the expense of oxygen efficiency and transfer economy.

Hines et al. also show how the necessary drive head for liquid circulation is related to the total friction loss through the shaft plus the difference between the total voidage head in the downcomer and the total voidage head in the riser. When the head is provided by the injection of air into the downcomer shaft the driving force arises from the difference in total bubble volume in the two legs. The section of downcomer above the injection point is not aerated and thus exerts a greater hydrostatic pressure than the corresponding section of the riser. An expression can be developed relating the necessary injection depth to the total head required for circulation and the theoretical gas voidage at the head of the downcomer limb. This voidage can be determined from the chosen values of superficial gas velocity and liquid velocity. Optimization studies have shown that the system is mechanically most efficient for superficial air velocities in the range of 0.1 to 0.2 m/s. For shafts having a low friction loss the necessary injection depth has been found to be 20 m or more.

Process optimization shows that deep narrow shafts are most suitable for weak liquors requiring high quality treatment while shallower broader shafts are more economical for the roughing treatment of high strength industrial waste. For the treatment of domestic sewage to give a 30:20 effluent given a design retention time of two hours it has been found that the optimum depth is about 150 m.

PERFORMANCE of the PILOT PLANT

Bolton and Ousby [22] give typical performance figures for the pilot plant at Billingham. *Table 3* indicates the type of average performance achieved.

They also report phosphorus levels in the effluent to have been consistently less than 1.0 mg/ℓ and that partial nitrification occurred.

Total sludge production from the plant has been compared with the notional amount that would be produced from a conventional plant treating the same waste. The authors estimate that sludge production from the plant is about half that to be expected from conventional processes. They also indicate that the sludge can be readily de-watered using polyelectrolytes or aluminium chlorohydrate. However, these results are based on small samples and relate to a limited variety of wastes.

It would be unwise to draw general conclusions from the limited operational data available at present and, in particular, more information is required concerning the degree of purification of a variety of wastes under a variety of operating conditions, and concerning the quantity of sludge produced and its response to treatment. If borne out by subsequent operational experience the low sludge production which has been reported could be of key importance in the success of the system as this factor will have an important effect on the total cost of the process.

TABLE 3 : Performance of Deep Shaft Operating at 1.3 hrs Retention Time and Treating Average Domestic Sewage Only
(Solids Separation by Floatation, Mechanical Degasing and Final Sedimentation (Boulton & Ousby [22]).

	Conditions	MLSS	4700 mg/ℓ
		MLVSS	3100 mg/ℓ
		Sludge Age	4.2 days (estimated)
			Concentration mg/ℓ
Influent	Raw Sewage BOD		195
	Settled sewage BOD		152
	Raw sewage solids		210
	Settled sewage solids		95
	Raw sewage COD		411
	Settled sewage COD		353
	Settled sewage TOC		78
Effluent	Suspended solids		36
	BOD		28
	Filtered BOD		4
	Filtered COD		65
	Filtered TOC		27

APPLICATIONS of the PROCESS

The Deep Shaft process is suitable for the treatment of a wide variety of wastes and the choice of the main design parameters will be a balance between the overall oxygen requirement and the kinetics of substrate removal.

Domestic sewage has a low organic content compared with many industrial wastes but must normally be treated to a high standard. The economics of treating waste with a relatively low oxygen requirement and low rate of substrate removal favour designing for high oxygen absorption efficiency and power economy, with a transfer intensity just adequate to satisfy the microbiological demand. From these considerations it would appear that shaft depth should be maximized to an extent consistent with the intensity require- ment.

The process appears especially suitable for the roughing treat- ment of strong industrial wastes. For such wastes the ability system to achieve large oxygen transfer intensities is most impor- tant and optimization results in shorter and wider shafts. Oxygen efficiency and power economy will be less than for treatment of domestic sewage but savings will be made in the volume of shaft to be provided.

The Deep Shaft can also be used for the treatment of sludge by aerobic digestion and shaft design would be based on high oxygen absorption efficiency and power economy. The process might be run at high temperature, (e.g.,45°C) and with residence times in the order of 4 days.

COST COMPARISON with ALTERNATIVE PROCESSES

For the treatment of domestic sewage to a 30:20 standard the complete treatment process incorporating the use of the Deep Shaft is - screening and maceration, biological oxidation in the Deep Shaft, Vacuum degasification and sedimentation of suspended solids and sludge treatment either by conventional methods or perhaps in a separate section of the Deep Shaft. Thus the Deep Shaft process replaces the primary settlement and biologial oxidation stage of a conventional works. Assuming that the claims of the innovators re- garding oxygen transfer effectiveness, degree of purification and sludge production are correct then the process will show capital savings compared with conventional processes on land requirement, primary settlement tanks, biological oxidation tanks and air supply equipment, and on sludge treatment provision. Also, there should be a saving in running expenses due to the efficiency of oxygen transfer. To set against these savings will be the cost of the shaft and its associated air supply equipment and the facilities for degasing.

The innovators consider that the Deep Shaft will show substan- tial savings when applied to the treatment of domestic sewage from large populations of 100,000 or more. It is said to remain cheaper than conventional treatment until a break even point somewhere in the range of 5 to 20,000 people. Experience of full-scale practice must now be awaited to establish the economics of the process in a range of situations.

NOTATION

a	=	microorganism growth yield coefficient	(mg/mg)
A	=	concentration of ammonia in the liquid phase	(mg/ℓ)
A_O	=	concentration of ammonia in influent	(mg/ℓ)
b	=	microorganism decay coefficient	(d^{-1})
C	=	concentration of *Nitrosomonas*	(mg/ℓ)
D	=	dissolved oxygen concentration	(mg/ℓ)
D_O	=	initial dissolved oxygen concentration	(mg/ℓ)
D_S	=	saturation concentration of oxygen	(mg/ℓ)
G	=	superficial air velocity	(mg/s)
h	=	Henry's law constant	(mg/ℓ atmos)
H	=	depth of shaft	(m)
k	=	maximum rate of substrate utilization	(mg/mg d)
k_n	=	maximum growth rate constant for *Nitrosomonas*	(d^{-1})
K_S	=	Michaelis constant	(mg/ℓ)
K_C	=	mass transfer coefficient for oxygen	(mg/ℓ)
L	=	BOD$_5$	(mg/ℓ)
M	=	required methanol concentration	(mg/ℓ)
M_C	=	required methanol concentration expressed as COD	(mg COD/ℓ).
N	=	effluent nitrate - nitrogen concentration	(mg N/ℓ)
N_i	=	influent nitrite - nitrogen concentration	(mg N/ℓ)
N_O	=	influent nitrite - nitrogen concentration	(mg N/ℓ)
N	=	proportion of nitrate removed	
p	=	ratio of flow of returned sludge to that of sewage	
P	=	pH value	
P_p	=	partial pressure of oxygen	(atmos)
R	=	rate of oxygen supply	($k_g O_2/hm^3$)
S	=	total concentration of substrate	(mg/ℓ)
S_{ci}	=	effluent substrate concentration expressed as COD	(mg COD/ℓ)
S_{co}	=	influent substrate concentration expressed as COD	(mg/ℓ)
T	=	solids retention time	(d)
T_h	=	hydraulic retention time of mixed liquor	(d)
T_s	=	hydraulic retention time of sewage	(d)
T_{min}	=	theoretical minimum solids retention time	(d)
X	=	concentration of microorganisms	(mg/ℓ)
X_n	=	concentration of microorganisms produced during denitrification	(mg/ℓ)
X_r	=	concentration of microorganisms in return flow	(mg/ℓ)
Θ	=	temperature	(°C)

REFERENCES

[1]   World Health Organization, *European Standards for Drinking Water, 2nd Ed.*, WHO Geneva,]970.

[2]   Painter, H.A., and Jones, K., The Use of the Wide-Bore Dropping Mercury Electrode for the Determination of Rates of Oxygen Uptake and of Oxidation of Ammonia by Micro-Organisms, *J. appl. Bact* 26, 471, 1963.

[3]   Ministry of Technology, *Water Pollution Research*, HMSO, 1964.

[4]   Wuhrmann, K., and Mechsner, K., *Path. Microbiol.* 28, 99, 1964.

[5]   Wuhrmann, K., Objectives, Technology and Results of Nitrogen and Phosphorus Removal Processes. In *Advances in Water Quality Improvement* Ed. Gloyna E.F., and Eckenfelder, W.W., Univ. Texas Press. 1968.

[6]   Hunnerberg, K., and Saffert, F., *Gas-u. Wassfach*, 108, 966, 1967.

[7]   Wuhrmann, K., *Schweiz. Z. Hydrol.*, 19, 409, 1957.

[8]   Bailey, D. A., and Thomas, E.V., The Removal of Inorganic Nitrogen from Sewage Effluents by Biological Denitrification. *J. Inst. Wat. Pollut. Control* 74, 5, 1975.

[9]   McCarty, P.L., Beck, L., and Amant, P.P. St., *Proc. 24th Ind. Waste Conf. Purdue Univ.* 1271, 1969.

[10]   Barth, E.F., Brenner, R.C., and Lewis, R.F., *J. Water Pollut. Control Fed.* 40, 2040, 1968.

[11]   Amant, P.P. St., and McCarty, P.L., *J. Am Wat. Wks. Ass.* 61, 12, 659, 1969.

[12]   English, J. N., Carry, C. W., Masse, A. N., Pitkin, J. B., and Dryden, F. D., *J. Wat. Pollut. Control Fed.* 39, 10, R71, 1967.

[13]   Monod, J., La technique de culture continue: Theorie et applications. *Annls. Inst. Pasteur, Paris* 79, 390, 1950.

[14]   Downing, A. L., Jones, K., and Hopwood, A.P., Proc. Symposium on New Chemical Engineering Problems in the Utilization of Water. *Am. Inst. Chem. Eng. and Inst. Chem. Eng.* 1965.

[15]   Stensel, H.D., Loehr, R.C., and Lawrence A. W., Biological Kinetics of Suspended-Growth Denitrification. *J. Water Pollut. Cont. Fed* 45, 2, 249, 1973.

[16]   Lawrence, A.W., and McCarty, P.C., Unified Basis for Bio-
        logical Treatment Design and Operation. *J. San. Eng. Div.
        Proc. Amer. Soc. Civ. Engr.* 96, 757, 1970.

[17]   Lewandowski, T.P., The Use of High-Purity Oxygen in Sewage
        Treatment, *J. Inst. Wat. Pollut. Control,* 73, 647, 1974.

[18]   Boon, A.G., Technical Review of the Use of Oxygen in the
        Treatment of Waste Water, *An. Conf. Inst. Wat. Pollut.
        Control,* 1975.

[19]   Sidwick, J. M., Lewandowski, T. P., and Allum, K.H., An
        economic Study of the Unox and Conventional Aeration Sys-
        tems, *J. Inst. Wat. Pollut. Control,* 74, 6, 1975.

[20]   Banks, N., U.K. Work on the Use of Oxygen in the Treatment
        of Waste Water, Associated with the CCMS Advanced Waste Wa-
        ter Treatment Project, *An. Conf. Inst. Wat. Pollut. Con-
        trol,* 1975.

[21]   Hines, D. A., Bailey, M., Ousby, J.C., and Roesler, F.C.,
        The ICI Deep Shaft Aeration Process of Effluent Treatment
        *Inst. Chem. Eng. Symposium on the Application of Chemical
        Engineering to the Treatment of Sewage and Industrial Li-
        quid Effluents,* York 1975.

[22]   Bolton, D.H., and Ousby, J.C., The ICI Deep Shaft Effluent
        Treatment Process and its Potential for Large Sewage Works,
        Int. Ass. Wat. Pollut. Research. Sewage Treatment Workshop.
        Vienna, 1975.

REUSE OF MUNICIPAL AND DOMESTIC WASTEWATER

S. J. Arceivala

Sanitary Engineer, W.H.O., Regional Office for Europe,
Copenhagen

Unintentional reuse of wastewater has been practiced since time immemorial. For example, rivers have almost always served as public water sources and as receiving water bodies for wastewaters. Drainage has often percolated into underground waters and has been later used. It is, however, the international reuse of municipal and domestic wastewater that is proposed to be discussed in this paper as it is currently increasing in scope and extent of application in many countries. The discussions presented below are also equally applicable to several industrial wastewaters which are similar in character to municipal and domestic wastewaters. In fact, many municipal wastewaters contain a fairly large proportion of industrial wastes in them (sometimes even over 50% of the total), often without any appreciable impairment of their treatability. *Fig. 1* shows how intentional and unintentional reuse of wastewater is made. Some of the more important intentional uses are: industrial, agricultural, recreational and non-potable municipal. Reuse for potable purposes has also begun (Windhoek, SW Africa).

Domestic use of fresh water adds inorganic and organic matter in suspended and dissolved form to that already contained in the fresh water supply. *Table 1* shows the additions likely to result from one cycle of domestic use. While degradable organic matter is readily removable by treatment, conservative substances are not. The latter thus determines the suitability of the wastewater for reuse.

It is, therefore, advantageous whenever industrial reuse is contemplated to 1., segregate as far as possible the less polluted fractions for direct reuse in an industry without treatment, (see *Fig. 1*), and 2., tap the public sewer as far as possible upstream

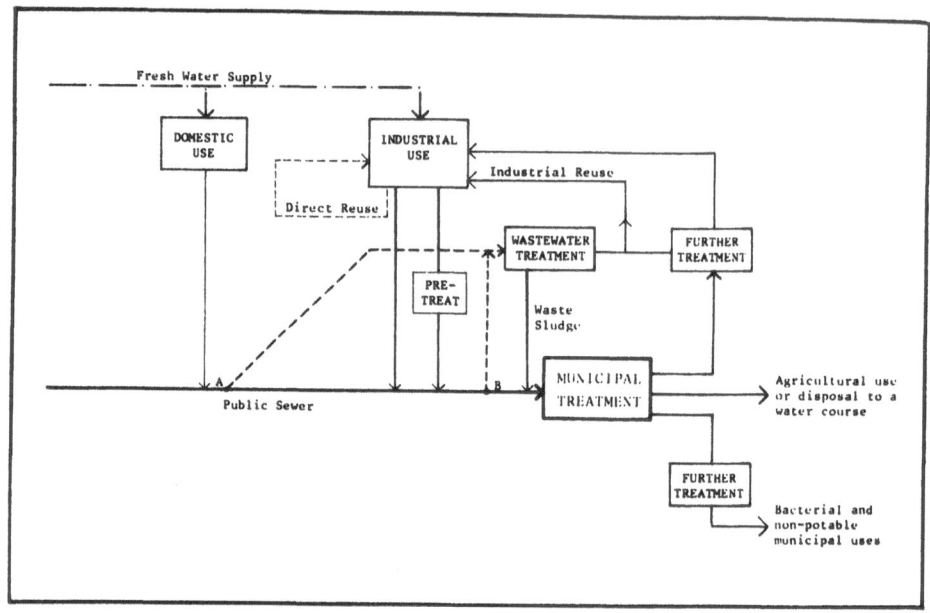

Fig. 1 :   Some Wastewater Reuse
Possibilities

of industrial discharges as their inflow into a public sewer can
add high amounts of both degradable and conservative pollutants.
Thus, in *Fig. 1*, the tapping point A would be preferrable to point
B, other conditions being equal. Often, the wastewater discharge
from a residential area can be turned into a good sources of "raw
water" for a treatment plant for reuse, and it may even be feasible
pumping it over a few kilometers by a separate pressure line laid
from the sewer tapping point to the industrial site.

Sewage cannot be stored for long (as it will turn septic), but
treated water can be stored and used as needed.   Thus, the pumping
rate from a public sewer must correspond to the usually fluctuating
flow in the sewer depending on its size and population served. The
treatment plant at the other end must also be capable of receiving
what may sometimes be widely fluctuating flows. On the other hand,
if the public sewer is large and the quantity to be tapped is equal
to or less than the minimum night time flow in the sewer, pumping
can be done at a steady rate. The waste sludge from the treatment
plant can also often be conveniently discharged back to the public
sewer (see *Fig. 1*) without any sludge treatment at all, thus econo-
mizing in the treatment costs for reuse of water.

Treatment methods used for reuse of water are often not different in any manner from those normally used for water and waste

TABLE 1 : Waste Water Quality and Irrigation Requirements

Item	Secondary Effluent (average of 33 U.S. cities) (mg/ℓ)	Average increment added from domestic water (USA) (mg/ℓ)	(g/c/d)	Increment in sewage (average of 62 Israeli municipalities) (g/c/d)	IRRIGATION STANDARDS California, USA Agricultural	Parks	India
BOD	8 - 24 - 80	8 - 22 - 45					
COD	44 -111 - 218	36 -111 - 218					
ABS	1 - 6 - 10	2 - 6 - 10					
Na$^+$	29 -134 - 384	8 - 69 - 115	10 - 20	0.6			
K$^+$	9 - 15 - 30	7 - 10 - 15	2 - 4 as N	2.12			
NH$_4^+$	3 - 19 - 44	3 - 19 - 50					
Ca^{++}	20 - 62 - 156	1 - 17 - 44					
Mg^{++}	0 - 25 - 94	0 - 7 - 24					
Fe^{++}	-	0.2					
Cl$^-$	49 -131 - 450	14 - 75 - 200	10 - 15	0.54 (low)	355	250	600
NO$_3^-$	0 - 13 - 66	0 - 9 - 26	5 - 8 as N	5.18 as N			
NO$_2^-$	0 - 1 - 2	0.1 -0.9 - 2.0					
HCO$_3^-$	110 -310 - 548	(-44)to 104- 265					
CO$_3^=$	0	0 -(-1)- 10					
SO$_4^=$	20 -103 - 358	10 - 28 - 57				250	1000
SiO$_3^=$	22 - 50 - 100	13 - 16 - 22					
PO$_4^=$ (total)	2 - 25 - 91	2 - 24 - 50	1 - 3 as P	0.68 as P			
PO$_4^=$ (ortho)	8 - 25 - 40	7 - 22 - 34	3 - 5				
Hardness.CaCO$_3$	8 -274 - 915	10 - 69 - 185		2.5			
Alkalinity.CaCO$_3$	90 -254 - 450	(-16) - 85 - 217					
TDS, mg/ℓ	350 -734 -1362	128 -123 - 541	50 - 150	40.0	2100	1500	2100
Boron, mg/ℓ				0.04 as B	2	2	2.
EC, μmhos/cm				600		x	3000 (25°C)
S.A.R., meq/ℓ				2.0	10	x	60
Percent Sodium							
RSC, meq/ℓ					2.5		

TDS: Total Dissolved Solids ; EC: Electrical Conductivity; S.A.R.: Sodium Absorption Ratio; RSC: Residual Sodium Carbonate

treatment. Basically a reuse plant is like a typical sewage treatment plant followed by a water treatment plant which may then be followed by additional units typically used in industrial water treatment depending upon the end use.

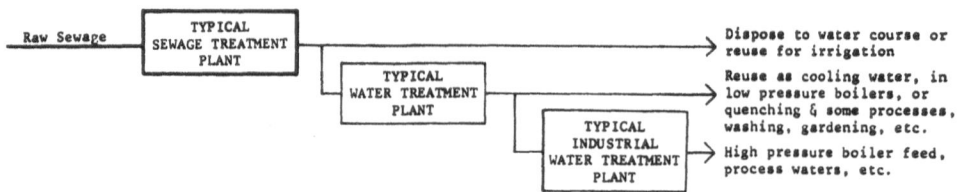

Fig. 2 :   A General Arrangement for
Treatment for Reuse of
Water

Fig. 2 shows a general arrangement for treatment for reuse of water. The first part of the plant is like a typical sewage treatment plant with settling, biological treatment and disinfection, except that sludge treatment may not be essential as stated earlier. The units selected may sometimes be different from those in municipal plants, owing to the need to economize on land requirements and use the higher level of skills often available in industries (e.g., one may use high-rate units, more mechanized methods, etc.). In this step, suspended and floating solids are removed, including any oils and grease, the carbonaceous and nitrogenous BOD satisfied and most bacteria killed off. Some phosphorus removal may also be achieved.

The second part of the plant assists in further removal of those constituents which could not have been sufficiently removed in the earlier stage. Removal of turbidity, hardness and phosphorus may also be emphasized here. Finally, in the third part, activated carbon may be employed to remove refractory materials and even partial or complete demineralization used for removing dissolved salts. Most of the methods likely to be used in this part of treatment are expensive and add considerably to the cost of reusable water.

TABLE 2:   Expected Efficiencies of Removal in
Different Degrees of Treatment

PROCESS	PERCENT REMOVAL (based on raw waste concentration)						
	BOD	Phosphates	Nitrogen	ABS	Suspended Solids	TDS	Coliforms
1. Conventional Sewage Treatment	90	40-50	40-50	50	90	5	90-99
2. 1, + Lime/Alum Coagulation, Settling, Filtration	93-95	95	50	50-55	99	10	99-99.9
3. 2, + Activated Carbon	99	95	50-55	95	99	15	99-99.9
4. 3, + Electro-Dialysis	99	97	75	98	99	50	99-99.9
5. 3, + Partial Demineralization	Reduction possible practically to any desired level						

Table 2 gives the expected percentage of removal of various constituents in different degrees of treatment. Of course, complete demineralization and distillation can give practically 100% removals, but that may generally not be necessary unless the end use is in high-pressure boilers.

REUSE as COOLING WATER

One of the major uses to which treated wastewater can be put is as cooling water in various industrial processes and in industrial and commercial air conditioning. A large number of industrial cooling operations can be conveniently and economically carried out with wastewater which has been reasonably well, but not too highly, treated. Based on studies in 45 different industrial and commercial establishments, Arceivala [1], [2], [3] has pinpointed various possible uses of treated wastewater for cooling and other purposes in textile industries, automobile and related manufacturing industries, in paper, chemical, petrochemical, and even in food and pharmaceutical industries where water quality requirements are more stringent. Several uses have also been shown in railway yards, city bus depots, docks, and harbors.

Arceivala [2] has also described the use of treated wastewater in several tall buildings in Bombay, India, based on his pilot plant studies. These uses are primarily as make-up water in cooling systems and are described further below.

An open recirculating system is generally adopted for cooling water used in air-conditioning systems in these buildings. The amount of water kept recirculating in the system is approximately 11 liters per minute for every ton of refrigeration capacity. For every 5°C temperature drop in the cooling tower, the water loss is about 1% of the quantity recirculating. Windage loss in cooling towers is of the order of 0.1 to 0.3% of the recirculating quantity when mechanical draft towers are used. Higher losses occur with atmospheric towers and still higher where spray ponds are used. The blowdown requirement is estimated as follows:

$$B = \frac{E + W(1 - C)}{(C - 1)} \qquad \ldots \ldots (1)$$

where;

B = blowdown requirement, ℓ/min
E = evaporation loss, ℓ/min
W = windage loss, ℓ/min
C = cycles of concentration (build-up ratio for salts)

The build-up of solids in the system depends on the relative proportions of B, E and W. For trouble-free operation, and minimum use of water quality control chemicals in the recirculating water, the cycles of concentration are generally kept at less than 4.0, depending on feed water quality, temperature, etc. Thus, for a 500 ton industrial plant, with temperature drop of 100°C in a mechanical draft cooling tower, we may estimate:

$$\text{Recirculating flow} = 11 \times 500 = 5500 \ \ell/\text{min}$$
$$E = 2\% \times 5500 = 110 \ \ell/\text{min}$$
$$W = 0.2\% \times 5500 = 11 \ \ell/\text{min}$$

$$B = \frac{110 + 11(1 - C)}{(C - 1)}$$

If C is to be kept at, say, 3.0 cycles of concentration, we get B = 44 ℓ/min.

$$\text{Total make-up}$$
$$\text{water} = B + E + W = 44 + 110 + 11 = 165 \ \ell/\text{min}$$

If the cycles of concentration are 3.0, all the conservative substances (salts) contained in the make-up water will be increased threefold by recirculation in the cooling system. Thus, it is possible to estimate the concentrations of various substances in the recirculating water, and knowing the temperature, determine whether the water quality will be corrosive or scale forming. Recirculating waters have to meet certain quality guidelines (see Appendix I). The quality of the treated wastewater used as make-up water should be such that the recirculating water is of the desired quality.

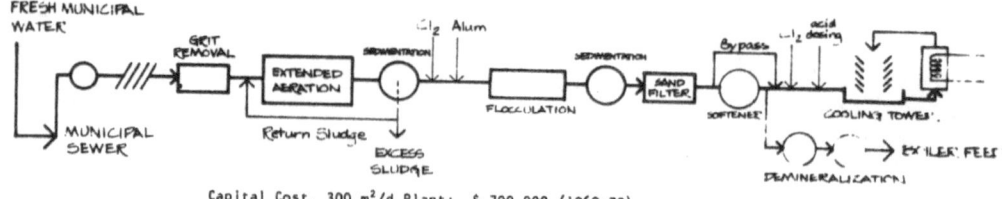

Capital Cost, 300 m²/d Plant: $ 700,000 (1969-70)
Operating Cost: $ 0.09 per 1000 liters

Fig. 3 : Flow Sheet & Expected Efficiencies
in Treatment of Wastewater for
Reuse in Cooling Systems

Fig. 3 gives the flowsheet developed for reuse of treated wastewater in tall buildings and several industrial cooling operations [3][2]. The quality of water at each treatment step is indicated in *Table 3* to show the progressive renovation achieved. In each building, the wastewater is drained directly to the treatment plant

TABLE 3 : Water Quality at Different Treatment Steps

ITEM	WATER QUALITY AT DIFFERENT TREATMENT STEPS					
	Fresh Municipal Water	Raw Domestic Sewage	After Extended Aeration & Settling	After Filtration	After Softening +CL$_2$	After Demineralization
pH	7.6-7.8	7.15-7.65	7.2-7.8	7.1-7.3	7.1-7.2	8.75
Total Hardness, mg/l, as CaCO$_3$	35-40	120-160	120-160	120-170	4.0*	NIL
M.O Alkalinity, mg/l, as CaCO$_3$	40-45	125-200	125-200	110-180	110-180**	5.0
Chlorides, mg^3/l, as Cl$^-$	15-20	60-130	60-130	60-130	60-130	NIL
Sulphates, mg/l, as SO$_4$	1.5-2.5	10-20	10-15	15-25	15-25	NIL
Phosphates, mg/l, as PO$_4$	traces -0.1	6-16	3-5	0.2-0.5	0.2-0.5	NIL
Nitrates, mg/l, as NO$_3$	1-2	1-3	13-19	13-19	13-19	NIL
Silica, mg/l, as SiO$_2$	8-24	10-24	10-24	10-20	10-20	NIL
Total Solids, mg/l	80-90	500-800	300-500	300-450	320-480	5.0
Suspended Solids, mg/l	5-10	150-250	15-30	NIL	NIL	NIL
Turbidity (SiO$_2$ units)	5-10	turbid	10-20	2-3	2-3	0.2
BOD$_5$, mg/l	0.5-1.5	200-250	6-10	1-2	1-1.5	NIL
COD, mg/l	1-2	250-350	16-40	4-6	3.5-5.0	NIL
Bacteriological Quality	safe	unsafe	unsafe	safe	safe	--
Specific Conductance	-	-	-	-	-	10 micromhos

* Softened water is blended with unsoftened water to give a final hardness of 40 mg/l as in fresh municipal water.

** Alkalinity is reduced by acid treatment just prior to use in cooling towers, by adding H$_2$SO$_4$, which increases sulphates somewhat.

located in its basement. Arrangement is provided for supplementing the flow by tapping an adjacent municipal sewer.

In these plants, the extended aeration process was selected to be used as it was likely to be least odorous (in a warm climate) and capable of removing up to 95% BOD. The large aeration volume would also be better able to withstand fluctuations. Further treatment was given by coagulation and sedimentation followed by rapid sand filtration to polish up the effluent since a consistently high performance could not be assured under local conditions with only coagulation and sedimentation. The effluent was softened and acid treated as necessary to meet make-up water quality requirements.

Similar installations have been made for industrial purposes and have been reported to be working satisfactorily [5]. Additional treatment by partial demineralization has been given for the fraction of water to be used in high pressure boilers in the industry.

About 90% of the raw sewage inflow is thus converted into re-usable water. Operating costs including repayment of capital and interest charges have been such as to give the product water before demineralization at less cost than buying fresh municipal water. Demineralization would have been necessary even if municipal fresh water supply was used, but would have been a little less costly owing to lower dissolved solids load.

IRRIGATIONAL USE of WASTEWATER

Irrigational use of wastewater is the common practice in many countries, especially in warm arid regions. In India, the first sewage farm was established in 1895. Today, there are over 130 farms covering 12,000 ha and using over 500 million m³/year. In Germany, the first sewage farm was established even earlier (1880) and in 1962 about 120 to 300 million m³ of sewage was used in this manner per year [9]. In the USA, over 1300 industries and 950 municipalities practice land disposal [7], and one of the best examples in recent times is the wastewater disposal project for Muskegon Country, Michigan, U.S.A., designed for disposing of 43 million gallons per day by 1992 over 371,000 acres of land after treatment in aerated lagoons.

Interest in land disposal of wastewater is being revived once again currently, owing to the observation that land disposal actually constitutes primary, secondary and tertiary treatment, all in a single operation! Not only is the water reused, but nutrients are effectively removed from it. Thus, land disposal becomes an effective method of tertiary treatment. The major disadvantages, however, are the large land requirements (5-40 ha/1000 m³ per day flow) and the seasonal nature of irrigation demand requiring alternative methods of disposal or of holding the wastewater until such time as it can be used again for irrigation. Indiscriminated use can also damage soil condition, pollute ground waters and affect the health of those who work on the farm or consume its products. Of course, none of these problems should exist in a well-designed and properly managed farm.

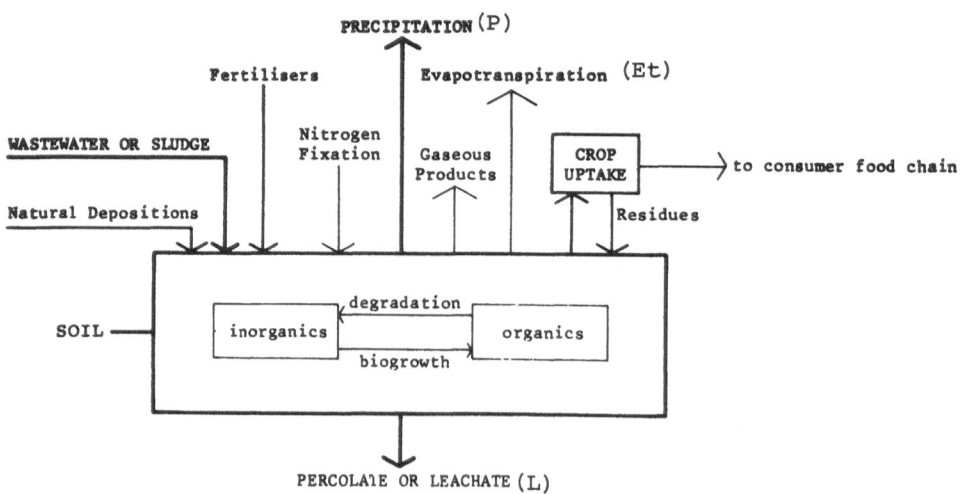

Fig. 4 : A Typical Soil System

Fig. 4 shows the various inputs and outputs of a typical soil system [4]. In summary, the various factors affecting land disposal of wastewaters and sludges are the following:

1. Water application rates depending on climate, crop and soil,
2. Nutrient requirements of crops and nutrients available in the wastewater,
3. Organic load on the soil,
4. Pathogen transmission,
5. Toxic substances and transmission in the food chain,
6. Build-up of salts in the soil,
7. Long term effects on soil and plants; in this regard several traditional irrigation criteria exist covering such items as electrical conductivity, total dissolved solids, sodium absorption ratio, boron, and other items,
8. Possibilities of ground water pollution, and
9. "Treatment" efficiency.

Generally, fodder grasses (forage crops) are preferred to be grown on sewage farms since they have a high nutrient demand, a greater tolerance to wastewater quality and a relatively longer growing season. In colder, wetter climates a combination of agricultural and woodland irrigation may be preferred. Land application rates during the growing season have been reported from the USA by Pound and Crites [10] to range from 1.27 to 10.16 cm/week. Typical land requirements are summarized in *Table 4*.

TABLE 4 : Typical Land Requirements for Irrigation [4]

Wastewater	Climate	Overall Land Requirement (ha/1000 m^3/day)
Domestic Sewage	warm temperate	5 - 10 10 - 40
Food Industry Wastes (milk, breweries, sugar, canning, etc.)	warm to temperate	5 - 25

Requirements vary depending on climate, type of soil, type of crops, and the need to have additional land for use when water is not needed on the main farm during harvesting.

ORGANIC LOADINGS on SOIL

As an example, if the crop watering practice is to give about 20 cm of water every 8 days during the growing season, the wastewater applied on an average per day over a hectare of land is

$$0.2 \text{ m} \times 10,000 \text{ m}^2 \times \frac{1}{8} = 250 \text{ m}^3/\text{ha/day}$$

If the wastewater is raw sewage or some industrial waste with a BOD_5 of 500 mg/ℓ, the BOD load applied to land is

$$250 \times 10^3 \times 500 \times \frac{1}{10^6} = 125 \text{ kg/ha/day}$$

Table 5 gives some typical BOD loadings applied in some cases. They are seen to vary widely.

Considerably high loadings are tolerated under favorable climatic and soil conditions. Excessive organic loadings can damage the cover vegetation and soil permeability. Often, the hydraulic load on a soil controls rather than the organic load (e.g., when treated effluents of low BOD are used).

TABLE 5 : Organic Loading on Land

Waste	Organic Loading kg BOD_5/ha/day	Reference
Dextrose	1026	[7]
Pulp and Paper	225	[6]
Horse Manure	160	[7]
Raw Domestic and Industrial Wastes (India)	25 − 150	[4]
Milk Wastes	12 − 125	[8]
Canning Wastes	100 − 2000	[8]
Protein Waters from Potato Starch Plants	600	[8]

PATHOGEN TRANSMISSION

The many variables affecting survival rate of different organisms make it difficult to apply available data from one sewage farm to another. Generally, however, non-edible crops are preferred as stated earlier. Some essential oil-bearing plants of commercial value may also be grown without fear of pathogen transmission. *Table 6* gives some recommendations regarding the crops and wastewater treatment suitable for them [12]. Stoppage of wastewater irrigation about 2 weeks prior to harvesting is often recommended, so as to give enough time for natural die-off to occur. Farm workers have an occupational hazard in this respect and need to be given protective clothing and encouraged to follow hygienic methods of working to avoid worm and other infections.

TRANSMISSION THROUGH the FOOD CHAIN

Certain persistent substances contained in the irrigational waters may first accumulate in the soil, and then be taken up by the crops, thus entering the food chain. More serious consideration is being given at the present time to the fate of cadmium, copper and zinc in the food chain. Other metals do not seem to accumulate in the edible portions of crops and are mainly phytotoxins. Though wastewater farms have been in operation for a number of years, little is known concerning biological accumulations of persistent substances and hence no limits for such substances have been prescribed in most irrigation water standards (*Table 1*).

Municipal wastewaters containing many industrial wastes may show the presence of substances like Cd, Hg, Zn, Cu, etc. in the range of 0.5 to 4.0 mg/ℓ. But their concentrations in digested sludges may be several times more. However, recently University of Illinois agronomists studying the application of Chicago sludge on land concluded that the land in question had the ability to accept accumulations of heavy metals in Chicago's sludge for at least 100 years without any adverse effects on crop productivity or usefulness [7].

Plant uptake depends on plant species and soil chemistry. Generally, in alkaline soils, heavy metals are precipitated and are therefore not available. Metals are adsorbed on hydrous oxides of Fe and Mn. Organic matter in soil also tends to chelate and fix heavy metals. Chemical tie-up of metals is promoted in clayed soils with higher cation exchange capacity. Various synergistic and antagonistic activities are also observed. For example, presence of phosphates tends to reduce the availability of metals to plants.

Although many soils have considerably capacity for retaining heavy metals and other toxic substances, their entry into wastewa-

394

TABLE 6 : Recommended Pre-Treatment for Irrigation [12]

	CROPS EATEN RAW	CROPS EATEN AFTER COOKING/PROCESSING	FODDER, FIBER AND SEED CROPS	ORCHARDS AND VINEYARDS
U.S.A.:				
Surface Irrigation	Coliforms 22/100 ml	Primary Effluent	Primary Effluent	Primary Effluent
Spray Irrigation	Coagulation, Filtration, Disinfection; Turbidity, 10 units	Secondary Effluent, Disinfection; Coliforms 23/100 ml	Primary Effluent	NOT ALLOWED
ISRAEL:	Wastewater irrigation NOT ALLOWED EXCEPT for fruits that have to be peeled	Coliforms 1000/100 ml in 80% samples	Secondary Effluent (seed crops for producing edible vegetables NOT ALLOWED)	Secondary Effluent
SOUTH AFRICA:		Tertiary Effluent	Tertiary Effluent	Tertiary Effluent, heavy Cl_2 (NO SPRAY IRRIGATION)
FEDERAL REPUBLIC OF GERMANY:		Irrigation allowed up to 4 weeks before harvesting (potatoes and cereals irrigated through flowering stage only)	Surface: Pre-Screen and Sedimentation / Spray: Biological Treatment, Cl_2	NO SPRAY IRRIGATION

ters proposed to be used for irrigation is best controlled at the source.

BUILD-UP of SALTS in SOIL

The soil-crop system acts as a concentrating mechanism for salts contained in the irrigant. This can be explained using a simple mathematical model [4]. Referring to *Fig. 4*, one can state the water balance as follows, assuming that the soil moisture content is unchanged:

$$I + P = Et + L \qquad \qquad \dots \quad (2)$$

where,

$$
\begin{aligned}
I &= \text{volume of irrigant used, mm/year} \\
P &= \text{volume of precipitation, mm/year} \\
Et &= \text{volume lost by evapotranspiration, mm/year} \\
L &= \text{leachate volume passing beyond the} \\
&\quad \text{root zone, mm/year}
\end{aligned}
$$

Similarly, a salt balance can be drawn up as follows:

$$(I \cdot c_i) + (P \cdot c_p) = (E_t \cdot c_e) + (L \cdot c_s) + \text{plant uptake} \quad .. \quad (3)$$

in which,

$$
\begin{aligned}
c_i &= \text{concentration of substance in irrigant} \\
c_p &= \text{concentration of substance in precipitation} \\
c_e &= \text{concentration in evapotranspirant} = \text{zero} \\
c_s &= \text{concentration of substance in the soil saturation} \\
&\quad \text{extract (also in leachate, by assumption of uniform} \\
&\quad \text{mixing in soil element).}
\end{aligned}
$$

Plant uptake accounts for some loss of the substance from the system when the plant is harvested off. Neglecting plant uptake, however, gives a sort of upper limit to what can accumulate in the soil. Further, the terms c_i and c_p can be replaced by $c_{i(av)}$ and the above equation can be rewritten as

$$(I + P)c_{i(av)} = L \cdot c_s$$

or the salt build-up ratio can be expressed as inversely proportional to the leaching fraction as shown below:

$$\left[\frac{C_s}{C_{i(av)}} \right] = \frac{1}{\left[\frac{L}{(I + P)} \right]} = \frac{1}{\frac{(I + P - Et)}{(I + P)}} \quad \dots \quad (4)$$

From Eq. 4 it is evident that the greater leaching, the less the build up of salts in a soil. Thus, sandy soils show less build-up. Wet areas also show less build-up (both L and P increase). In hot, dry climates, evapotranspiration E_t is high, and consequently the build-up ratio increases. Artificial methods of increasing the leachate can be adopted where conditions permit. Salt build-up in soils often ranges between 2.0 and 10 times the original concentration, more in tight, clayey, soils and less in sandy, well-drained soils.

Plant growth is affected by actual salt concentrations present in the soil (root zone) rather than in the irrigant itself. Thus it is worthwhile to remember that an irrigant as such has no inherent quality; it must be evaluated in terms of the local soil, climate, and crops proposed to be grown. A wastewater that may be suitable in one region may not be suitable in another. Crop selection must also be guided by the tolerance limit of the particular plant to the concentration of various substances in the soil after salt build-up occurs. In this regard, several water and soil quality criteria have been traditionally used by agriculturists for evaluating irrigation waters. It is not possible to discuss them at length in this lecture. Some standards have been included in Table 1.

Mineral pickup in domestic sewage above the background value accounts for an increase in Total Dissolved Solids of 100 to 300 mg/ℓ, or electrical conductivity of 150 to 450 µmhos/cm. Similarly, for domestic sewage, the sodium absorption ratio (SAR) increases by 3 to 5 meq/ℓ above the background value in the water supply. Chloride pickup in domestic sewage averages 4 to 8 g/cap-day. Higher values may be observed for some of these constituents in industrial wastes. In some cases, a commonly used corrective measure is the addition of CaO or $CaSO_4$ (not $CaCO_3$) to reduce the percent sodium proportion, the SAR value, the residual sodium carbonate, etc. An alkaline soil is generally preferred by agriculturists. Metals contained in the wastewater also stay precipitated in the soil at higher pH values. Generally, domestic sewage does not present any special problem in its usage as an irrigant. Some industrial wastes, however, may need pretreatment of simple dilution with fresh water before being used for irrigation.

NUTRIENT REMOVAL THROUGH IRRIGATION

All soils effectively remove P, K, Ca and Mg in percolation through them. Coarse textured soils are good at removing P and Ca. Nitrogen removal, on the other hand, is mainly through crop harvesting. Without crops, nitrogen tends to leach through the soils as nitrates. Thus, phosphorus is effectively removed in the soil, while nitrogen is taken up by plants and must be harvested out. Erosion of soil returns the phosphorus back to the environment.

Nitrogen contained in the soil is partly absorbed by plant growth, partly lost by denitrification and partly carried away in the leachate. Mineralization of organic nitrogen to nitrates in the soil progresses at around 1% per year, with higher and lower values depending on various factors. These nitrates are then carried down to the groundwaters by the rainfall seeping through. For example, if the top soil has, say, 4000 kg N_2/ha and the leachate is, say, 800 mm/year, natural mineralization at an assumed rate of 1% per year would give nitrogen concentration in the leachate as

$$\frac{4000 \text{ kg/ha} \times 0.01 \times 10^6}{800 \text{ } \ell/m^2 \times 10,000 \text{ } m^2/ha} = 5.0 \text{ mg}/\ell$$

Thus, the nitrogen-nitrate concentration in the groundwater depends on the nitrogen content in the soil, extent of mineralization and the quantity leaching through. Drinking water standards require that nitrates should be less than 50 mg/ℓ as nitrates in the water.

Excessive addition of nitrogen to the soil in the form of wastewater or sludge would only increase the losses and perhaps the nitrate content in the ground water. If the aim of irrigation is nutrient removal along with wastewater disposal, it would be advantageous to optimize for nitrogen and phosphorus removal by diluting the waste if necessary and spreading it over a larger area. Crops known for their higher nutrient uptake (e.g., certain grasses) would also be advantageously cultivated in such cases.

TREATMENT EFFICIENCY

It was stated earlier that land treatment of wastewater is, in fact, comparable to primary, secondary and tertiary treatment in its efficiency. This can be seen below for spray irrigation which gives very high efficiencies of removal in passage through soil [11]. The natural purifying capacity of soil is indeed high. The renovated water can be captured for reuse if under-drains are provided or if other arrangements are made to pump out the ground water flow.

TABLE 7 : Treatment Efficiency in Spray Irrigation

ITEM	% Efficiency in Spray Irrigation
BOD	99
Suspended Solids	99+
Nitrogen	80 - 90
Phosphorus	99
Heavy Metals	99
Bacteria	99
Viruses	99+
Total Cations	0 - 75
Total Anions	0 - 50

Renovation of water can thus be achieved to a high degree if land disposal systems are properly designed and operated. Land disposal systems can be incorporated into a wastewater treatment strategy to enable renovation at least cost. Both recreational and potable use of renovated water can be more assuredly made if natural land treatment is included in the system rather than depending on solely artificial methods. Public acceptance of renovated water is also better assured when land treatment is incorporated.

CONCLUSION

In this paper, an attempt has been made to show how municipal and domestic sewage (and industrial wastewater of similar characteristics) can be treated for intentional reuse in industrial, agricultural and other activities.

The gradually changing pattern of water usage in industries is reflected in less consumption of fresh water and more recirculation of used water, with or without treatment. Technological changes are affecting water utilization rates and quality requirements. Consideration is beginning to be given to optimum water usage with built-in mechanisms for encouraging water conservation and reuse. Often municipal or domestic sewage constitutes the most dependable sources of raw "water" for an industrial or commercial user, particularly for use as cooling water, boiler-feed water, or even as process water. Methods discussed in this paper can often be readily and economically applied.

Land disposal must be considered as yet another method, not only for disposal of wastewater or sludge when a water course is unavailable nearby, but as a method capable of giving a high degree of treatment along with reusable water. Land disposal, wherever feasible, can often be a more ecologically sound solution than that afforded by only artificial treatment. Various factors involved in the design of land disposal systems have been discussed. They must be taken into account in preparing a properly engineered design rather than just leaving the wastewater to the wits of the local farmer.

Water reclamation and reuse provides a unique opportunity to sanitary engineers to project a better image on the public mind, because the end results are more readily assessable in terms of costs and benefits than possible in their traditional work.

REFERENCES

1. Arceivala, S.J., "Water Conservation and reuse in Industries in Bombay", reports submitted to Bombay Municipal Corporation and Associated Industrial Consultants, Bombay, India, 1967-68.

2. Arceivala, S.J., Kapadia, J.R., and Wadekar, V.R., "Economy and Reuse of Water in Textile Mills", *Proceedings,* 26th All-India Textile Conference, Bombay, India, 1969.

3. Arceivala, S.J., "Reuse of Water in India", in *Water Renovation and Reuse,* edited by Hillel Shuval, under publication by Academic Press, New York.

4. Arceivala, S.J., *Wastewater Disposal Technology,* under publication, Marcel Dekker, New York.

5. Bannerji, "Purified Wastewater for Industrial Use", Journal Indian Water Works Association, Vol. IV, Oct. - Dec. 1972.

6. Blosser and Owens, Water and Sewer Works Journal, Vol. 111, p. 424.

7. Kneeland, Godfrey, Jr., Civil Engineering, ASCE, Sept. 1973.

8. Koziorowski and Kucharski, *Industrial Waste Disposal,* Pergamon Press, 1972.

9. Kruse, H., W.H.O. Bulletin, Vol. 26, p. 542.

10. Pound, C., and Crites, E., Report prepared on behalf of Metcalf and Eddy, Boston, 1973.

11. U.S. Army Corps of Engineers, "Assessment of the Effectiveness of Land Disposal Methods", Report 72/1, 1972.

12. World Health Organization, "Reuse of Effluents", Technical Report Series No. 517, Geneva, 1973.

APPENDIX I : COOLING WATER QUALITY GUIDELINES

It is not desirable to specify absolute values for different constituents in the make-up water used for cooling systems of the open recirculating type. The quality is largely dependent on the cycles of concentration likely to take place in the cooling system as a result of the relative proportion of evaporation, windage and bleed-off. The make-up water characteristics also should not fluctuate much over short periods of time.

1. Total Dissolved Solids (TDS) in the make-up water should preferably be within the following limits and should not vary by more than 25% above or below the 8-hour average:

Total Dissolved Solids in Make-Up Water mg/l	Recommended Limits for Cycles of Concentration
500	6.0
1000	3.5
3000	2.0

2. pH Value of make-up water should be preferably between 6.8 and 7.0 and the pH variation less than 0.6 units in 8 hrs.

3. The water should be non scale-forming at the skin temperature of the heat exchange surface. The calcium and alkalinity may be adjusted if necessary so as to give the desired value of Langelier Index (around zero).

4. Caustic Alkalinity should be absent.

5. Methyl Orange Alkalinity should be less than 200 mg/ℓ as $CaCO_3$ depending on Langelier Index desired. Polyphosphates may be used if necessary. Acid may be added to reduce alkalinity.

6. Silica (SiO_2) in recirculating water should not be more than 150 mg/ℓ as SiO_2, or else silica scale may be formed at temperatures $\geq$ 60°C in heat exchanger.

7. Phosphates, Carbonate, and Sulphates should be such that their solubility limit is not exceeded (i.e., no precipitation will occur) in the recirculating water at the given pH and temperature.

8. Foaming (due to ABS) in the recirculating water should be such that no foam will persist for more than 1 minute after 10 seconds of vigorous shaking. Otherwise, anti-foam agents must be used.

9. Low BOD (Biochemical Oxygen Demand) values are desirable (preferably less than 5 mg/ℓ).

10. No appreciable amounts of Ammonia should be present.

11. Residual Chlorine should be about 0.3 mg/ℓ, if chlorine is used on a continuous basis for control of slime and algae. However, residual chlorine should not be more than 1.0 mg/ℓ for intermittent shock dosing method.

12. Chlorides ($C\ell^-$) should be preferably below 175 mg/ℓ.

USE OF MODELS IN BIOLOGICAL TREATMENT

İ. İlkin Esen

Associate Professor, Civil Engineering
Boğaziçi University, Istanbul, Turkey

INTRODUCTION

In the design of biological treatment units, the selection of the reactor type is one of the important considerations. Although specific design criteria have been established on many types of treatment units, increasing usage of biological treatment in recent years makes it economically feasible to conduct a laboratory model study to investigate the effects of a number of operational factors on the effluent quality. A laboratory investigation is also justified since both the design criteria and the effluent quality are both dependent on the particular waste under consideration.

In this lecture, a number of laboratory model studies whose results have been successfully applied to the design of biological treatment units will be briefly described. The main emphasis will be on the reaction kinetics governing the treatment process.

REACTION KINETICS

Operational factors that must be considered in the selection of the type of reactor that will be used in the biological treatment processes are listed by Metcalf and Eddy, Inc. [1] as follows:

1. The reaction kinetics governing the treatment process;

2. Oxygen-transfer requirements;

3. Nature of wastewater to be treated; and,

4. Local environmental conditions.

As an introduction to reaction kinetics, consider the mass balance equation for a reactor where the hydraulic flow pattern can be represented by a completely mixed flow condition; here the rate of substrate inflow minus the rate of substrate outflow minus the rate of substrate removal in reactor equals rate of change of substrate in the reactor, and the governing equation is;

$$Q \; C_{n-1} - Q \; C_n - k \; C_n V_n = V_n \frac{dC_n}{dt} \qquad \dots \dots \; (1)$$

where;

Q = flow rate
C_{n-1} = inflow substrate concentration
C_n = outflow substrate concentration
V_n = volume of reactor
k = substrate removal rate
$\dfrac{dC_n}{dt}$ = rate of change of substrate in reactor

At steady state, time rate of change of substrate concentration becomes zero and equation (1) can be written as

$$\frac{C_n}{C_{n-1}} = \frac{1}{1 + k \; V_n/Q} \qquad \dots \dots \; (2)$$

For the case of n reactors in series, application of equation (2) successively gives

$$\frac{C_n}{C_o} = \frac{1}{(1 + kV/nQ)^n} \qquad \dots \dots \; (3)$$

where;

V = volume of all reactors in series
n = number of reactors
C_o = influent substrate concentration

In the limiting case as $n \to \infty$, equation (3) becomes

$$\frac{C_e}{C_o} = e^{-k\frac{V}{Q}} \qquad \dots \quad (4)$$

where; C_e = effluent (outflow) substrate concentration

Equation (4) characterizes a chemical reactor with plug flow.

Equation (3) clearly indicates that the total volume required for a series of complete-mix reactors is less than that required for a single complete-mix reactor. Use of a series of reactors becomes more favorable with respect to volume requirements with increasing reaction order [1]. Thus, it is clear that unless the reaction order is specifically known, a model study is in order.

In cases of reactors with considerable axial dispersion, the equation developed by Wehner and Wilhelm [2] can be employed:

$$\frac{C_e}{C_o} = \frac{4a \ \exp(1/2d)}{(1+a)^2 \exp(a/2d) - (1-a)^2 \exp(-a/2d)} \qquad \dots \quad (5)$$

where;

a	=	$\sqrt{1 + 4ktd}$
d	=	dispersion factor = D/UL
D	=	axial dispersion coefficient
U	=	fluid velocity
L	=	characteristic length
k	=	first order reaction constant
t	=	detention time

In this equation, d = 0 would imply an ideal plug-flow reactor, and d → ∞ would yield the equation for a complete-mix reactor. Again, it would be difficult to predict values of d in the solution of equation (5) and laboratory studies would be in order.

SELECTED LABORATORY INVESTIGATIONS

1. COMPLETELY-MIXED SYSTEMS

Busch [3], presents a procedure for the establishment of design criteria for completely-mixed bio-oxidation systems encompassing integral clarification and continuous solids recirculation

406

with positive control of mixed liquor suspended solids (MLSS).

The basic mathematical relationships used in this study are:

$$E = K'L + b \qquad \dots \quad (6)$$

where;

L = BOD loading (weight of 5-day BOD added per day per unit weight of MLSS)
E = effluent BOD
K' = a coefficient
b = a constant

and,

$$Q(C_0 - C_t) = V\left(-\frac{dc}{dt}\right)C_t$$

or,

$$\frac{V}{Q} = \frac{(C_0 - C_t)}{\left(-\frac{dc}{dt}\right)C_t} \qquad \dots \quad (7)$$

where,

C_0 = initial BOD concentration

C_t = BOD concentration at time t

$\left(-\frac{dc}{dt}\right)C_t$ = rate of reaction at a concentration of C_t

The other terms are as previously defined.

With a proper laboratory model, the constants K' and b in equation (6) can be determined if effluent BOD is plotted against BOD loading through a continuous test; and the reaction rate in equation (7) can be determined from a batch test using organisms from a continuous unit.

Fig. 1 shows the experimental set-up used by Busch [3], and Figures 2 and 3 show the results obtained by the experiments that are used in the computation of K', b and $(-dc/dt)C_t$.

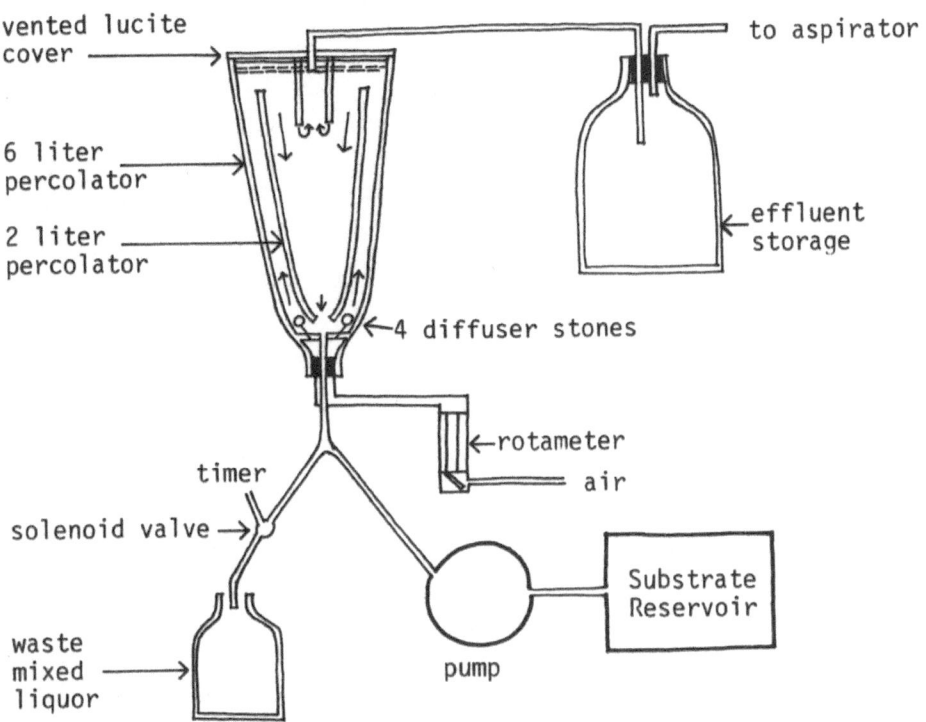

Fig. 1 : Experimental Set-Up of the
Bio-Oxidation Unit [3]

408

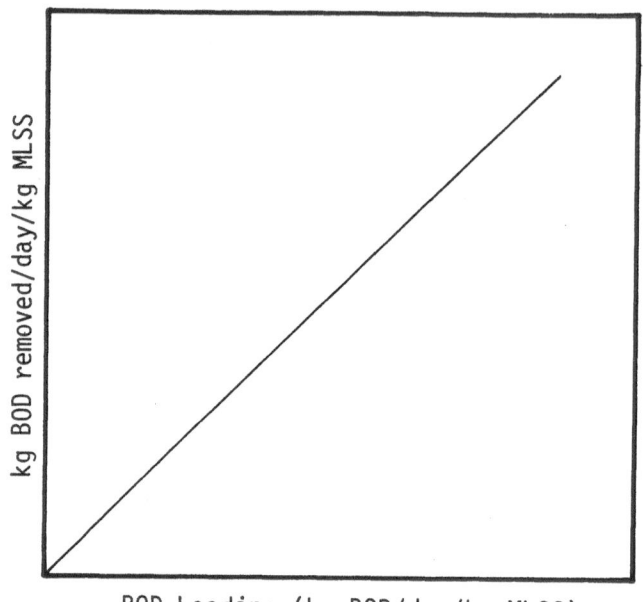

Fig. 2 : BOD Loading vs. Removal
        Curve [3]

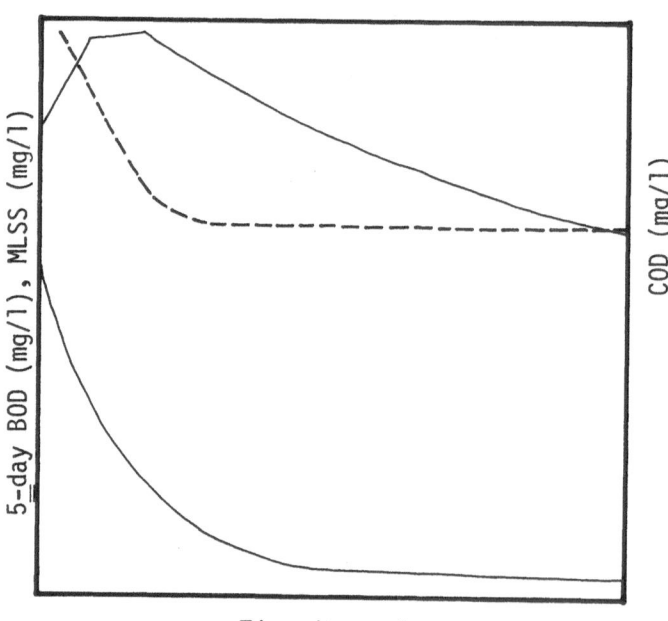

Fig. 3 : Batch Aeration Test [3]

2. GENERAL BIO-OXIDATION UNITS

The bio-oxidation unit reported by Ford, Gloyna and Yang [4] and Ford [5] can simulate aerated lagoons, activated sludge or extended aeration systems.

A schematic diagram of their bio-oxidation units is shown in *Fig. 4*. The main unit consists of an aeration chamber (capacity about 8 liters) and a clarification chamber (capacity about 2 liters) separated by a sliding baffle. The activated sludge and extended aeration processes can be simulate by sliding the baffle all the way to the bottom, then raising it approximately 0.6-1.3 cm. By adjusting the flow rate, the unit could simulate an activated sludge or extended aeration process. The aerated lagoon system was approximated by raising the baffle and resting it on top of the effluent weir to stabilize the tube. The suspended solids would then flow over the weir as in actual operation.

Basic analyses for such a system are outlined in *Table 1* as given by Ford [5]. The data obtained could then be plotted to show the effect of detention time or organic loading on process efficiency (*Figs. 5a* and *b*); to compare the efficiencies of the activated sludge and aerated lagoon systems (*Fig. 5c*); to determine the system oxygen requirements (*Fig. 5d*); to estimate the detention time of the extended aeration system (*Fig. 5e*); and to show the effects of the mixed liquor volatile suspended solids concentration of activated sludge process efficiency (*Fig. 5f*).

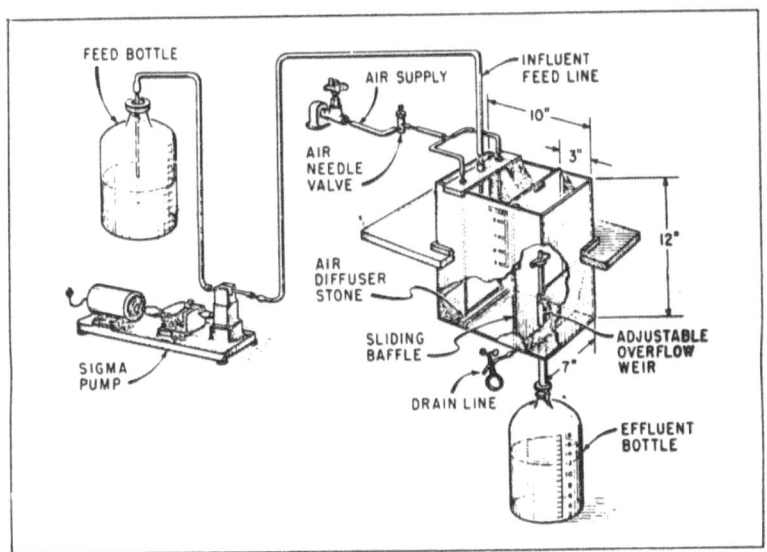

Fig. 4 :   Bio-Oxidation Unit [5]

TABLE 1 : Analyses for the Bio-Oxidation Unit
(After Ford [5])

ANALYSIS	RAW WASTE (Sample to be withdrawn from influent feedline or raw waste containers)	MIXED LIQUOR (Sample to be withdrawn from aeration chamber)	EFFLUENT (Sample to be withdrawn from effluent bottle)
COD, BOD or organic carbon, mg/l (filtered samples)	X		X
pH	X	X	
Suspended solids mg/l	X	X	
Volatile suspended solids, mg/l	X	X	
Oxygen uptake, mg O_2/day		X	
Dissolved oxygen		X	
Significant ions, compounds, etc., in wastewater	X		X

(a)

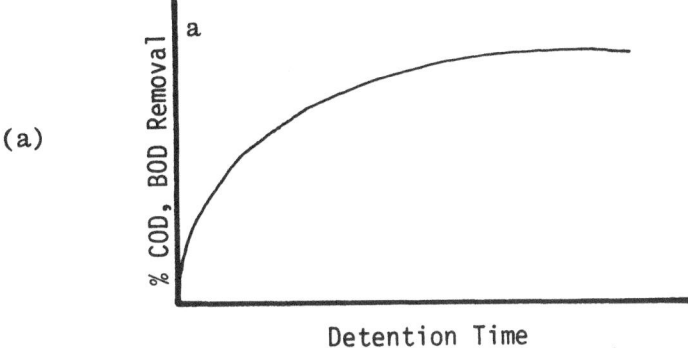

(b)

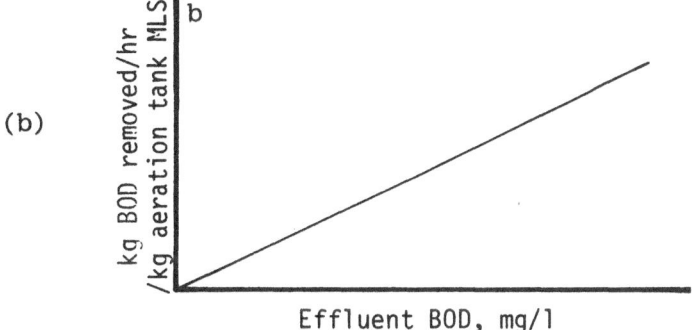

(c)

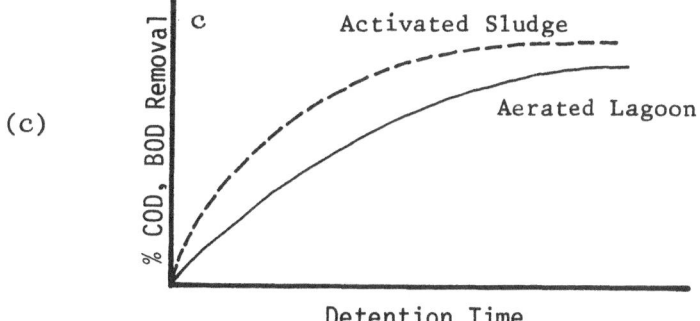

Fig. 5 : Test Results from the
Bio—Oxidation Unit
(Ford [5]).

(continued on next page)

412

(continued from previous page)

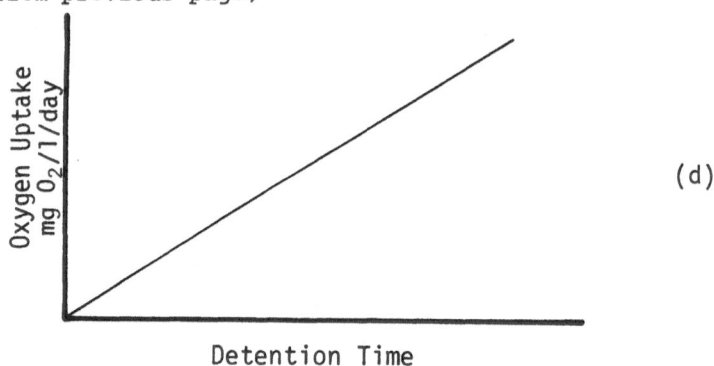

(d)

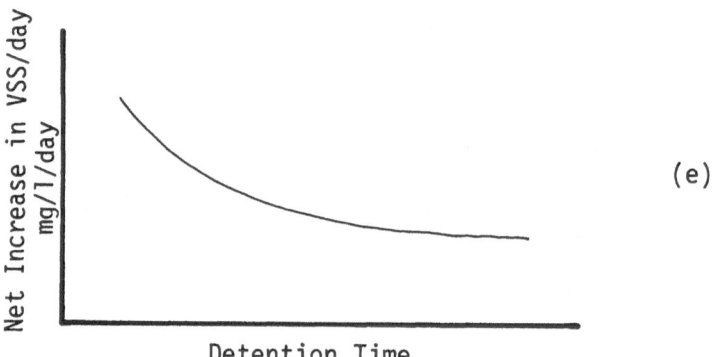

(e)

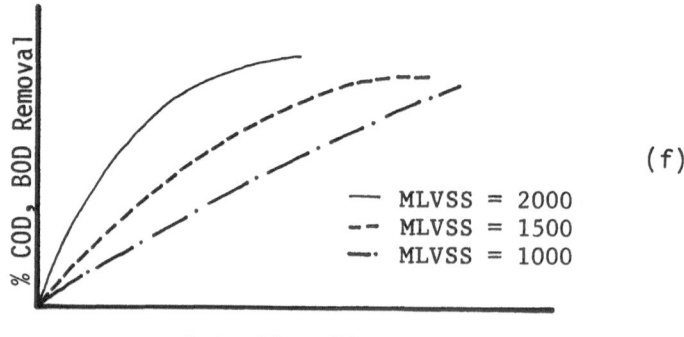

(f)

Fig. 5 :   Test Results from the
Bio-Oxidation Unit
(Ford [5])

3.   WASTE STABILIZATION PONDS

Thirumurthi (6) have used the Wehner and Wilhem equation, equation (5), to develop design criteria of waste stabilization ponds by studying the BOD removal and diffusivity characteristics of a laboratory model unit. The unit served the purposes of determining the parameters kt and d in equation (5), which are considered unreliable to predict analytically.

A waste stabilization pond was simulated by using a glass rectangular tank (45 cm length, 25 cm width and 20 cm depth) and a fluorescent lamp. The lamp was turned on for 8 hrs per day. Thirumurthi (1969) has used a synthetic sewage of known and constant qualities instead of domestic sewage since a domestic waste of constant BOD could not be obtained over a period of three months.

A number of tests were run by using different organic loads and pond temperatures. In these tests, routine measurements of BOD, COD, pH and temperature of influent and effluent samples were conducted for about 50 days. A plot of percent BOD remaining vs. time is plotted in *Fig. 6*.

The dispersion factor, d, defined by equation (5), was determined by employing the method suggested by Levenspiel and Smith (7). For this purpose, a tracer study using sodium chloride as the tracer material was conducted. Concentration vs. time measurements reported by Thirumurthi (6) are shown in *Fig. 7*. The mean detention time, t, and dispersion factor, d, are then computed from the following equations:

$$\sigma \quad t = \frac{\Sigma tc}{\Sigma c} \qquad \qquad \cdots \cdot (8)$$

$$\sigma t^2 = t^2 \{2d - 2d^2 \ (1-e^{-1/d})\} \qquad \cdots \cdot (9)$$

where

$$t^2 = \frac{\Sigma t^2 c}{\Sigma c} - \{\frac{\Sigma tc}{\Sigma c}\}^2 \qquad \cdots \cdot (10)$$

Thirumurthi (6), also developed a nomograph for design purposes which gives the relationships between BOD load and first order BOD removal coefficient and the influent BOD concentration.

414

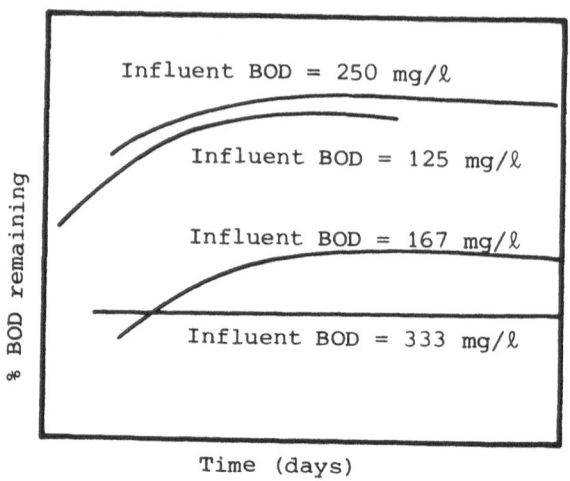

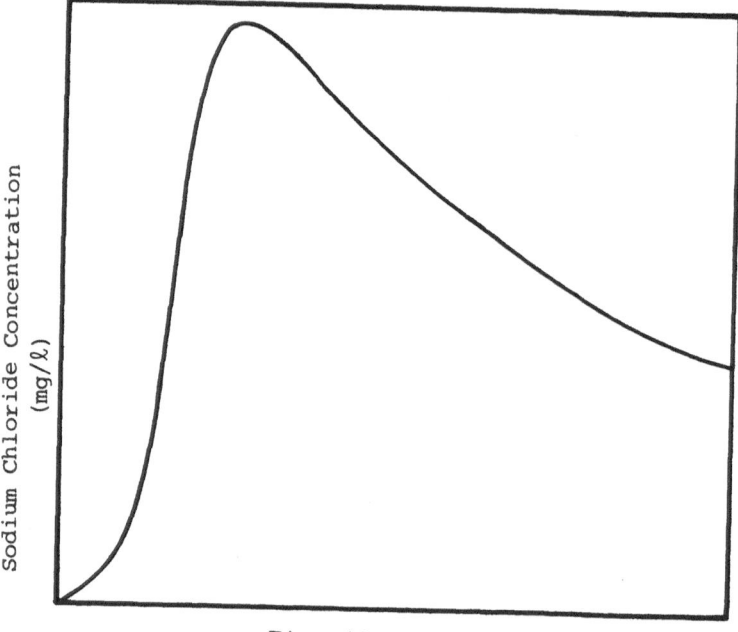

Fig. 6 : BOD Data for the Model Waste Stabilization Pond (Thirumurthi (6)).

Fig. 7 : Results of Tracer Studies (Thirumurthi (6)).

MODEL LAWS

In general, model laws leading to the similarity criteria between a model and the prototype can be directly obtained if the mathematical models representing the physical phenomena are known. In this respect, consider the first order BOD equation as an example:

$$\frac{dL}{dt} = - K_1 L \qquad \qquad \dots \text{(11)}$$

where

L = BOD at any time t;

K_1 = rate constant for deoxygenation

When the following substitutions are made

$$L' = \frac{L}{L_o} \qquad \qquad \dots \text{(12)}$$

$$t' = \frac{t}{t_o}$$

where

L_o = ultimate first stage BOD;

t_o = a reference time

equation (11) can be written in dimensionless form as

$$\frac{dL'}{dt'} = -(K_1 t_o) L' \qquad \qquad \dots \text{(13)}$$

Equation (13) implies that similarity between two systems would exist if the values of $(K_1 t_o)$ were the same in both systems.

A main disadvantage of laboratory studies of biological phenomena is that the duration of the experiments is often very long. Therefore, it would be helpful to develop methods which require less testing time. For example, if the BOD bottle tests were made

416

at 40°C instead of at 20°C, the model deoxygenation constant would be

$$(1.047)^{(40-20)} = 2.51$$

times greater than that of the prototype, since a generally accepted empirical equation giving the value of the deoxygenation constant as a function of temperature and value of the deoxygenation constant at 20°C is (1)

$$K_{1_\theta} = K_{2_\theta} (1.047)^{\theta-20} \qquad \qquad \dots\dots (14)$$

where

K_{1_θ} = value of deoxygenation constant at θ^0C

K_{2_θ} = value of deoxygenation constant at 20°C

θ = temperature in °C

Thus, a BOD test made for the purpose of determining the 5-day BOD value would require 5.00/2.51=1.99 days if performed at 40°C instead of at 20°C.

However, empirical relationships similar to equation (14) are not available for many kinetic models, and even if they were, they would not be very reliable; and methods as suggested above should be used with care in biological systems.

REFERENCES

1. Busch, A.W., "Treatability vs. Oxidizability of Industrial Wastes, and the Formulation of Process Design Criteria", Proceedings of the 16th Industrial Waste Conference, Purdue University, 1961.

2. Ford, D.L., Gloyna, E.F., and Yang, Y.T., "Development of Biological Treatment Data for Chemical Wastes", Proceedings of the 22nd Industrial Waste Conference, Purdue University, 1967.

3. Ford, D.L., "Laboratory Methodology for Developing Biological Treatment Information", in Manual of Treatment Processes, ed. W. Wesley Eckenfelder, Environmental Science Services Corporation, 1970.

4. Hermann, E.R., and Gloyna, E.F., "Waste Stabilization Ponds, III Formulation of Design Equations", Sewage and Industrial Wastes, Vol. 30, Aug. 1958.

5. Levenspiel, O., and Smith, W.K., "Notes on the Diffusion-Type Model for the Longitudinal Mixing of Fluids in Flow", Chemical Engineering Science, Vol. 6, 1957.

6. Metcalf and Eddy, Inc., Wastewater Engineering, McGraw-Hill Book Company, New York, 1972.

7. Thirumurthi, D., "Design Principles of Waste Stabilization Ponds", ASCE, Journal of the Sanitary Engineering Division, April 1969.

8. Wehner, J.F., and Wilhem, R.H., "Boundary Conditions of Flow Reactor", Chemical Engineering Science, New York, Vol. 6, 1958.

ECONOMICS OF BIOLOGICAL WASTEWATER TREATMENT

Klaus R. Imhoff

Director of Ruhrverband and Ruhrtalsperrenverein

Biological processes are applied in general for wastewater treatment.This is tne most certain aspect of the topic. Already the dimensioning of the plant is depending on the receiving water. The economic picture is even more complicated since capital cost, personnel cost, cost of maintenance and repair and energy cost are involved.

It also must be considered if we talk of a developing country or a highly industrialized area. In developing countries the a-mount of capital investment is of great concern. There is a high demand for everything and wastewater treatment stands at the end of the priority list. For wastewater treatment simple and inex-pensive solutions are required with little mechanization and more handlabour. In highly industrialized areas personnel cost are of greatest importance. In consequence there is a tendency to cen-tralize wastewater treatment and to highly mechanize the plants to save on labour.

From these few statements it may be concluded that the ques-tion of economics of biological wastewater treatment can be ana-lyzed only for a certain area. If a clear statement is required it becomes necessary to define a certain task, to make several de-signs and to ask for several bids like it has been done for the advanced wastewater treatment plant for the city of Zürich [1].

Because of the limited value of special cost analyses for an international gremium it is the purpose of this paper to show some trends which influence the cost picture of biological wastewater treatment based on West German experience.

1. DEGREE of TREATMENT

As pointed out by Müller-Neuhaus [2] construction cost is
very much dependent on the degree of treatment. As can be read
from *Fig. 1* 80% of BOD removal by biological treatment is less expensive in terms of kilograms of BOD removed than the application
of only mechanical treatment. Above an 80% of BOD reduction treatment cost rises considerably. In more recent curves a smaller increase of cost is shown. But there is no doubt about cost prog-
ression if the residual pollution has to be abolished.

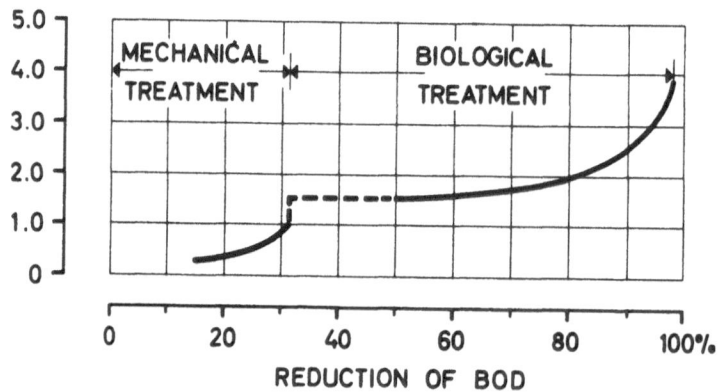

Fig. 1 :   Construction cost for Biological
          Treatment Plant in Relation to
          the degree of Purification according to
          Müller-Neuhaus [2].

Cost figures which are given later in the paper refer to an
overall BOD removal of about 90%.

2. SPECIFIC CONSTRUCTION COST

There is a definite trend that specific construction cost decrease with increasing size of plant.

A plant of 10,000 population equivalents requires only 50% of
the specific construction cost than a plant of 1,000. This law
may be extrapolated even to bigger plants if conventional sludge
treatment (digesters and sludge lagoons) can be applied. Due to
lack of land in many cases mechanical sludge dewatering and incineration cannot be avoided in plants of more than 100,000 popula-

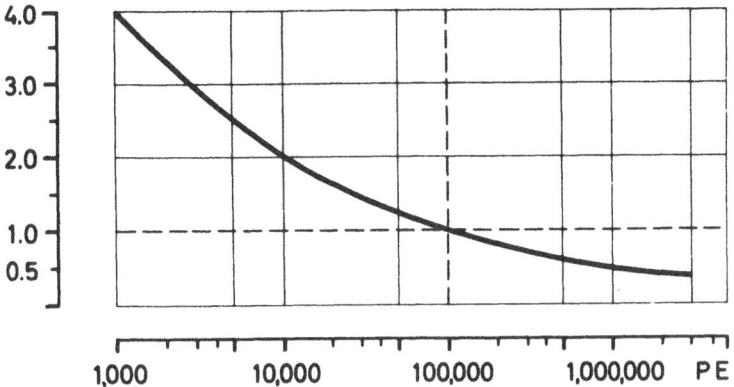

Fig. 2 : Relation between specific construction
cost for Biological Treatment Plants
and population equivalents served
[3]. Specific cost for a plant
for 100,000 population equivalents = 1.0.

tion equivalents. In such cases there is no longer a degression
of construction cost. The curve then will swing out more or less
horizontal.

From *Fig. 2* it has to be concluded that central waste-water
treatment may be more economical than the treatment in separated
plants. Depending on the specific case it may be justified to
conduct wastewater to a central treatment plant for several kilometers. Bigger plants have the advantage of a more equalized load
distribution and of better operational results. Pumping stations
with screw pumps are able to lift screenings and sand. Consequently this material should be eliminated only at the central works.

3. CAPITAL COST

The interest rate for capital is always higher than the inflation rate. In the Federal Republic of Germany the inflation rate
presently (1976) equals 5.5% per year and a community may borrow
money over a period of five years for 8% per year. Communities are a
very safe debtor because according to their legal status they cannot turn bankrupt. Since inflation may increase in future, money
becomes more expensive if it is borrowed for longer periods like
15 or 25 years.

Concrete structures of the wastewater treatment plant may last
over a period of 30 years while mechanical equipment like scrapers,

screens, pumps and motors will last for only 10 years. About 75%
of the investment is for concrete structures and 25% of investment
for mechanical equipment. The average depreciation time is therefore 25 years. For such a long time in the Federal Republic of
Germany money is only available for an interest rate of 9% per
year, and if there is a considerable change, interest rates will
be adapted. From *Fig. 3* it may be read that 9% interest rate
and 25 years of depreciation time result in annual capital cost of
10%.

Estimating the construction cost one has to look at the design
population and the design flows. Depending on the growth rate of
the area one will construct works with an extra capacity of 20, 30
or 40%. This extra capacity has to be financed by the living population.

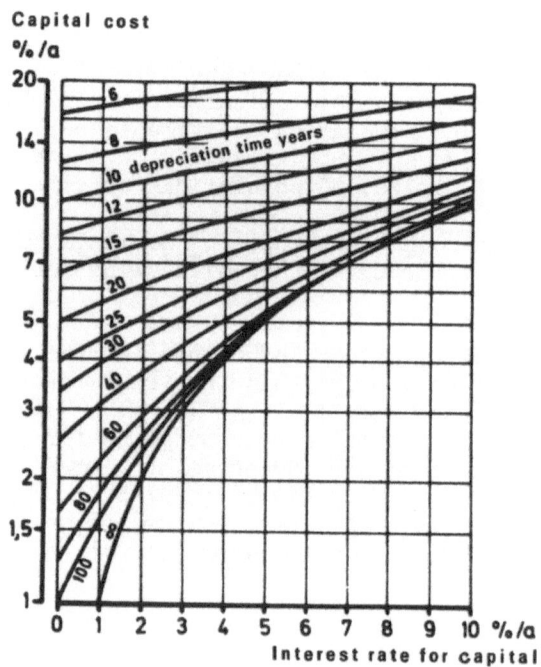

Fig. 3 : Capital cost in dependency of
interest rate and depreciation
time.

The population equivalent mostly is based on the BOD_5- load of
60 g/per capita-day. Another basis may be the flow rate. This
may be assumed to be 3.5 ℓ/s per 1,000 capita. If there are considerable differences between the BOD and the flow population e-
quivalents one can refer to the arithmetic average of the two figures because some structures are sized on the basis of flow and

others on the basis of BOD volume loading or the amount of sludge loading.

A breakdown of construction cost for activated sludge plants with heated digesters is shown in *Table 1*.

TABLE 1: Activated Sludge Plants with Heated Digesters: Breakdown of Construction Cost in % according to Bucksteeg [5]

	Minimal	*Average*	*Maximal*
Design and Construction	5	7	9
Land	2	3	5
Access, fence	4	6.5	9
Equipment of Construction Site	3	4	6
Soil Movement	4	7	10
Retaining Water	2	3.5	6
Pipes and Manholes	2	4	7
Screen and Grit Channel	2	3.5	10
Primary Clarification	4	5.5	8
Activated Sludge Tanks	7	10	14
Secondary Clarification	4	6.5	9
Digestion	11	14	19
Sludge Drying Beds	2	4	7
Gas Container	2	3	5
Operation Building & Pumping Station	14	16	22
Electrical Equipment	2	2.5	4
		100 %	

4. OPERATION COST

Three main influences have to be considered:

Personnel cost
Cost for upkeep and repair &
Cost for power

424

Fig. 4 gives average results for the personnel demand according
to Sickert and Londong. If labour is expensive and if there is
an incentive to save on labour cost one can estimate one plant o-
perator per 10,000 population equivalents connected.

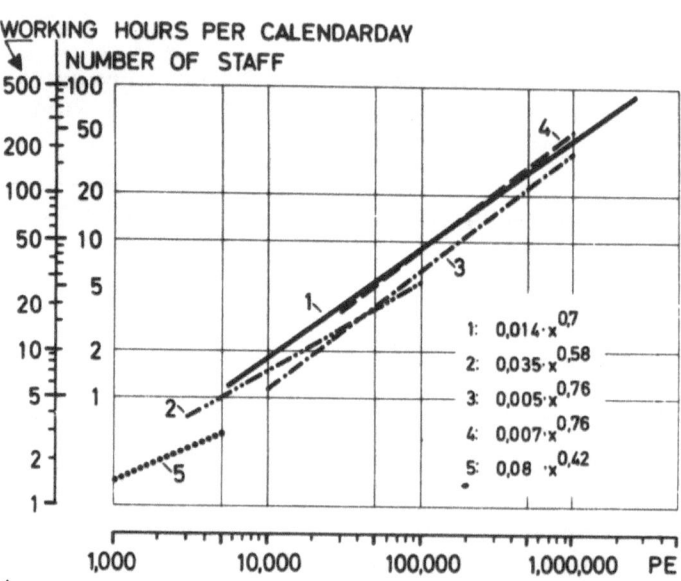

Fig.: 4   Demand of personnel for municipal treatment plants in Germany.
1.   Activated sludge plants with separate sludge treatment, but without power generation
2.   Trickling filter plants with separate sludge treatment, but without power generation.
Personnel being used on treatment plants of the organizations 'Emschergenossenschaft' and
'Lippeverband'
3.   Activated sludge plants with separate sludge treatment, but without power generation
4.   Activated sludge plants with separate sludge treatment and with power generation
5.   Activated sludge plants with aerobic sludge stabilization.

    Over the last 25 years in the Federal Republic of Germany la-
bour cost was increased by 500%. Construction cost increased for
the same period by 280% and energy cost (electricity) increased by
66%. In general energy cost is, not so decisive as may be concluded comparing the energy requirement of different types of plants:

Trickling filter   0.7 kWh/kg BOD removal
Activated sludge process   1.0 kWh/kg BOD removal
Activated sludge process
    with aerobic sludge
    decomposition   1.8 kWh/kg BOD removal

Personnel demand is much more important as at present time in the Federal Republic of Germany 35,000 DM have to be paid per plant operator per year.

The relation between personnel cost, cost for upkeep and repair and for power is demonstrated by *Fig. 5*. Since mechanization is increased with increasing size of plant the percentage of personnel cost is diminished while the cost for upkeep and for power slightly increases.

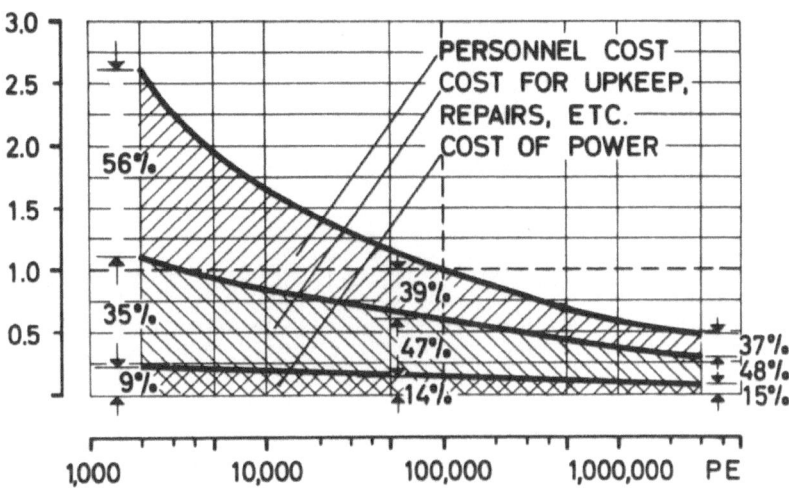

Fig. 5 : Relation between Specific Operation cost for Biological Treatment Plants and Population Equivalents (PE) served according to Sickert [3].

5. TOTAL COST

Total cost consists of capital cost and operation cost. Mönnich investigated annual cost for different types of plants without mechanical sludge dewatering. In *Fig. 6* his results are presented.

For German conditions the oxidation ditch is feasible up to 5,000 population equivalents. Between 5,000 and 25,000 population equivalents the trickling filter still may be applied. For plants with more than 25,000 population equivalents mostly the activated sludge process is selected because it provides a better degree of treatment and is more flexible.

426

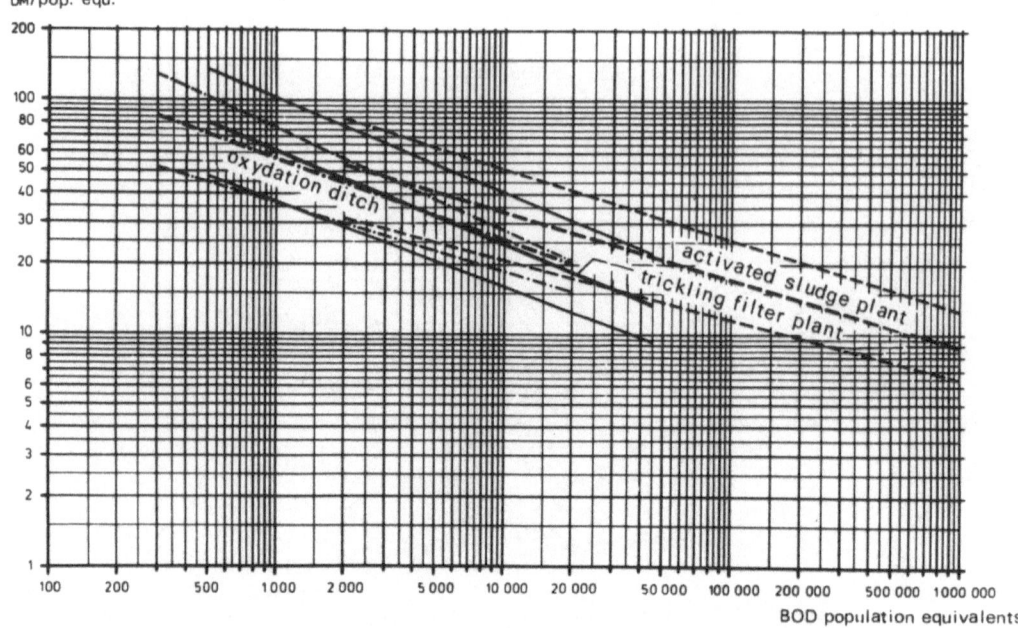

DM/pop. equ.

BOD population equivalents

Fig. 6 : Annual Cost (1974 prices) for
different types of Wastewater
Treatment Plants according to
Münnich [7]

Type 2: Imhoff tank, Trickling filter, Dortmund tank
Type 3: Schreiber package trickling filter
Type 4: Activated sludge plant with heated digester
Type 9: Oxidation ditch with aerobic sludge decomposition

Table 2 has been compiled from Ruhrverband figures for 90%
BOD removal and conventional sludge treatment [6].

TABLE 2: Cost for Biological Wastewater Treatment in 1974

Capacity	Construction Cost		Annual Cost (DM/E/a)			Total Cost
(E)	(DM/E)	(DM/m³/d)	Operation	Capital	Total	DM/m³
1 000	600	3 000	23,00	72,00	95,00	1,30
5 000	330	1 650	14,40	39,60	54,00	0,74
10 000	260	1 300	11,50	31,20	42,70	0,58
25 000	180	900	9,60	21,60	31,20	0,43
50 000	150	750	7,80	18,00	25,80	0,35
100 000	130	650	6,60	15,60	22,20	0,30

The total cost well compare with the diagram of Mönnich. In Table 2 also specific costs per m³ of wastewater treated are shown. It should be pointed out that the cost for the sewerage system is not included in that figure.

SUMMARY and CONCLUSIONS

The economics of biological wastewater treatment are very much related to the local situation and to the time you look at it. The Federal Republic of Germany may be an example of how the situation has changed during the last 25 years. In 1950 there was a tremendous shortage of capital. Consequently the high loaded biological treatment processes with only some 80% of BOD removal and relatively short sedimentation times were applied. Since the sewerage system requires about 3 to 5 times the capital investment of the wastewater treatment plant for local developments frequently small plants were chosen. At that time the higher personnel cost could be paid.

Nowadays a high level of industrial production and a good standard of life has been achieved. Manpower has become extremely expensive. To sanitate the rivers 95% of BOD removal is required. In consequence the sewerage systems have been expanded and central wastewater treatment plants are applied. The treatment volumes per capita roughly have been doubled to provide for sufficient stormwater capacity and to upgrade the process. Many sites for wastewater treatment have become too small and they are surrounded by houses. Starting wastewater treatment, from the point of view of the author, it is more economical to buy land of sufficient size far out of the city. In the first period inexpensive treatment processes like lagoons can be applied. Later on more sophisticated processes have to be chosen.

The best economical effect can be reached by a basin wide water pollution control. The money, which is collected from all polluters, can be invested according to a priority list. That means, that most feasible projects will be constructed first and that selfpurification of the receiving water can be utilized in the desired limits [8].

REFERENCES

1. Wiesmann, J. and Roberts, P.V.: Results of an International Competition for the Advanced Wastewater Treatment in Zürich-Werdhölzli, *Berichte der Abwasser-technischen Vereinigung* Nr. 28, Bonn 1976, 259.

428

2. Müller—Neuhaus, G.: Possibilities and Limits of Wastewater Treatment from the Technical and Economical Point of View, *Gas- und Wasserfach* (GWF) 104, 1214, (1963).

3. Sickert, E.: The Development and Effect of Construction and Operation Cost in Biological Sewage Treatment Plants, 6th Int. Conference on Water Pollution Research, B/7/13, *Pergamon Press,* (1972).

4. Londong, D.: Operation and Maintenance of Wastewater Treatment Plants, *Gewässerschutz - Wasser - Abwasser,* Band 4, 109 (1971).

5. Bucksteeg, K.: Construction Cost of Wastewater Treatment Plants, *GWF 112,* 163, (1971).

6. Imhoff, K. and K. R.: Pocket Book of Town Drainage, 24th edition, *Verlag R. Oldenbourg,* München 1976.

7. Mönnich, K. H.: Construction and Operation Cost of Wastewater Treatment Plants, *ATV - Fortbildungskurs Ottilienberg 1976.*

8. Rinche. G.: Economic Considerations of Regional Wastewater Treatment in a River System, *GWF 109,* 1229, (1968).

INSTREAM AERATION

Klaus R. Imhoff

Director of Ruhrverband and Ruhrtalsperrenverein

1. INTRODUCTION

After the second world war the construction of wastewater treatment plants did not keep pace with pollution from settlement and industry. This caused increasing oxygen difficulties in flowing and stagnant waters, resulting not only from the discharge of organic substances but also from eutrophication due to fertilizing salts.

Sufficient oxygen contents can be achieved in flowing water by three different methods: Dilution water from reservoirs, biological wastewater treatment or instream aeration. Roughly estimated their relative costs are in the ratio of 16: 4:1. This shows that instream aeration is the cheapest procedure. Moreover it can be applied immediately. On the other hand aeration does not remove the polluting substances from the water. Therefore it cannot be regarded as a perfect method or final solution. Instream aeration can be compared to the oxygen tent of the physician, as transitional solution in case of emergency.

This paper is restricted to the most promising procedures from the technical and financial point of view. It is not a summary on all possibilities of instream aeration by mechanical means.

2. DEVELOPMENT & TASK of INSTREAM AERATION

2.1 HISTORICAL VIEW

Already in 1929 Bach proposed to improve the oxygen content

in small receiving waters by fine-bubble aeration known from the activated-sludge process. In the same year Nolte picked up this suggestion and got installed a corresponding device at a creek loaded with wastewater of a sugar factory. Due to the small capacity of the system only a limited success could be achieved. In 1938 the Wupperverband used a floating aerator on a lagoon [2]. This promising start was interrupted by the war. In 1943 intensive investigations were started in the USA, to aerate Flambeau River. At first Tyler installed fine bubble aeration at the turbine outlet of a power station. Later on the turbine inlet was aerated. In the meantime Wagner developed a technique of venting the turbine, which has been successfully used in the Flambeau River and at other places [1].

Due to the increasing importance of this field, several associations 1965 founded a working committee. In 1969 this committee submitted recommendations for instream aeration which involve more than 135 references [3]. Besides, the American report on "water quality behavior in reservoirs" by J.M. Symons with 616 pages is of special importance. In 1967 the first technical conference mainly dealt with the different aspects of instream aeration.

2.2 TASK

Aerobic conditions in receiving waters can be achieved by instream aeration, if other measures are not sufficient or have not yet been realized. This is an important economic aspect, as the decomposition of organic pollutants in aerobic water takes place more rapidly than without the presence of elementary oxygen.

The daily decomposition rate at 20°c may amount to 20.6% for aerobically digested matter and to 8.4% for the anaerobic process [5]. Aerobic water therefore will decompose 50% of the organic pollution at 20°c within 3 days, whereas 7 days would be necessary in the anaerobic environment. Moreover, digestion processes generate toxic decomposition products (e.g., hydrogen sulphite). Oxygen-deficient water may also dissolve iron and manganese from the sediments.

For aerobic self-purification a minimum oxygen content of $1.5 \ g/m^3$ is necessary. This requires higher values for the average low flow conditions. There must be a security for influences which cannot be estimated in detail. This leads to the oxygen range required for fish life, i.e., $3-4 \ g \ O_2/m^3$. Simultaneously the fish serves as indicator for sufficient water quality. Water pollution control therefore should ensure oxygen levels in the order of $3-4 \ g/m^3$.

Also after completion of additional wastewater treatment plants instream aeration may be useful to compensate for the effects of algal dissimilation, unusual rain pollution or other pollution events.

3. POSSIBILITIES of INSTREAM AERATION

3.1 AERATION of STAGNANT SURFACE WATERS

From the point of view of aeration technique one has to distinguish between flat and deep waters.

3.1.1 Deep Stagnant Waters

In deep waters during summer a temperature stratification occurs which prevents the exchange of water and the transport of oxygen taken up at the surface. This causes high oxygen contents in the warm epilimnion, whereas the deeper and colder layers of the hypolimnion have a lack of oxygen due to absorption processes. By nature, stratification is only destroyed during the winter period.

Therefore the oxygen balance of stratified waters can be essentially improved by blowing compressed air or by pumping water and thus effecting an artificial turnover [3], [4]. By a stability equation the necessary energy can be calculated [4]. It depends on the geometrical shape of the water body, the season, the meteorological situation, and the quality requirements. In the average of 8 different reservoirs 1.6 Wh/m^3 water content were spent as mixing energy [4]. The necessary mixing time varied between 20 and 90 days.

For drinking water reservoirs a complete mixing has the disadvantage that the temperature of the abstracted drinking water essentially rises. This undesired effect can be avoided by the aeration device of the Wahnbachtalsperrenverband (Fig. 1).

This equipment consists of 2 vertical tubes being connected by an open channel at the top. The longer tube serves as uptake pipe for the air/water mixture. The degased water is discharged back into the hypolimnion by the gravity tube with horizontal distribution arms [6], [7], [31].

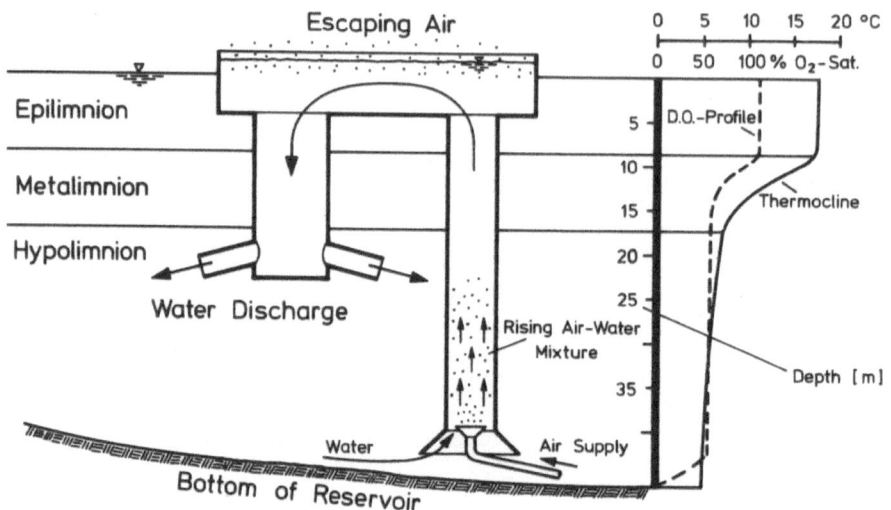

Fig. 1 :   Diagram of the Wahnbach
            Aeration Device [3].

3.1.2 Flat Stagnant Waters

These involve surface waters like shallow reser‑
voirs or harbours up to a depth of about 6 m in which temperature
stratification does not occur. Here the sufficient distribution
of oxygen becomes difficult.

Hoses or tubes with orifices may be used which are
kept on the bottom by weights [8], [9]. There is a close depen‑
dency between the infiltrated amount of air, the air pressure, and
the number of orifices. The losses are indicated by a diagram [9].
Large bubble aeration involves the danger of unequal air distribu‑
tion. This must be considered especially in large systems with
differing depth. Oxygen utilization varies between 0.9 and 1.3%
per m injection depth [10]. The air flow amounts to 5-15 Nm^3 per
m of hose and hour. *Fig. 2* shows the operational results of a hose
with orifices.

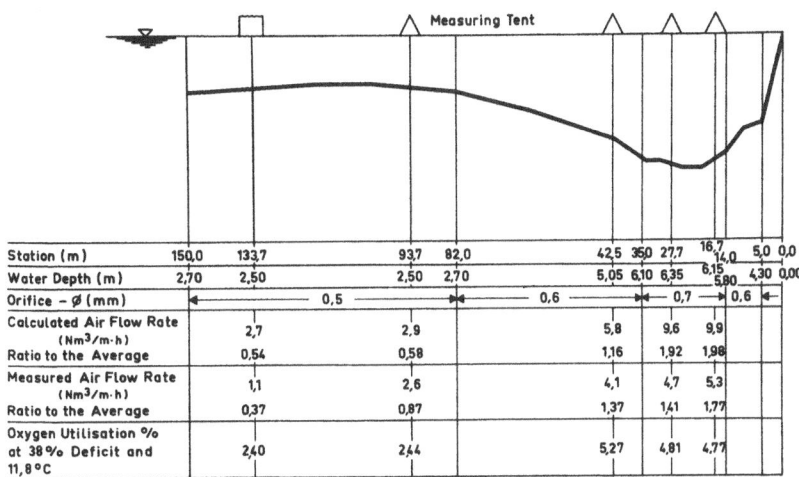

Station (m)	150,0	133,7		93,7	82,0		42,5	35,0	27,7	16,7 / 14,0	5,0	0,0
Water Depth (m)	2,70	2,50		2,50	2,70		5,05	6,10	6,35	6,15 / 5,80	4,30	0,00
Orifice – ø (mm)			0,5			0,6			0,7		0,6	
Calculated Air Flow Rate (Nm³/m·h)		2,7		2,9			5,8	9,6	9,9			
Ratio to the Average		0,54		0,58			1,16	1,92	1,98			
Measured Air Flow Rate (Nm³/m·h)		1,1		2,6			4,1	4,7	5,3			
Ratio to the Average		0,37		0,87			1,37	1,41	1,77			
Oxygen Utilisation % at 38 % Deficit and 11,8 °C		2,40		2,44			5,27	4,81	4,77			

Fig. 2 : Operational Results of
Compressed Air Aeration
Hoses [9].

Nozzle diameters of 0.6 and 0.7 mm have proved to be advantageous. The orifices of 0.5 mm diameter are highly susceptible to clogging. Air hoses are also used to retain oil in harbours [11]. This double function makes the process especially interesting for stagnant waters. Hoses can be stored on drums and installed within a few hours.

Flat stagnant waters can also be oxygenated by *surface aerators*. Due to the restricted range of influence the units should not be too great and carry out a self motion [12], [13]. Otherwise an oxygen pillow occurs strongly impairing the effect of the aerator due to the low deficit.

The applied aerator consists of a submerged motor with propeller for horizontal water transport. The water stream flows through a Venturi pipe where air enters at the narrow passage. The following characteristics have been determined: Injection depth about 50 cm, water transport 1 160 m³/h, air entrainment 206 Nm³/h, energy consumption of the motor 5.1 kW, oxygen input 5.5 kg O_2/h, oxygenation efficiency 1.08 kg O_2/kWh (at 100% deficit). With corresponding position the repulsion from the horizontal water transport effects a moving power Thus the turbo-oxyders on the lagoon rotate on a circle of 30 m diameter.

Fig. 3 : Rotating Turbo Oxyders

3.2 AERATION of SMALL RIVERS

Most experiences are available with these waters. It turned out that the oxygen balance of small rivers can be influenced successfully by instream aeration. But the installed aeration capacity must be great enough and a careful control is necessary. A calculation method is shown in reference [2].

3.2.1 Weir Aeration

Aeration efficiency is very high and covers the whole overflowing water. Londong made investigations the results of which are shown in *Fig. 4*.

The smaller the head the better the energy is utilized. As weirs with great heads mostly are connected to hydropower stations, turbine aeration becomes economical at heads of more than 6 m.

The greatest portion of the oxygen is taken up in the downstream water where air is entrained. Dams with a vertical fall therefore achieve a better effect than inclined weirs or underflown roller gates. In *Fig. 5* the results of 7 different

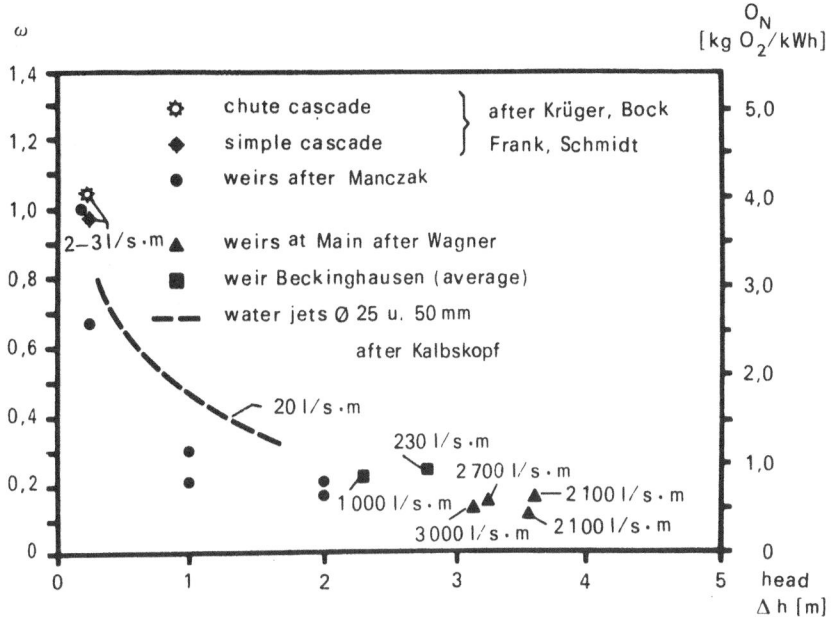

Fig. 4 :   Oxygenation Efficiency
in Relation to the Fall
Head and Overflow Rate [10].

weirs are compared.

As the increase of oxygen may vary considerably,
oxygenation capacity*and efficiency are related to an average oxygen deficit of 50% between head-water and tail-water. With other
average deficits of X (%) the oxygen input varies according to the
ratio (OC: 50). x. By weir aeration the oxygen balance can be especially improved when in critical times the turbine discharge is
diminished and the weir overfall is opened.

3.2.2 Turbine Aeration

With turbine aeration two procedures are applied.
According to Wagner-Voith the air penetrates through numerous o-
penings in the underpressure zone of the turbine case [1]. The air
quantity is regulated by a valve. The decrease of generator efficiency corresponds to the energy amount which is necessary to inject the air. By the procedure of Wolff the air is mechanically
compressed and supplied above the impeller level in the zone of
overpressure [15]. For newly installed turbines the system of

---

*
OC= Oxygenation Capacity

	Weir Spillenburg (Ruhr)	Weir Kettwig (Ruhr)	Weir Kahlenberg (Ruhr)	Weir Baldeney (Ruhr)	Weir Werne**) (Lippe)	Weir Beckinghausen (Lippe)**)	Weir Duisburg (Ruhr)
Type of weir	two stage slope weir 1:5; 1:2.8	slope weir (sector) 1:1	underflown roller gate and two steps	underflown roller gate	slope weir 1:10	overflown step weir	overflown step weir
Scheme	1,2 m 1,6 m	5,9 m	4,5 m	8,75 m	1,8 m	2,6 m	5,5
Oxygen input [kg O_2/h] at 50 % average deficit, 20°C and 20 m^3/s discharge Q*)	151	495	339	570	101	323	635
Energy [kW] $N = \eta \cdot 9.81 \cdot Q \cdot H$ ($\eta = 1,00$)	549	1159	883	1718	353	510	1080
Oxygen yield [kg O_2/kWh]	0,28	0,43	0,38	0,33	0,29	0,63	0,59
Length of weir crest at 20 m^3/s discharge [m]	300	90	430	33	100	30	15
Discharge per meter of weir crest at 20 m^3/s [l/s · m]	67	222	47	606	200	667	1.333

**)after Londong

*)maximal oxygen input under these conditions = 635 kg O_2/h

Fig. 5 : Oxygenation Capacity and Efficiency of Different Weir Systems at a flow of 20 m^3/s [14].

Wagner-Voith is more economical, for existing turbines that of Wolff.

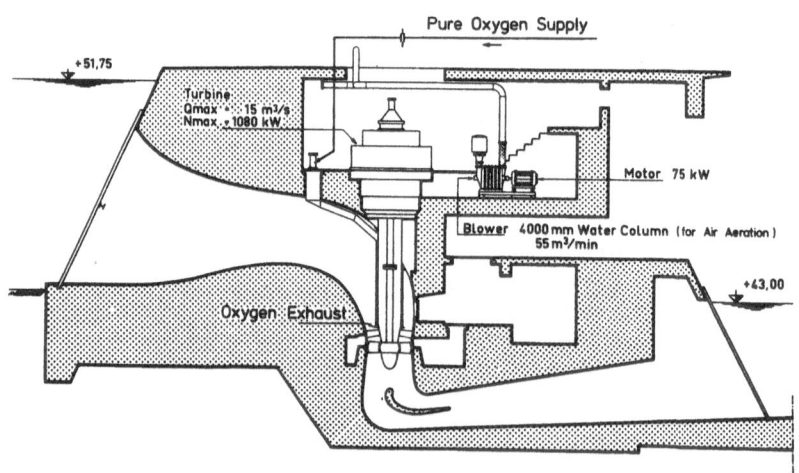

Fig. 6 : Baldeney Turbine Pump with Aeration Equipment. The Superstructure is not shown.

The Ruhrverband installed a blower and pipelines to inject the air immediately above the impeller level *(Fig. 6)*. The optimal air quantity has been determined by tests with respect to the oxygenation capacity and efficiency for a volume ratio of relaxed air to water of 1:14. At a supply of 3,500 Nm^3 of air per hour, 50% initial oxygen deficit and 18°C water temperature an oxygen input of 150 kg O_2/h and an oxygenation efficiency of 0.90 kg O_2/kWh could be obtained. The utilization of air oxygen was 15% for 50% oxygen deficit of inflowing water [9].

3.2.3 Surface Aerators

Surface aerators known from wastewater treatment plants may also be used for instream aeration. Here they are installed in a floating arrangement. They may get motor capacities between 10 and 220 kW and are mostly equipped with a reduction gear. But there are also models of high speed. The electricity supply requires a good anchoring. Existing bridges can be used.

In 1965 the Ruhrverband developed a system driven by a 150 hp diesel engine. The total length is 19 m, the inside width between the pontoons 7 m. By the speed of the motor the revolution of the 2.20 m diameter Simcar impellers can be adjusted between 30 and 50 rpm. The unit is equipped with position lights. It can be anchored at any suitable place. The diesel oil tank has sufficient capacity for a 10 days continuous operation. From tests results the oxygen input for 50% average oxygen deficit, 10 m^3/s runoff and 20°c water temperature was estimated to be 47 kg O_2/h [14].

Fig. 7 : Floating Aerator of
Ruhrverband.

To compensate for the oxygen demand of discharged wastewater from a paper mill, in the Thames estuary, the largest known floating aerator with a motor capacity of 150 kW and a wheel diameter of 3 m has been installed [18]. The unit is supported by 4 cylindrical pontoons. It covers a surface of about 10x10 m. Concrete blocks were chosen for anchoring. The energy is supplied by a submerged cable. According to preliminary investigations the oxygenation efficiency shall be 1.13 kg O_2/kWh, corresponding to 170 kg O_2/h. The location of the unit is marked by position lights. A further device with a motor capacity of 220 kW has later been taken into operation.

3.2.4 Compressed Air Systems

Beside of hoses and tubes with orifices as mentioned under 3.12 there are experiences with a floating aeration bridge [10], [16], [17]. This equipment has been constructed with barrels and other simple material. It creates a large bubble aeration zone across the river. The aeration system is attached to 14 pontoons. It consists of tubes with 6 mm holes which are 1 m below the water surface. The blower which has been installed at the banks supplies 12,000 m^3 air/h. For 50% oxygen deficit of water the oxygen supply amounts to 30 kg O_2/h [10]. The bridge must be removed before a flood appraoches.

3.2.5 Experiences at River Ruhr

Based on a research grant of the State of Northrhine-Westfalia in 1964 the first instream aeration tests were carried out in the lower Ruhr. Since that time Ruhr aeration has been systematically practiced and extended [9], [12], [14].

Oxygen Control

Because of the great velocity of flow a lack of oxygen does not occur in the upper and medium part of the Ruhr River. Only in Lake Baldeney and the downstream reaches the effects of algal blooms and the resulting secondary pollution can be observed. Over the last 40 kilometers of the river a careful control of the oxygen content is necessary in order to enable corresponding measures during critical times.

After a heavy fish kill in 1963 and several oxygen difficulties of short duration in the following years the Ruhrverband installed 5 continuously recording oxygen meters (*Fig. 9*). The location of the meters is especially important to achieve good results. Hydropower stations and weirs are very suitable, as here.

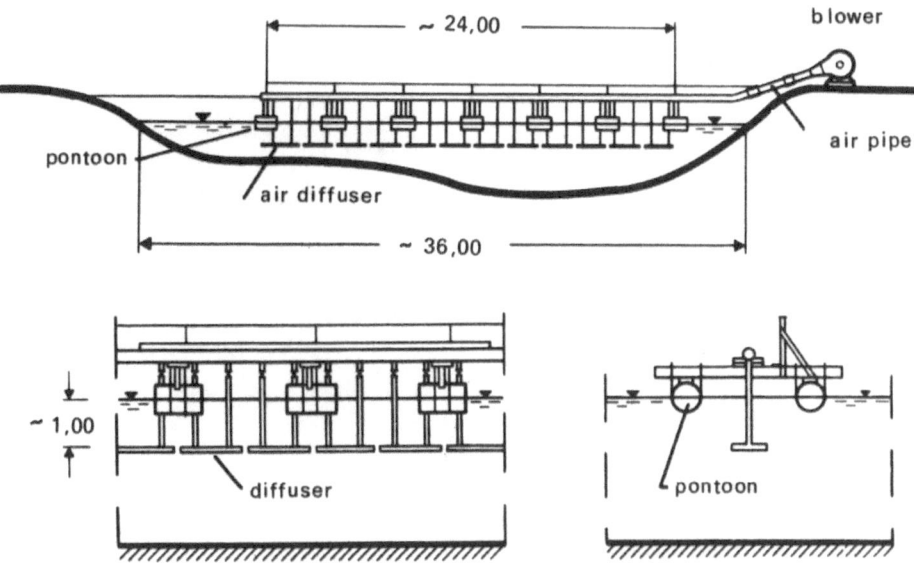

Fig. 8 : Floating Aeration Bridge of
of Lippeverband

1. the flow is mixed and concentrated at the
   turbine inlets and the abstraction of representative water samples renders no difficulty.

2. the measuring devices can be easily installed
   and protected.

3. the staff of the power station and weir can
   frequently control the device.

Only one of the oxygen recorders is not located
at a power station as the oxygen balance of Lake Baldeney shall
be controlled by measurements in its influent and effluent.

The results of the stationary meters are controlled twice per week by a portable oxygen electrode. During
low run-off and critical times in addition longitudinal oxygen
profiles are taken everyday.

The control station Baldeney can be regarded as an
example for one of the oxygen meters [13].

The water sample is abstracted from the influent
of the operating turbine by a submerged pump and is conducted in
a closed pipe system to the oxygen electrode, which can be seen
in the lower third of the figure. The flow of electric current
of the membrane-covered gold-silver electrode indicates the oxygen content of the water. The value is adjusted to the given tem-

perature and recorded. The installed electrodes are very reliable.

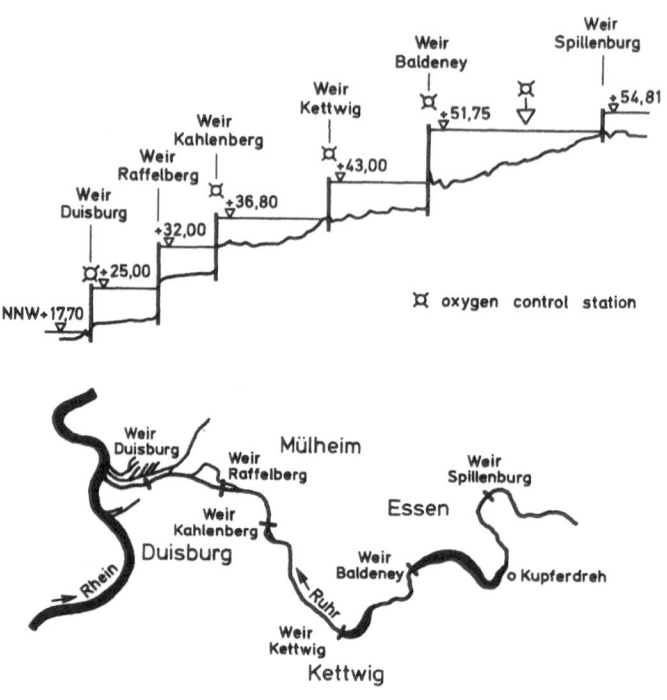

Fig. 9 : Longitudinal Section of the
lower Ruhr with Location of
Oxygen Meters [14].

Aeration Devices

Table 1 shows operating times and costs of the
different aeration devices of the Ruhr in the years 1970 through
1973. In 1971 the considerable increase in operation time can be
traced back to the very low runoff and to the accumulation of bottom sludge. In 1971/72 a scouring flood did not occur. Apart
from the air hoses which have the fourfold energy costs due to local circumstances, 0.06 to 0.16 DM per kg of oxygen are spent.The
total costs involve the capital and the energy costs, assuming 30
days of operation per year and 10% for interest rates and amortization. The examples of turbine and weir aeration at Baldeney indicate the favourable energy costs of turbine aeration. But only
in case of more than 30 days of operation time per year the total
costs of turbine aeration offer a certain advantage.

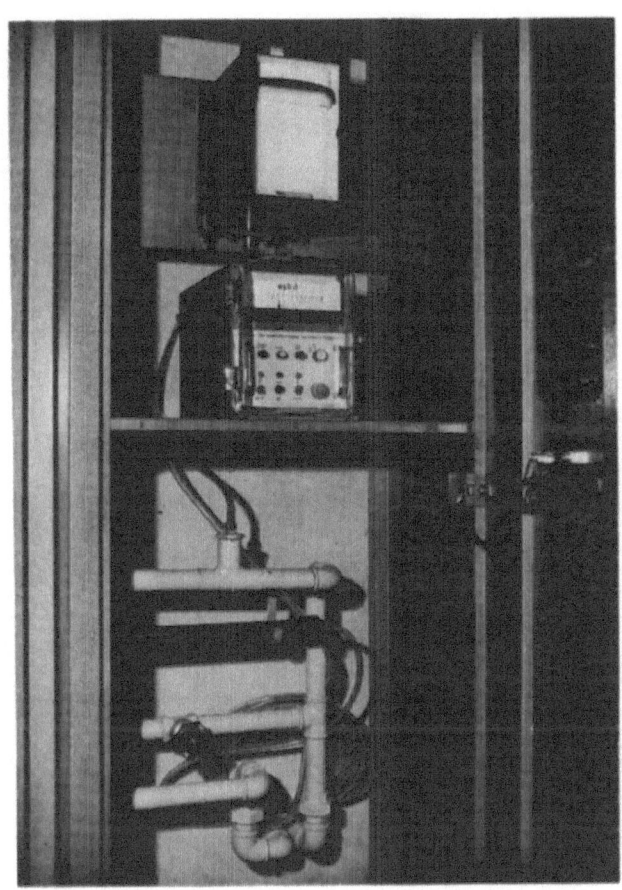

Fig. 10 : View of the Oxygen Meter

TABLE 1 : Operation Times and Relative Costs of Aeration Systems at the Lower Ruhr

Plant	Ruhr-km	Days of operation				Energy costs		Total costs
		1970	1971	1972	1973	DM/d	DM/kg O_2	DM/kg O_2
Weir Spillenburg	42.7	365	365	366	365	—	—	—
Compressed air aeration	32.9	15	27	10	—	377	0.56	1.00
Floating mechenical aerator	30.7	15	35	10	22	147	0.13	0.50
Turbine aeration Baldeney	29.3	10	43	13	33	241	0.06	0.15
Weir Baldeney	29.3	—	10	—	—	2,180*	0.16*	0.16
Weir Kettwig	21.6	10	51	21	42	1,460*	0.12*	0.12
Weir Kahlenberg	12.6	7	11	10	35	1,130*	0.14*	0.14
Weir Duisburg	2.7	365	365	366	365	—	—	—

*) Calculated for 20 m^3/s of river flow and 50 % average D.O. deficit. Since only 75 % of the theoretical energy of the water has to be paid to the owners of hydro power stations, in the case of weir aeration energy costs are very low.

Results

In *Fig. 11* the average daily oxygen values and related data are plotted for two weeks of May and June 1971.

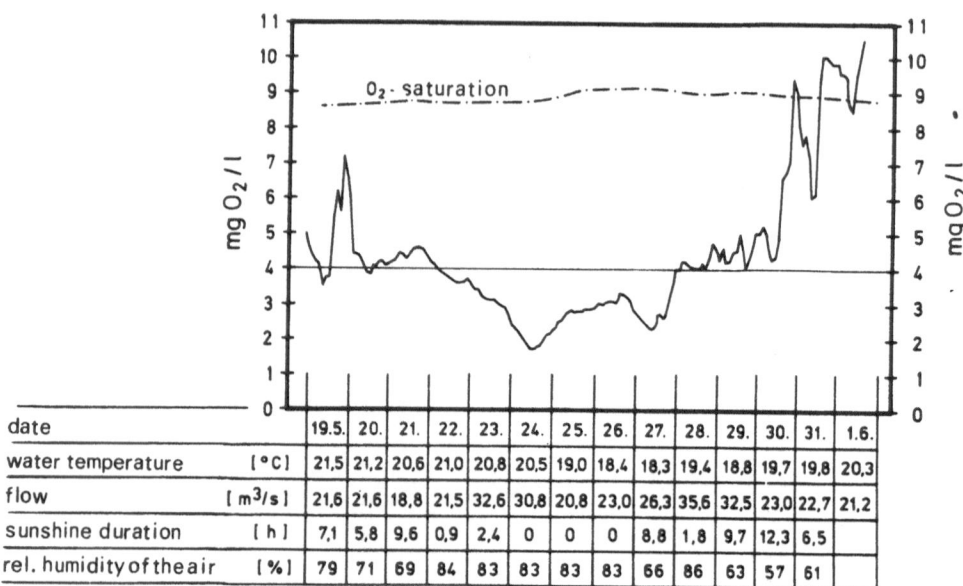

date		19.5.	20.	21.	22.	23.	24.	25.	26.	27.	28.	29.	30.	31.	1.6.
water temperature	[°C]	21,5	21,2	20,6	21,0	20,8	20,5	19,0	18,4	18,3	19,4	18,8	19,7	19,8	20,3
flow	[m³/s]	21,6	21,6	18,8	21,5	32,6	30,8	20,8	23,0	26,3	35,6	32,5	23,0	22,7	21,2
sunshine duration	[h]	7,1	5,8	9,6	0,9	2,4	0	0	0	8,8	1,8	9,7	12,3	6,5	
rel. humidity of the air	[%]	79	71	69	84	83	83	83	83	66	86	63	57	61	

Fig. 11 : Oxygen Contents of the Ruhr
Upstream of Weir Baldeney [14].

This figure clearly shows development, depletion and recovery of the oxygen content. Despite of relatively high run-offs of more than 30 m³/s already in the middle of May low oxygen contents were measured. This was caused by a sudden lack of sunshine and increasing humidity of the air. Due to a layer of clouds there was only little algal activity. After a six days oxygen depression a rapid improvement could be observed on the 27th May, caused by decreasing humidity and increasing duration of sunshine. During the following 2 days photosynthesis developed again and finally a distinct assimilation could be measured. Some longitudinal oxygen profiles are plotted in *Fig. 12*.

On Friday, 21st May, the oxygen meter in the effluent of Lake Baldeney showed a value of 4.5 mg/ℓ *(Fig. 11)*, at a sunny warm weather. Over the weekend thunder-storms developed and a drop in temperature occurred. The sky remained clouded during the following days. Until Monday the 24th the oxygen content

had dropped to 1 mg/ℓ *(Fig. 12)*. All available aeration devices
were immediately taken into operation. Although the oxygen content of the effluent of Lake Baldeney did not show a substantial
increase, satisfactory conditions have been achieved by instream
aeration up to the 26th May. Fortunately no fish kill had hap-
pened, since zones of high oxygen contents were provided in each
impounded section.

In summary it can be stated that by the different
aeration devices of the lower Ruhr it is possible to increase the
oxygen content from 1 to 3.5 g O_2/m^3. But a continuous control
is of decisive importance.

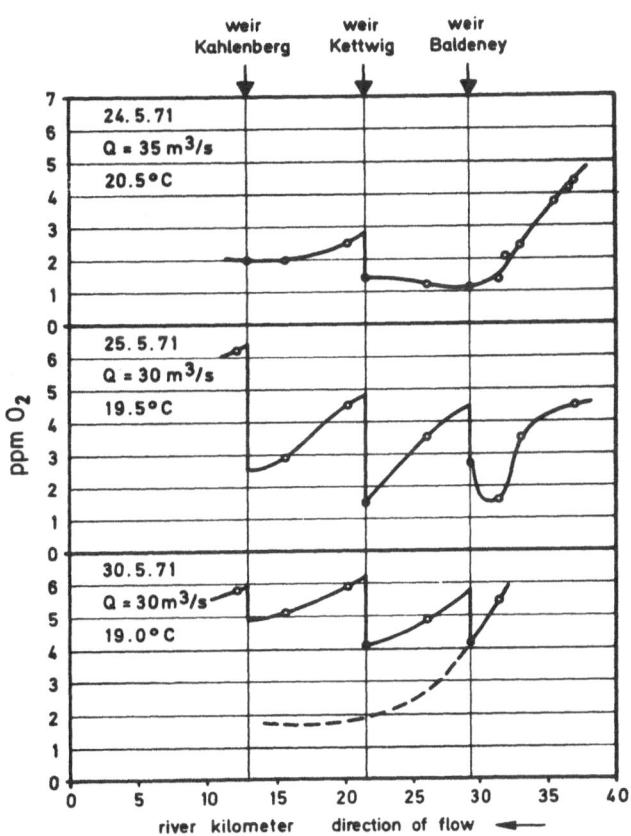

Fig. 12 : Oxygen Profiles of the
Lower Ruhr [14].

3.3 AERATION of a LARGE RIVER [Rhine]

There are no aeration experiences with large rivers, as in the past the self-purification capacity was sufficient for aerobic decomposition of the introduced organic pollutants. With the Rhine this is no longer guaranteed.

3.3.1 Required Aeration Capacity

In order to get a basis for further reflections, the necessary aeration capacity has to be estimated [2, p. 355].

Critical Oxygen Profile

At first a critical oxygen profile with respective runoff and temperature must be assumed. The critical oxygen curve on the one hand shall imply a certain extreme that the corresponding aeration capacity covers many events. On the other hand economic boundaries must be considered. The critical oxygen profile has a similar importance for instream aeration as the selected rainfall for dimensioning a sewerage system. The expediency of the choice must be proved in detail by frequency investigations. At the Rhine, a critical oxygen profile of 0.5 g O_2/m^3 for $1,000\,m^3/s$ and $22°c$ was chosen. By application of instream aeration the oxygen content shall not fall below 3 g O_2/m^3.

The necessary aeration capacity can be distributed to many small devices and thus may be regarded uniform. But it is also possible to concentrate great aeration capacities to a limited flow section. The calculation of the distributed aeration is carried out at first.

Distributed Aerations

The critical oxygen profile is the equilibrium of oxygen consuming and oxygen increasing processes. In the considered range of concentration the oxygen consumption is not changed by instream aeration. Therefore the artificial oxygenations can be compared to the loss of natural reaeration which results from the D.O. increase of the water from 0.5 to 3.0 g O_2/m^3 and a corresponding decrease of the oxygen deficit. Natural reaeration is depending on velocity of flow, water depth and surface area. Assuming a medium width of water surface of $bm= 300$ m, a depth $tm= 3.00$ m and a velocity of flow $vm= 1.05$ m/s for the Rhine with $Q= 1,000$ m^3/s, the natural reaeration rate for 100% deficit amounts to $w= 18$ g $O_2/m^2 \cdot d$ *(Fig. 13)*.

At $22°c$ 0.5 g O_2/m^3 correspond to a deficit of 94% and 3.0 g O_2/m^3 to a deficit of 65%. With artificial aeration

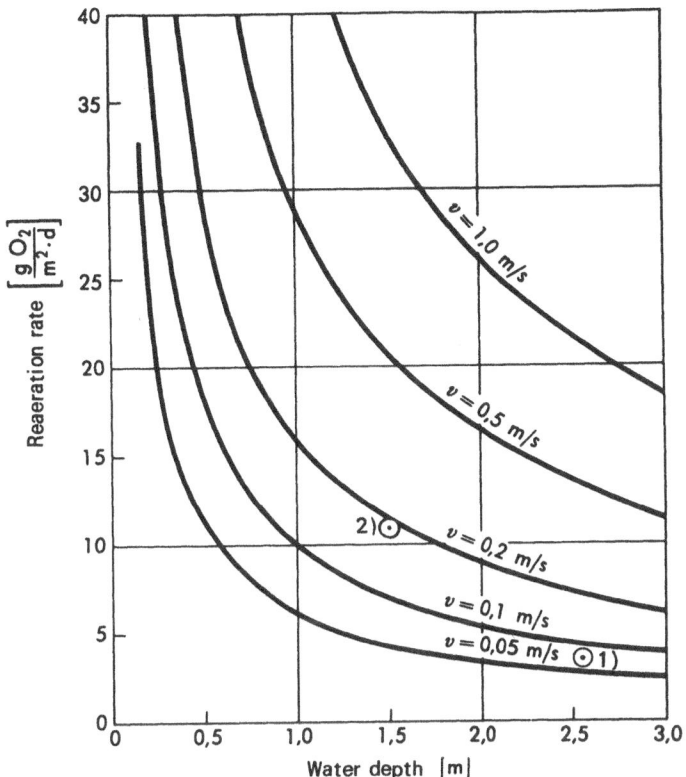

Fig. 13 : Reaeration Through the Water
           Surface depending on Velocity
           of Flow and Water depth at
           22°C and 100% deficit   [2]

the deficit decreased by 95-65 = 30 %. The loss of natural reaeration tion amounts to   18•0.29 = 5.2 g O_2/m^2•d. The necessary artificial cial oxygen input is

$$\frac{5.2 \text{ g } O_2 \cdot 300 \text{ m}^2 \cdot 1000 \text{ m} \cdot \text{kg} \cdot \text{d}}{\text{m}^2 \cdot \text{d} \cdot \text{m} \cdot \text{km} \cdot 1000 \text{ g} \cdot 24 \text{ h}} = \frac{65 \text{ kg } O_2}{\text{km} \cdot \text{h}}$$

Calculation of the Effect of
Locally Concentrated Aeration
Devices in Flowing Waters

The calculation is preferably based on a measured critical oxygen profile, which shall be improved by one or by several aeration installations.

The D.O. increase at the aeration point equals:

$$\Delta\text{Co} = \frac{B_L \cdot D}{Q \cdot 3.6 \cdot 50} \cdot K \qquad \dots \quad (1)$$

ΔCo (g/m^3)   Oxygen increase at the aeration point.

B_L (kW)   Spent aeration energy. For weirs: 9.81 x Flow (m^3/s) x head (m)

D (%)   Average oxygen deficit

Q (m^3/s)   Water discharge

K (kg O$_2$/kWh)   Coefficient

The coefficient describes how much oxygen is added per kWh with 50% average oxygen deficit (of upstream and downstream values) and 20°c water temperature. If detailed measurements cannot be made, the following coefficients can be assumed:

Cascade	1.5	(kg O$_2$/kWh)
Weir with vertical overflow	0.6	(kg O$_2$/kWh)
Weir with inclined overflow	0.4	(kg O$_2$/kWh)
Vented turbine	1.0	(kg O$_2$/kWh)
Surface aerators	0.5	(kg O$_2$/kWh)
Compressed air	0,4	(kg O$_2$/kWh)

The new oxygen profile can be calculated out of the measured critical oxygen profile in finite steps. For this purpose the measured curve is substituted by straight lines (Fig. 14).

After B_L has been chosen, Δ Co can be calculated by equation (1). a' is a+ ΔCo, b' can be calculated by the following concept:

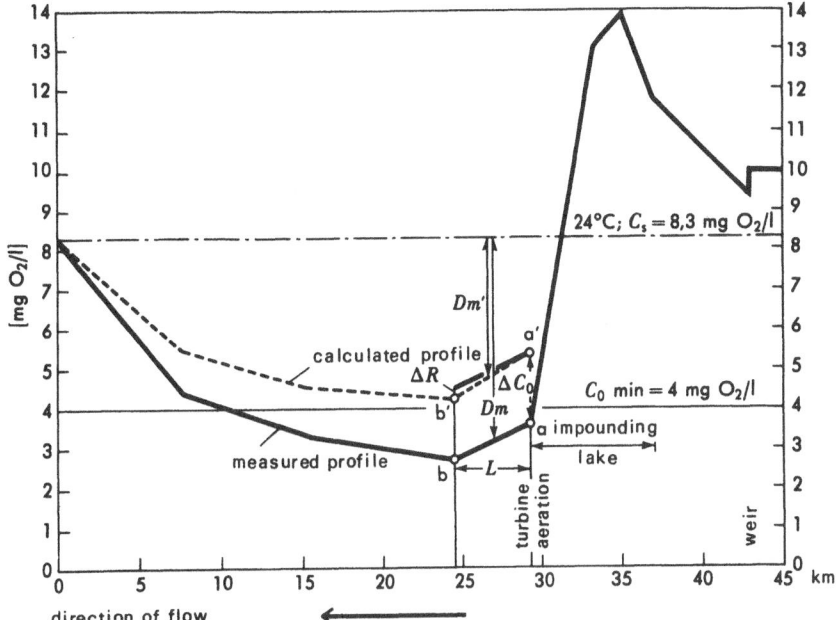

Fig. 14 : Calculation of the
Aerated Oxygen Profile
[2].

$$b' = b + \Delta\ Co - \Delta R \qquad\qquad \cdots\cdots\ (2)$$

Flow, water temperature, respiration, oxygen consumption of bottom sludge and the influence of algae shall be the same for both profiles. Their different slope then only results from natural reaeration.

Because of the smaller average deficit Dm' there is a loss in natural reaeration. For this reason the aerated profile has an inclination to the measured profile of Δ R over the distance L.

$$\Delta\ R = (Dm - Dm')\ \frac{b_m \cdot L \cdot w}{C_s \cdot Q \cdot 24 \cdot 3600} \qquad \cdots\cdots\ (3)$$

Δ R (g/m^3)   loss of natural reaeration

$Dm = C_s - \dfrac{a + b}{2}$ average oxygen deficit of measured profile (g/m^3)

$$Dm' = C_s - \frac{a' + b'}{2} \qquad \text{average oxygen deficit of aerated profile} \quad (g/m^3)$$

bm (m)   average width of river section

L (m)   length of river section

w (g $O_2/m^2 \cdot d$)   natural reaeration rate at 100% deficit

C_s (g O_2/m^3)   Saturation value

Q (m^3/s)   Flow

Substituting the above parameters in equation (3) and (2) point b' can be expressed:

$$b' = \frac{C_s \cdot Q \cdot 48 \cdot 3600 (b + \Delta Co) + bm \cdot L \cdot w (b - \Delta Co)}{C_s \cdot Q \cdot 48 \cdot 3600 + bm \cdot L \cdot w} \qquad \ldots \ldots \quad (4)$$

For the River Rhine the following data have been chosen;

C_s = 8.53 (g O_2/m^3)   Saturation value

Q = 1000 (m^3/s)   Flow

bm = 300 (m)   Average width

L = 10000 (m)   Length of section

w = 18 (g $O_2/m^2 \cdot d$)   Reaeration rate

b = 0.5 (g O_2/m^3)

ΔCo = 3.0 (g O_2/m^3)

Equation (4) results in

$$b' \text{ km } 655 = 3.30 \text{ (g } O_2/m^3)$$

At km 655 b equals 0.5 g O_2/m^3 and Δ Co= 3.30-0.50 = 2.80 (g O_2/n^3), By repeated application additional points are obtained according to Table 2.

TABLE 2 :   Coordinates of the Stepwise Calculated Oxygen Profile

River - km	655	665	675	685		695
b' $(g\ O_2/m^3)$	3.30	3,12	2.94	2.77		2.61
River - km	705	715	725	735		745
b' $(g\ O_2/m^3)$	2.46	2.33	2.21	2.09		1.98
River - km	755	775	795	815	835	855
b' $(g\ O_2/m^3)$	1.88	1.69	1.53	1.39	1.26	1.15

Fig. 15 shows a diagram. The graphic interpola-
tion indicates that after 27 km an oxygen content of 3 g/m³ is
achieved. Here again aeration is necessary with

$$OC = \frac{0.5\ g\ O_2 \cdot 1000\ m^3 \cdot 3600\ s \cdot kg}{m^3 \cdot s \cdot h \cdot 1000\ g} = \frac{1800\ kg\ O_2}{h}$$

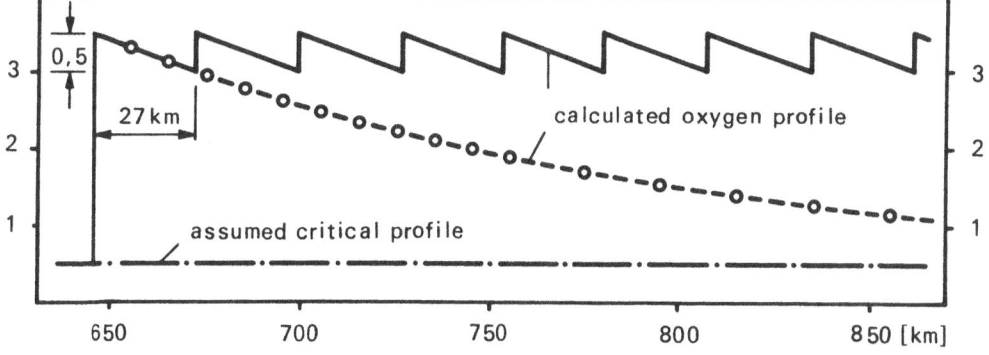

Fig. 15 :   Oxygen-Profile of the River
Rhine with Locally Concentrated
Aeration Devices

450

If it is assumed that the Rhine, due to above aerations, passes km 645 with 3 g O_2/m^3, in total 220: 27 = 8.2 aeration points must be selected over a river length of 865 - 645 = 220 km. The range can be extended, e.g., to 58 km, if a minimum oxygen content of 2.5 g O_2/m^3 is permitted. Even after 220 km the falling oxygen line has not yet reached the respective critical oxygen profile.

The difference of calculation results between the distributed aeration with 65 kg $O_2/km \cdot h$ and for the locally concentration with 1,800: 27 = 67 kg $O_2/km \cdot h$ is not yet important. The latter however requires the greater oxygenation capacity the greater the distance between the aerated sections becomes.

3.3.2 Discussion of Different Aeration Processes

As the Rhine is not regulated by dams, two promising aeration processes, weir aeration and turbine aeration, cannot be applied. Due to great required amounts of oxygen also the injection of compressed air is not advisable, as its distribution by tubes or hoses would be very expensive. Navigation experts, besides, suspect that the transition from aerated to unaerated water may cause shocks by which the gear of the ship could be damaged [21]. Therefore the reflections are restricted to aeration with ships, with surface aerators and with pure oxygen.

Ships

Already the 1964 report of the ministry of agriculture of Northrhine-Westfalia states that during the dry summers 1959 and 1961 a depletion of the oxygen content to a total fish kill was prevented by the additional aeration caused by the movement of ship propellers. This effect could be improved systematically.

Therefore it is very useful that the *Versuchsanstalt für Binnenschiffbau* in Duisburg started systematically investigations [21]. According to the reflections of Heuser, compressed air shall be injected behind the propeller in the range of the propeller tunnel.

Some proof for the expediency of the procedure can be derived from the turbine aeration according to von Wolff [paragraph 3.2.2] and the experiences with Turbo-Oxyders [paragraph, 3.1.2]. Thus for a maximum aeration effect the ratio of the amount of relaxed air and the transported water quantity might range between 1:5 and 1:10. The water flow of ships according to [21] amounts to 10 - 20 m^3/s. Therefore a good distribution of

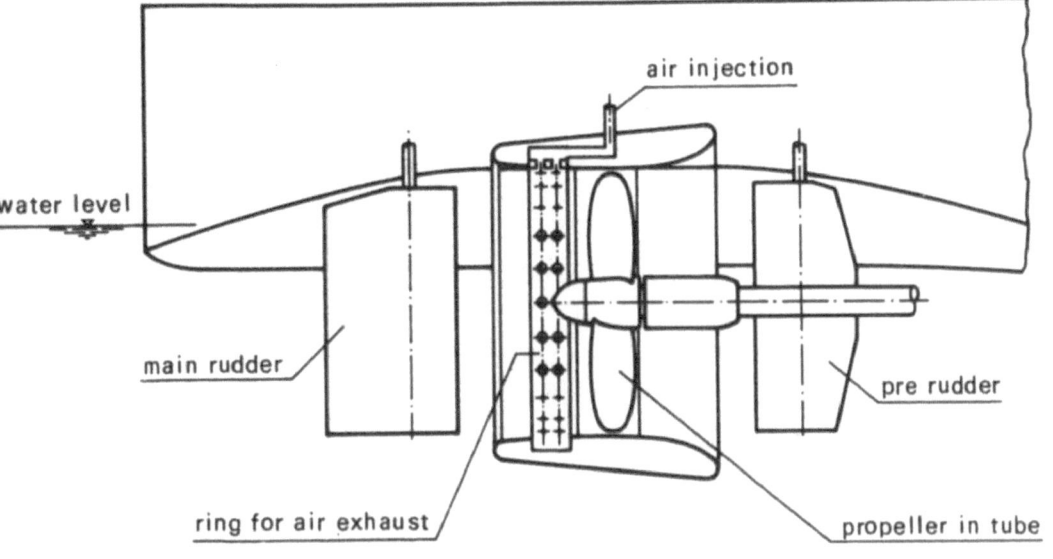

air injection

water level

main rudder

ring for air exhaust

pre rudder

propeller in tube

Fig. 16 : Aeration of the Ship
Propeller

the atmospheric oxygen is guaranteed. As for turbine aeration, also here the existing devices are utilized and only a few additional installations are necessary. Therefore the procedure is very economical.

Floating Surface Aerators

Floating surface aerators have already been mentioned as aeration process for stagnant surface waters and small rivers. The application on the Rhine would require especially great units with favourable hydraulic shape.

Holste proposed an *aeration ship* with rotating cylinders installed at the longitudinal sides. Transferring this reflection to a 60 m long Rhine ship, prepared for aeration purposes, about 80 running m of Mammoth Rotor can be installed at both longitudinal sides. About 6 kg O_2 per hour are transferred by 1 m rotor of $\emptyset$ 1 m. Corresponding to 6x0.65 = 3.9 kg O_2/m·h at 65% oxygen deficit of the river. Then the ship would have an oxygenation capacity of

$$\frac{3.9 \text{ kg O}_2}{\text{m} \cdot \text{h}} \cdot .80 \text{ m} = 312 \text{ kg O}_2/\text{h}.$$

According to the calculation in paragraph 3.3.1 this is sufficient for additional oxygen supply of a river section with a length of

$$(312 \text{ kg O}_2/\text{h}): (65 \text{ kg O}_2/\text{km} \cdot \text{h}) = 4.8 \text{ km}$$

For the drive power of the rotors about 350 kW must be installed. The aeration ships if possible should be anchored at the downstream side of bridge-piles. The drive by diesel motors, the storage of oil, the marking by signals, and the maintenance in general should not render difficulties.

In paragraph 3.2.3 the *floating surface aerator* of Thames Board Mills Ltd. used for the aeration of Thames Estuary is mentioned [18]. The unit transfers 110 kg O_2/h at 65% deficit. It would be sufficient for the oxygen supply of a (110 kg O_2/h): (65 kg O_2/km·h)= 1.7 km long Rhine section. With the same specific efficiency values, the 220 kW floating aerator would transfer 250 kg O_2/h, corresponding to 162 kg O_2/h at 65% deficit. This oxygen amount is sufficient for a Rhine section of 162:65=2.5 km.

Aeration with Pure Oxygen

- Fundamentals

The partial pressure of pure oxygen is 4.75 times higher than the partial pressure of the air oxygen. This results in a better solubility. By sufficient addition of pure oxygen to the water, oxygen contents of more than 40 g O_2/m^3 can be achieved.

- Procedure

There are three major possibilities for the addition of pure oxygen to water: Addition of oxygen gas to a partial stream of water, direct injection into the water and addition at a turbine.

In Japan the *addition of oxygen gas to a partial stream* was tested. By a pumping station a portion of the water which shall be aerated is abstracted and enriched with pure oxygen by means of an ejector. The ejector is a submerged stirrer of high velocity with a hollow shaft by which oxygen cavitates into the water. In a subsequent tank, water and non-utilized oxygen are separated. The water enriched with oxygen flows back into the river and is mixed with the untreated stream. The separated oxygen is discharged back to the ejector.

By the described procedure good results could be achieved with 10 kg oxygen input in a volume of 150 m^3 of water. On the one hand the oxygen concentration of the oxygenated partial stream with 23 g O_2/m^3 was high enough to enrich a large water body. On the other hand the oxygen utilization with 32% was still tolerable. These results are related to an oxygen/water ratio of 1:21.4.

As mentioned pure oxygen may also be directly added by pipes or hoses. American investigations at Pearl River have shown that with conventional distributors an oxygen utilization of maximal 22% could be achieved [26]. The most favourable oxygen yield was 0.23 kg O_2/kWh.

The paper of Cohen gives a promising hint in respect to the economical application of pure oxygen in open tanks [27]. Based on the observation that very small oxygen bubbles dissolve after a short rise, aerators were developed producing gas bubbles with diameters of less than 0.2 mm. By hydraulic shear it could be prevented that single bubbles unify above the aeration tubes. The oxygen was utilized up to 72%. Operational experiences with a large-scale plant are not yet reported.

As a further procedure *pure oxygen addition in a turbine* was tested [26]. In a Pelton Turbine, utilization rates between 33 and 62% were obtained. The low value applies to a volume of oxygen gas of 7.8 Nm^3/min, the high utilization to an addition of 0.28 Nm^3/min. Again it turned out that with additions of small oxygen volumes high utilization rates could be achieved at a low increase of dissolved oxygen concentration.

3.3.3 Costs

In paragraph 3.2.5 energy costs and total costs of the aeration plants at the lower Ruhr are mentioned. Table 3 gives a more detailed survey. It was developed from Table 4 of reference [3] and completed with additional data.

According to the given data and experiences it is difficult to estimate the costs for artificial aeration of the Rhine River. Nevertheless a rough indication shall be made. The costs of the large 220 kW surface aerator of Thames Board Mills Ltd. serve as basis [28]:

delivery	350,000.--	DM
anchoring	35,000.--	DM
energy supply	100,000.--	DM
T O T A L	485,000.--	DM

TABLE 3: Relative Costs of Different Aeration Systems (year 1972).

Aeration System	Oxygenation capacity [kg O$_2$/hr]^x	Capital costs [DM]	Relative capital costs [DM/kgO$_2$/hr]^x	Energy consumption [kWh/kg O$_2$]^x	Total costs [DM/kg O$_2$]^{xx}
Aeration Bridge, Lippe	30	63 000	2 100	1.62	0.45
Hoses with Orifices, Lippe	22	171 000	7 800	3.45	1.43
Hoses with Orifices, Baldeney	28	102 000	3 640	1.45	0.64
Wahnbach Aeration Device	23	96 000	4 200	1.20	0.70
Turbine Aeration Poppenweiler	115	10 300	90	1.41	0.15
Turbine Aeration Baldeney	120	49 000	410	0.98	0.16
Floating Aerator Baldeney	41	147 000	3 600	1.52	0.65
Turbo-Oxyder Ruhrverband	13	64 000	4 900	1.90	0.87
Floating Aerator Thames Estuary	125	485 000	3 900	1.75	0.73

^{x} With 50 % D.O. deficit.

^{xx} With 50 % D.O. deficit, operating 30 days per year, 10 % interest rate and 0.10 DM/kWh.

As according to paragraph 3.3.2   one unit would be sufficient for 2.5 km of the river, the State of Northrhine-Westfalia had to invest for its river section of 223 km

$$\frac{485\ 000 \cdot 223}{2.5} = 43.5 \text{ Mil DM}\quad (1972 \text{ cost level})$$

With operation of 100 days per year and 0.10 DM/kWh the energy costs amount to

$$\frac{220 \text{ kWh} \cdot 0.10 \text{ DM} \cdot 24 \text{ h} \cdot 100 \text{ d} \cdot 223 \text{ km}}{\text{h} \cdot \text{kWh} \cdot \text{d} \cdot \text{a} \cdot 2.5 \text{ km}} = 4.7 \text{ Mil DM/a.}$$

Including interest and amortization the total cost equals 0.26 DM/kg O_2 without service and maintenance.

4. SUMMARY

The most promising instream aeration processes have been described. Experience has shown, that within certain limits of overloading, stagnant surface waters and small rivers can be aerated effectively. Two methods of calculation and specific cost data are presented. At the Ruhr and other places instream aeration has become an indispensable process for water pollution control.

REFERENCES

1. Wagner, H., Die künstliche Belüftung kanalisierter Flüsse. *Besondere Mitteilungen zum Deutschen Gewässerkundlichen Jarhbuch Nr. 15,* Koblenz 1956.

2. Imhoff. K. und K,R., *Taschenbuch der Stadtentwässerung,* 23. Auflage, Verlag R. Oldenbourg, München 1972.

3. KfK, ATV, DVGW, Die künstliche Belüftung von Oberfläckengewässern, *Arbeitsblatt AW 161,* ZfGW – Verlag, Frankfurt 1971.

4. Symons, J.M., Water Quality Behavior in Reservoirs, U.S. Department of Health, Education and Welfare, *Public Health Service Publication No. 1930,* Washington 1969.

5. Imhoff, K. and Fair, G.M., Sewage Treatment, John Wiley and Sons, New York 1940.

6. Bernhardt, H., Belüftung stehender Gewässer, aufgezeigt am Beispiel der Wahnbachtalsperre. *Gewässerschutz-Wasser-Abwasser,* Aachen 1968, S. 129. Bd. 1.

7. Bernhardt, H., Aeration of Wahnbach Reservoir without changing the temperature profile, JAWWA 1967, 943.

8. Krolewski, H., Die Flussbelüftung der VEW in der Lippe. *Energie und Technik 1966,* 247.

9. Imhoff, K.R., Sauerstoffhaushalt und künstliche Wiederbelüftung im Bereich des Baldeneysees und der unteren Ruhr, *GWF,* 936, 1968.

10. Londong, D., Flusswasserbelüftungen an der Lippe. *Gewässerschutz-Wasser-Abwesser (GWA),* Bd. 1, S. 105, Aachen 1968.

11. Stehr, E., Geräte gegen Ölverschmutzungen, Pressluftsperre. *GWF,* S. 822, 1961.

12. Imhoff, K.R., Künstliche Belüftung für den Baldeneysee und die untere Ruhr, *GWA,* Bd. 1, S. 88, Aachen 1968.

13. Muskat, J., Abwasserteichbelüftung, Technik auf neuer Grund-
    lage, *Abwassertechnik 1969*, I.

14. Albrecht, D. and Imhoff, K.R., Erfahrungen mit der künstlichen
    Ruhrbelüftung, *GWF*, H. 3, 1973.

15. Eckoldt, M., Die künstliche Flusswasserbelüftung nach R. von
    Wolff, Ergebnisse der Versuche am Saarkraftwerk Mettlach, *DGM
    1966*, 1.; *GWF*, 1034, 1966.

16. Knop, E., Künstliche Belüftung von Flüssen und Klärteichen.
    *Technisch-Wissenschaftliche Mitteilungen der Emschergenossen-
    schaft*, H. 5, S. 5, 1962.

17. Bohnke, B., Möglichkeiten der künstlichen Fluss- und Gewässer-
    belüftung, aufgezeigt am Beispiel der Lippe, *Münchener Beiträ-
    ge*, Bd. 12, 318.

18. Anonymous,Thames aerator lessens board mill pollution, *Surve-
    yor*, 11 February 1972.

19. Albrecht, D., Belüftung des Ruhrwassers am Wehr Spillenburg.
    Die Wasserwirtschaft, Heft 11, 1968.

20. RP Düsseldorf, Schreiben an alle Oberkreisdirektoren und Ober-
    stadtdirektoren vom 22.9.71. 64.II.500.

21. Heuser, H.H., Mündliche Mitteilung anlässlich eines Besuches
    der Versuchsanstal für Binnenschiffbau in Duisburg am 21.8.
    1972.

22. RWE, Studie über die thermische Belastbarkeit des Rheins. Mo-
    tor-Columbus Ingenieur-Unternehmung AG, Dezember 1971.

23. Holste, D., Einrichtung und Verfahren zum Belüften von Wasser-
    läufen. Patent angem. A.Z.P 1654 069.5/84 a.

24. Reimann, H., Einsatz von Sauerstoff in Belebtschlammverfahren.
    Referat vor dem ATV Fachausschuss 2.6, Werkgruppe München der
    Linde AG, Juni 1971.

25. Tokyo Metropolitan Government: Tokyo Fights Pollution, *TMG
    Municipal Library No. 4*, 1971.

26. Amberg, H.R., Stream Reaeration Using Molecular Oxygen. 4. In-
    ter. Abwasserkonferenz Prag 1968, 1 - 17, *Pergamon Press*,
    London.

27. Cohen, D.B., A low cost open tank pure oxygen system for high
    rate total oxidation, 6. Intern. Abwasser-konferenz Jerusalem
    1972, Hall B - Paper No. 9, *Pergamon Press*, London.

28. O'Neill, O.C., Letter to the author dated 29.8.1972.

29. Tsivoglou and Wallace, Hydraulic Properties Related to Stream
    Reaeration, Symposium on the Use of Isotops in Hydrology,
    Wien, März 1970.

30. Speece, R.E. and Malina, J.F., Application of commercial oxygen to water and wastewater systems, Center for Research in Water Resources, University of Texas, Austin 1973.

31. Bernhardt, H., Ten years experience of reservoir aeration, Progress in Water Technology, Vol. 7, Nos. 3/4, pp. 483-495, *Pergamon Press,* 1975.

PHYSICAL AND MATHEMATICAL MODELLING FOR WASTE-WATER OUTFALLS DESIGN

George J. Balafoutas and
Themistocles S. Xanthopoulos

Hydraulics Department, School of Technology,
Aristotelian University of Thessaloniki, Greece

INTRODUCTION to SEWAGE DISPOSAL in a MARINE ENVIRONMENT

There is a cycle of water which is used by human communities
for their urban, industrial and agricultural needs: the human community draws water from the natural water environment and gives
back in the form of waste-water. Improvement of physical, biological and chemical properties of water on the causality 'b' was always considered as quite evident, though similar care in the opposite direction 'c' seemed not so obviously necessary   (See *Fig. 1*).

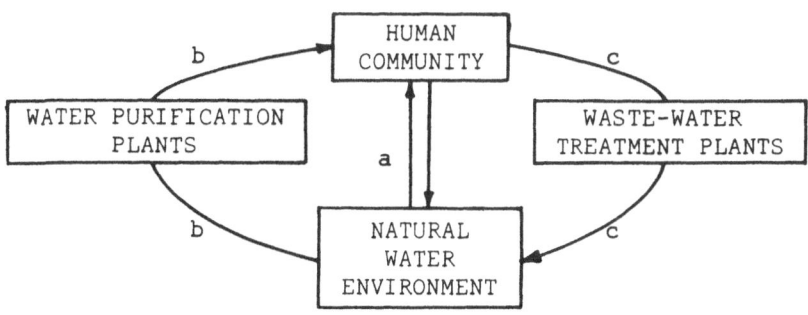

Fig. 1 : The Cycle of the Used Water

In a water (say marine) environment, where a certain self-
contained ecosystem is living, a physical equilibrium is maintained,
as far as the requirement in dissolved oxygen of each species is

fulfilled. A limited amount of organic sewage matter can be assi-
milated by this environment, if it doesn't exceed its capacity.
Therefore, the water body itself constitutes a quite powerful 'natural treatment plant', which can improve, complete or even in
some limited cases replace the manmade treatment plants; both of
them have to maintain the required quality in the causality 'b'
and in the interrelation 'a'.

Waste-water disposal into the sea is usually the best alternative for coastal communities. Sewage effluent treated preliminar-
ly by sedimentation or even biologically - and only in some limited
cases untreated - is transported through long submarine pipelines,
roughly perpendicular to the shoreline, several kilometers (5 to
10) offshore, up to depths of 50 to 100 meters and is dumped into
the sea water, through multiport diffusers. A simple diffuser system consists of a manifold (i.e. a pipe with many small holes)
which distributes the waste-water over a large distance. "Typically, the overall length of a diffuser is an order of magnitude larger than the depth, which in turn is an order of magnitude larger
than the pipe diameter and port spacing, which again are an order
of magnitude larger than the typical port diameters" (Koh & Brooks,
1975) [1].

Besides the problem concerning the hydraulic computation of the
long pipeline and the multiport diffuser [2, 1], the design of a
sea outfall is mainly based on a good knowledge of the turbulent
mixing process of the sewage effluent with the sea water.

In this paper, after a critical description of the sequential
phases of this mixing process, physical and mathematical modelling
techniques of this phenomenon are analyzed and some typical examples
are mentioned. Because of the scope of this presentation a good
part of the material is not new and rather general. For a deeper
insight into more specific subjects and extensive bibliography is
presented.

MIXING PROCESS and LOCAL MATHEMATICAL DESCRIPTION

Density currents are defined as movements of a fluid under,
through or over another fluid, being in a gravity field, and the
density of which differs by a small amount from that of the first
fluid. Such density differences may result from differences in
temperature as well as in salt, air and fine solid matter content.

The displacement and spreading of a liquid patch (named for convenience tracer) in the body of a different liquid (named for convenience ambient medium) are governed by a rather complicated momentum, mass and energy transport process. Thus, e.g. in wastewater disposal phenomenon physical and biochemical properties of

ambient medium and tracer influence the mixing process and constitute the main factors which fix the length of the outfalls and the
design of the diffuser system.

The 'mixing process' block in the relation diagram of *Fig. 2*

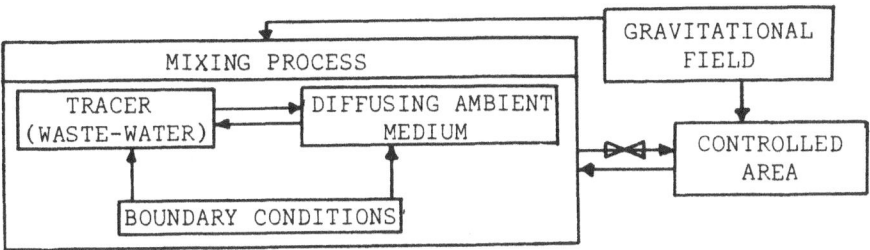

Fig. 2 :   Relation Diagram of a
              Waste-Water Disposal

could be characterized as a black box model, which creates a controlled causality for any affecting influence from the tracer on
the controlled area. The reciprocal reverse causality determines
the boundary conditions of the mixing process. Our aim is to de-
tect the internal structure of the mixing process block.

Tracer and ambient medium may be non-missible or missible fluids
and the whole phenomenon isothermal or thermodynamically irrever-
sible. Isothermal phenomena for non-missible fluids are determined
by momentum and buoyancy forces, while for missible fluids additional diffusion effects are required. Temperature difference between the two phases entails a corresponding density difference and
therefore cooling water disposal into the sea is similar to sewage
disposal in this medium. Being limited to sewage effluent we shall
refer, for those who are interested in cooling water outfalls design, to some recent works [3, 4, 5]. As biochemical properties
characterize, sewage effluent, the bacterial decay and organic matter degradation must be taken into consideration.

The mathematical description of density currents phenomena is
based on the Boussinesq approximation which can be summarized by
the statements [6]:

a.   The fluctuation in density which appears with the advent
      of motion results principally from thermal and/or salt con-
      tent effects (as opposed to pressure effects).

b.   In the equations of momentum and mass, density fluctuations
      may be neglected (density may be taken as constant) in all

terms except those describing the body forces field.

As a particular result of this approximation, the condition

$$\frac{\Delta\rho}{\rho_0} \ll 1 \qquad\qquad\qquad \dots \ (1)$$

is obtained, where ρ_0 = space average of density, and,

$\Delta\rho$ = maximum variation of ρ across the layer where the phenomenon is developed.

After this approximation the following mathematical description of the local mechanical situation in a mixing process results in: Equation is continuity

$$\frac{\partial u}{\partial x} + \frac{\partial v}{\partial y} + \frac{\partial w}{\partial z} = 0 ; \qquad\qquad \dots \ (2)$$

Equations of Motion

$$\frac{\partial u}{\partial t} + u\frac{\partial u}{\partial x} + v\frac{\partial u}{\partial y} + w\frac{\partial u}{\partial z} = -\frac{1}{\rho_0}\frac{\partial p}{\partial x} + \nu\Delta u + \Omega v, \qquad \dots \ (3)$$

$$\frac{\partial v}{\partial t} + u\frac{\partial v}{\partial x} + v\frac{\partial v}{\partial y} + w\frac{\partial v}{\partial z} = -\frac{1}{\rho_0}\frac{\partial p}{\partial y} + \nu\Delta v - \Omega u, \qquad \dots \ (4)$$

$$\frac{\partial w}{\partial t} + u\frac{\partial w}{\partial x} + v\frac{\partial w}{\partial y} + w\frac{\partial w}{\partial z} = -\frac{1}{\rho_0}\frac{\partial p}{\partial z} + \nu\Delta w + g(1+\frac{\rho}{\rho_c}) \ .. \ (5)$$

Equation of Turbulent Diffusion

$$\frac{\partial c}{\partial t} + \frac{\partial (uc)}{\partial x} + \frac{\partial (vc)}{\partial y} + \frac{\partial (wc)}{\partial z}$$

$$= \frac{\partial}{\partial x}(\varepsilon_x\frac{\partial c}{\partial x}) + \frac{\partial}{\partial y}(\varepsilon_y\frac{\partial c}{\partial y}) + \frac{\partial}{\partial z}(\varepsilon_z\frac{\partial c}{\partial z}) - Kc ; \ \dots \ (6)$$

where;

u, v, w = x-, y-, z-component of velocity
p = pressure
g = acceleration of gravity
ν = kinematic viscosity
c = concentration of a diffused component
(salt, dye, organic matter, bacteria, etc.)
K = generalized sink factor
(only for non-persistent components)
K_c = component decay per unit volume and per unit time
due to bio-chemical reactions,
$\Omega = 2\omega \sin\phi$ = Coriolis force coefficient
ϕ = earth latitude
ω = earth rotation velocity
$\varepsilon_x, \varepsilon_y, \varepsilon_z$ = local diffusion coefficients

It has to be emphasized that a mathematical model describing the mechanical situation of an N-component mixture is a superposition of N continuous media, even in the case of a discrete component (bacterial concentration). At any point in space occupied by the mixture we must consider that N material particles - one from each of the continuous media representing individual species —co-exist.

ANALYSIS of the WASTE-WATER MIXING PROCESS

WASTE-WATER DISPOSAL SCHEMATIZATION

In the usual schematization of the sewage mixing phenomenon in a marine environment three zones are distinguished.

In the first zone a buoyant jet or plume rises entraining the ambient medium and mixing with it. This zone is known as near field and the achieved dilution as initial dilution.

In the second zone, which is a transitional field between the two others, the sewage field tends to spread out horizontally.

In the third zone finally, tracer and ambient medium have the same density and so, there is no buoyancy effects, but only turbulent advection and diffusion due to the sea currents and turbulence. This zone is known as far field and the achieved dilution as subsequent dilution.

464

A short description concerning results of the near and transitional field is given further on. These results constitute the initial condition of the far field modelling; on this far field our main interest is focused here.

NEAR FIELD and INITIAL DILUTION

Diffusers are usually laid on the sea bottom, across the most unfavorable current of the site, and they have one row of holes in each site. They discharge the waste-water horizontally in order to achieve higher dilutions with the heavier bottom water and to create longer jet trajectories. However, outfalls during the longer period of their life bring much less flow rate than this one which is taken into consideration for their hydraulic calculation. Therefore, the outcoming jets are rather weak and as Liseth, 1970 [7] has experimentally found their horizontal momentum is quickly cancelled out and a single slot plume is formed. Wakes from strong currents in the downstream side of the diffuser may increase this initial mixing [1]. Cederwall, 1971 [8] has given diagram for various types of possible regimes.

An extensive theoretical and experimental investigation of buoyant jets and plumes has allowed a rather good knowledge of these phenomena [7, 8, 9, 10]. Koh & Brooks, 1975 [1] present as advisable for preliminary design the following formulation given by Brooks, 1972 [10].

Without Blocking A plume in a stably stratified medium reaches up to a certain level, where its buoyancy force is cancelled out, due to the locally existing thermocline. If the medium is motionless and the entrainment of it, throughout the slot rising plume, has a steady character (*Fig. 3*), the following formulae may be used:

Weight of Rise

$$z_{max} = 2.5 \ (q \ g')^{1/3} \ (- \ \frac{g}{\rho_1(z)} \ \frac{d\rho_a(z)}{dz})^{-\frac{1}{2}} \ ; \qquad \qquad \dots \ (7)$$

Centerline Dilution at Top

$$S_t = 0.36 \ (\frac{g'}{q} z)^{1/3} \ z_{max} \ ; \qquad \qquad \dots \ (8)$$

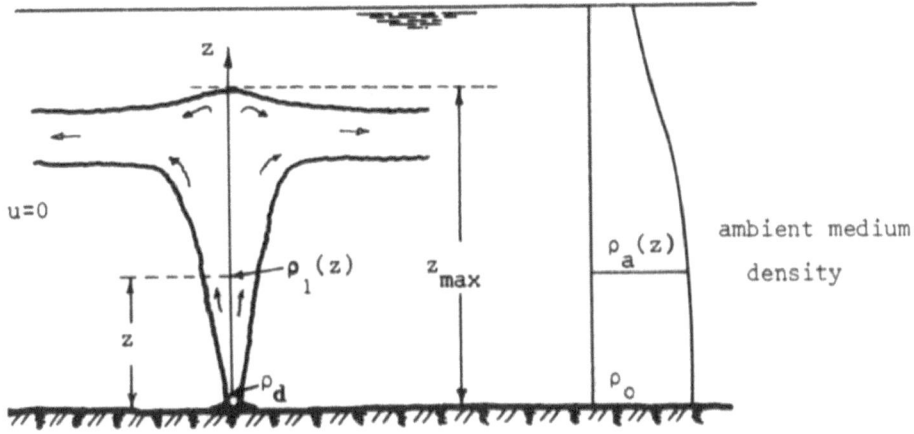

Fig. 3 : Slot Rising Plume in a Stratified
Medium without Blocking Effects [1]

where;

q = discharge per unit length

g' = $g(\rho_0 - \rho_a)/ \rho_0$

ρ_0 = reference density (ambient medium at source level)

ρ_d = density of discharge

ρ_1 = centerline density at the end of the zone of flow establishment

ρ_a = density of the ambient medium

Multiplication of the Eq. 8 by $\sqrt{2}$ gives the average dilution.

With Blocking   In the case that a light current exists in the ambient medium, the mixing must effectively cease for $z > z_b$ (Fig. 4), because of the unfavorable condition for the waste-water spreading.   In this case Eq. 8 overestimates initial dilution, and a corrective factor

$$P = \frac{\sqrt{2}\ q\ S_t}{u\ z_{max}} ,$$ .... (9)

characterizing the blocking effect has to be introduced (u = mean current velocity). The reduced value of the centerline dilution is

466

$$S_{t,b} = \frac{S_t}{1 + P} \qquad\qquad \dots (10)$$

multiplication by $\sqrt{2}$ gives again the corresponding mean value, and the thickness of the waste-water field, under the same assumption is

$$\dots (11)$$

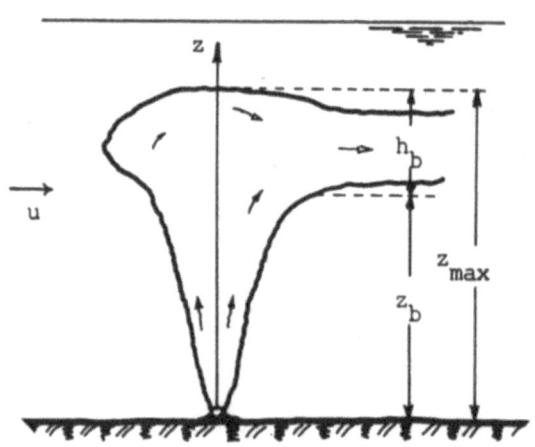

Fig. 4 : Slot Rising Plume with Blocking Effects [1].

TRANSITIONAL FIELD

For the transitional zone, some experimental results of a rather limited interest for our case, are available. Therefore, it is supposed that, for high initial dilutions (say 1:100), the rather limited influence of this zone is included to the near field results. Nevertheless, a more detailed approach is required for lower initial dilutions (say 1:10).

FAR FIELD and SUBSEQUENT DILUTION

In this phase of the mixing process we use as initial conditions the results of the two previous phases and as boundary conditions those imposed by the site in view; the circulation and stability structure of the ambient medium determine the transport process.

Our aim is to calculate the time interval required for the waste-water field to reach the controlled area, and the dilution and/or biochemical decay, which correspond to this interval.

The width of the waste-water field at the area where the initial dilution is given almost coincides with the length of the diffuser.

Further on we are limited to the far field investigation, through physical and mathematical models, and only some remarks concerning the near field investigation are given.

MODELLING and IN SITU INVESTIGATION

There are three main phases in any model technique:

(a) Schematization of the physical phenomenon, measurements in situ of the determining factors and their transfer to the model. Any physical constraint, starting from the most effectual, has to be taken into consideration.

(b) Exploitation of the model, i.e. for a physical model, build up and manipulation of the model, while for a mathematical model development and application of a computer program. It is advisable that any model have a general character with many facilities concerning input-output variability. Thus, it is better to construct a hydraulic basin, instead of an individual hydraulic model, or to develop a design system instead of a one-off mathematical model. Moreover the exploitation procedure foreseen has to be as simple and as cheap as possible.

(c) Interpretation of the obtained model results in a reliable solution for the physical phenomenon in view. In this phase restrictions and scale effects of the hydraulic model have to be considered or physical objectivity of numerical solutions has to be proved.

The investigation in situ, which is required before any model exploitation, may contain measurement on the following factors:

(a) General and local circulation of the sea water (tidal, wind shear stress, density and topography currents);

(b) Vertical gradient of salinity and temperature (these factors characterize the ambient medium stability);

(c) Wave characteristics;

(d) Biochemical characteristics (Measurement techniques and the significance of these factors have been analyzed in [11,12,13]).

468

This investigation in situ will have a duration of, at least, one year and from the combination of its results may be obtained typical circulation and stability structures of the ambient medium. Our study, through the model, has to be performed for all these typical structures.

Although in both cases investigation in situ is necessary, there is a difference in the planning of our activities. In a hydraulic model the results of this investigation belong to the design criteria of this model; so we need a large number of them before starting any design procedure. On the contrary, the development of a mathematical design system is completely independent of the above investigation, which may be realized at the same time with this development.

PHYSICAL MODELS

DIFFUSION MEASUREMENTS IN SITU

Besides hydraulic models, natural environment itself may be used for diffusion measurements of some detectable tracer release (dye, cooling water, radioactive material, floating particles). Okubo, 1971 [14] has given a survey of such experimental data.

Although this technique has already been used many times, we cannot characterize it as a physical model technique, but only as a technique for individual measurements campaign, into the prototype ambient medium, because,

a. We cannot repeat the same experiments changing input-output data as we wish;

b. We need expensive and rather complicated measuring equipment (aerophotography, tracer measurement in situ etc);

c. Such experiments are usually undesirable to local communities; etc., etc.

Such measurements can mainly been used for estimation of diffusion coefficients by fitting a prescribed mathematical model to experimental data. Usually this fitting is based on the Einstein formula [15]

$$\varepsilon = \frac{1}{2} \frac{\partial \sigma^2}{\partial t} \qquad \qquad \dots . \ (12)$$

where;

ε = diffusion coefficient,
σ = standard deviation of the concentration c,
t = time

Practically a typical length b (size of a dye patch or distance
between two floating particles) is measured in successive time intervals and an empirical relation is accepted between b and σ, e.g.
that which Cannerson, 1968 [16] has accepted for his experiments

$$b = 4\sigma ,$$ .... (13)

which corresponds, for a normal distribution, to the 95.5% of the
initial dye release.

HYDRAULIC MODELS

Hydraulic models are scale representations of hydraulic phenomena.

The scale of a parameter is defined as ratio between the prototype value and the model value of this parameter [17]. In order
to transfer a physical phenomenon in a hydraulic model the scale
relations have to be investigated and in this investigation dimen-
sional analysis is a powerful tool [18]. Scale relations may be
obtained by keeping in the model constant the value of some dimensionless parameters, which characterize the physical phenomenon.
These relations are known as scale laws or scale conditions [17].

We must point out that an absolute similarity between nature and
model is possible only in some limited cases; scale effects are
interfered with and for this reason calibration tests have to pre-
cede any model use [17].

The hydraulic models for sewage effluent have many similarities
with hydraulic models for air pollution control [19, 20]. These
models may be distinguished in

a.   Models for large extent phenomena, and
b.   Models for limited extent phenomena.

The first category concerns mainly far field phenomena [21] and is
rather limited because of the induced scale effects. The second
concerns near field phenomena [22] and special local problems, as
diffuser's design [1], etc. Through the second category experimental calculations of the dispersion coefficients may be performed
[23]. A critical discussion of dispersion studies in hydraulic models is given by Fisher & Holley, 1971 [24].

SIMILARITY ARGUMENTS

In order to reproduce the mechanical situation of the ambient
medium in a hydraulic model, Froude, Reynolds, Weber and Rossby con-

conditions, corresponding to the dimensionless products

$$Fr = \frac{u^2}{gH} \ , \quad Re = \frac{uL}{\nu} \ , \quad We = \frac{\rho u^2 L}{\sigma} \ , \quad Ro = \frac{u}{\Omega L} \ , \qquad \dots \ (14)$$

have to be fulfilled (σ= surface tension, H and L = typical vertical and horizontal lengths). Moreover roughness condition [17]
has to be fulfilled too. But, as a matter of fact the simultaneous
fulfillment of all these conditions is impossible, and to overcome
this impasse some simplifications are required.

a. Reynolds condition is not important here as long as the
    model is not to small. Although there are techniques to
    maintain the turbulent character of the flow artificially,
    it is advisable to avoid them because they usually create
    intervals of decreasing turbulence.

b. Weber condition may introduce scale effects due to surface
    tension and arises in the study of capillary waves and of
    breakup of liquid droplets (eg. dye droplets) in the free
    surface. When wave effects are studied this condition imp-
    lies a minimum wave length in the model (say 0.2 m).

c. Rossby condition may be ignored in small extent phenomena,
    but in models of large regions Coriolis forces may in-
    fluence the circulation of the ambient medium and therefore
    the mixing process. The best alternative is to construct
    the model basin on a rotating disc, but this technique be-
    comes very expensive for large models. In this case Corio-
    lis tops [17] may be used, but special care is required for
    secondary effects due to the Magnus effect [25].

After this simplification the scale conditions are specified by
the Froude condition.

In a stratified medium Richardson number

$$Ri = \frac{g \ \frac{d\rho_a}{dz}}{\rho_a \ (\frac{du}{dz})^2} \qquad \dots \ (15)$$

leads to a generalized Froude condition which constitutes a stability criterion for the ambient medium. A critical evaluation of
this criterion is given by Long, 1972 [26].

In density currents with sharp interfaces the stability of the
interface and the buoyancy effects are characterized by a quite

simpler generalization of the Froude condition corresponding to the internal or densimetric Froude number

$$Fr_i = \frac{u^2}{\frac{\Delta\rho}{\rho_a} gH}$$ .... (16)

The fulfilment of a generalized Froude condition, instead of the normal one, permits, up to a certain scale, the simultaneous fulfilment of the Reynolds condition [27].

In order to investigate diffusion phenomena in a model the scale condition which corresponds to the Peclet number

$$Pe = \frac{\varepsilon}{uL}$$ .... (17)

has to be fulfilled. The three-dimensional character of the diffusion phenomenon shows that realistic results concerning the vertical diffusion may be obtained only in undistorted models. However, if we are obliged to use distorted models, reduction factors must be provided. In our case, where the length and velocity scales have already been chosen from the Froude condition, the dimensionless product of the Eq. 17 cannot be correctly scaled, giving model diffusion coefficients 30 to 75 times higher than those corresponding to the prototype. In this case the Schmidt number

$$Sc = \frac{\nu}{\varepsilon}$$ .... (18)

offers a partially corrective alternative.

DESIGN CRITERIA for HYDRAULIC MODELS

Concerning the construction technique of the hydraulic models the following remarks may be pointed out:

a. The scaling depends on the significance of the problem concerned and the facilities of the laboratory involved. De Vries, 1973 [17] has mentioned that routine model studies hardly do exist and it is necessary to start each model with an open mind.

b. For medium extent phenomena a hydraulic basin containing special facilities is a fruitful idea. A peripheral chan-

472

nel, separated in chambers, supplying or evacuating water individually through regulated gates and weirs in the inner wall, may be used for the reproduction of currents. Wind shear stress may be introduced by a series of ventilators. A sophisticated measurements circuit may be provided.

c. Sea bottom topography may be reproduced by vertical aluminium-sheets for isodepth lines, sand for filling the space between and concrete or plaster for covering the top, so as to keep the sand in place. Another technique is the reproduction of the sea bottom by plastic blocks [19, 20].

d. In distorted models roughness condition implies the use of additional roughness elements, such as gravel, concrete cubes and bars. Secondary effects due to these roughness elements have to be detected.

e. Choice criteria for the used liquids are: (1) the dissolubility between them, (2) their price, (3) their easy handling and safe use, (4) the simple measurement of their concentrations, (5) their ability to reproduce sufficient stratification gradient into a model of a small vertical dimension, etc. Thus, the use of the combination of cold-warm water is rather limited and liquids of different densities, such as fresh water and salt, alcohol and sugar solutions, are preferred. Alcohol dilutions in fresh water ambient medium is an expensive procedure with limited detection possibilities (photography). Fresh water in saline ambient medium is a better alternative (salt is not expensive and measurements can be performed through dynamic conductivity probes). In the last case the salinity of the ambient medium has to be checked and corrected during experiments. Many times the combination fresh-fresh water (say simple dye solutions) may be used. Hydrostatic pressure differences may replace the required buoyancy effects.

f. For local studies (near field, transitional spreading in stratified medium, diffuser's design etc) the inverse similarity technique [28, 29, 30] is the most favorable. The model ambient medium is motionless, fresh or slightly saline (in the case of stratification) water; the tracer is a heavier saline water. Local currents may be reproduced by displacement of the tracer source. The turbulence of the ambient medium is of course nullified, but its effects on local phenomena are rarely significant.

MATHEMATICAL MODELS

HYDRODYNAMIC MODEL of the AMBIENT MEDIUM

We have already seen that far field is locally described by the

hydrodynamic model of the ambient medium (Eq. 2 to 5) and a diffusion equation for each diffusing species (Eq. 6). Our basic assumption is that the continuity and momentum equations of the ambient medium are not influenced by the diffusion equation of the
tracer. However, the solution of this diffusion equation requires
a previous hydrodynamic description of the ambient medium, which
may be obtained either through a solution of the above hydrodynamic system or through field data for velocities.

In hydraulic engineering we normally use integrated forms of
the local description (Eq. 2 to 5).

Nearly horizontal flow is a similar schematization, described
[31] by:

Equation of Continuity

$$\frac{\partial h}{\partial t} + \frac{\partial(\bar{u}h)}{\partial x} + \frac{\partial(\bar{v}h)}{\partial y} = 0 \quad ; \qquad\qquad \dots (19)$$

Equation of Motion

$$\frac{\partial \bar{u}}{\partial t} + \bar{u}\frac{\partial \bar{u}}{\partial x} + \bar{v}\frac{\partial \bar{u}}{\partial y} + g\frac{\partial h}{\partial x} = -g\frac{\bar{u}\sqrt{\bar{u}^2+\bar{v}^2}}{C_x^2 h} + \frac{\tau_{wx}}{\rho h} - gi_x + \Omega\bar{v} \quad , \dots (20)$$

$$\frac{\partial \bar{v}}{\partial t} + \bar{u}\frac{\partial \bar{v}}{\partial x} + \bar{v}\frac{\partial \bar{v}}{\partial y} + g\frac{\partial h}{\partial y} = -g\frac{\bar{v}\sqrt{\bar{u}^2+\bar{v}^2}}{C_y^2 h} + \frac{\tau_{wy}}{\rho h} - gi_y - \Omega\bar{u} \quad , \dots (21)$$

where bars refer to the average value over the depth h, i_x,

i_y = local bottom slope
C_x, C_y = local Chezy coefficients
τ_w = ρCW^2 = wind shear stress [32],
ρ_w = air density
W = wind velocity
C = resistance coefficient.

This system may be solved numerically [33] or by the method of characteristics [34].

For a stratified ambient medium a multilayer system [35] may be provided. A solution of the two-dimensional two-layer flow, based on the method of characteristics is given by Balafoutas, 1976 [35].

Eddy Diffusion and Dispersion Coefficients

In the diffusion equation, besides the new variable (the concentration c), some new parameters have also been introduced

a.   the diffusion coefficients ε_x, ε_y, ε_z, and

b.   the generalized sink factor κ, especially for non-persistent species.

Factor κ depends on the local biochemical characteristics of the ambient medium and may be experimentally calculated [12, 13].

The term of eddy diffusion coefficient has been proposed by Taylor in 1915 [36] and it is based on the assumption that the turbulent flux vector of a diffusion admixture is proportional to the gradient of the mean admixture concentration

$$\overline{c'u'_i} = -\varepsilon_{ij} \frac{\partial \bar{c}}{\partial x_j} \qquad \qquad \dots (22)$$

This formula is similar to that proposed by Boussinesq in 1887 for turbulent shear stress. Two other semi-empirical hypotheses of the turbulence theory are also transferred to the mass diffusion phenomena:

a.   Prandtl's theory of mixing length [36]

$$\varepsilon_x = \ell^2 \left| \frac{d\bar{u}}{dz} \right| \qquad \qquad \dots (23)$$

where ℓ = mean dimension of the turbulent fluctuation.

b.   Taylor's theory of vorticity transfer [37]

$$\varepsilon_x = \frac{1}{2} \overline{u'^2} \int_0^\infty R(\tau) \, d\tau \qquad \qquad \dots (24)$$

where;

R(τ) = correlation between the velocity
u of a particle in the direction of
x at one instant, t_o, and the velocity
of the same particle at the time
t_o + τ.

However such formulae have only a theoretical interest, because in hydraulic engineering integrated mathematical descriptions, instead of local ones, are usually used.

In order to integrate in the same manner the diffusion equation (Eq. 6), the term 'dispersion', which refers to a spatial averaging, is introduced. We can have an average over the depth (for unit width), an average over the width (for unit depth) or even an average over a more complicated two-dimensional area (say the cross section of a river). In each case a corresponding dispersion coefficient may be provided.

The dispersion equation, which corresponds to the nearly horizontal flow, is

$$\frac{\partial \bar{c}}{\partial t} + \bar{u} \frac{\partial \bar{c}}{\partial x} + \bar{v} \frac{\partial \bar{c}}{\partial y} = \frac{\partial}{\partial x}(D_x \frac{\partial \bar{c}}{\partial x}) + \frac{\partial}{\partial y}(D_y \frac{\partial \bar{c}}{\partial y}) + \frac{\partial}{\partial z}(D_z \frac{\partial \bar{c}}{\partial z}) - K\bar{c}$$

.... (25)

Concerning the dispersion coefficient,

1. Taylor, 1954 [37] has demonstrated that the dispersion coefficient in a uniform flow through a straight pipe is

$$D_x = 10.1 \, Ru_*$$ .... (26)

where;

R = pipe radius
u_* = $(\tau_o/\rho)^{\frac{1}{2}}$ = shear velocity
τ_o = shear stress on the wall

2. Elder, 1959 [38], for a flow of a logarithmic velocity profile in an infinitely wide open channel, has obtained

$$D_x = 5.9 \, hu_*$$ .... (27)

$$D_y = 0.23 \, hu_*$$ .... (28)

where h = depth of flow. The coefficient 0.23 has been corrected by Sulivan, 1971 [39] in 0.16.

3. Fisher, 1968 [40], for natural rivers, has introduced the formula

$$D_x = -\frac{1}{A} \int_0^b u\acute{}h(y)\,dy \int_0^Y \frac{dy}{\varepsilon_y\,h(y)} \int_0^Y dy \int_0^{h(y)} u\acute{}dz \quad .. \quad (29)$$

where;

    A   =   cross-sectional area

    $u\acute{}$ =   spatial variation of the velocities u from
              their cross-sectional mean value

    y   =   coordinate in lateral direction

  h(y) =   depth as a function of y

    b   =   width

    ε_y =   eddy diffusion coefficient in lateral direction

4. In a marine environment the horizontal eddy diffusion coefficient is usually estimated by Richardson's "four thirds power
law" [41]

$$D_y = \alpha\, L^{4/3} \qquad\qquad .... \quad (30)$$

where;

    α =   constant which takes values between
              10^{-4} to 10^{-3} $m^{2/3}.s^{-1}$

    L =   length scale of the diffusion.

5. Batchelor, 1950 [42] has extended Eq. 30 to

$$D_y = \alpha\, E^{1/2}\, L^{4/3} \qquad\qquad .... \quad (31)$$

where E = local flow energy dissipation per unit mass.

6. Orlob, 1959 [43] has detected this formula and proved that it
is valid only in the first phase of a point source diffusion
process; after a transitional zone where the diffusion coefficient depends linearly on L, this coefficient becomes constant.

7. Hadjianghelou, 1976 [44] has given the formula

$$D_y = \int_0^h \varepsilon_y(z)\ dz = \int_0^h \alpha z u_* \frac{z}{h} \left| f'(\frac{z}{h}) \right|\ dz \qquad \dots\dots (32)$$

where

$$f\ (\frac{z}{h}) = \eta\ \frac{\bar{u}_{max} - \bar{u}(z)}{u_*}\ , \qquad \dots\dots (33)$$

α = coefficient (using the Elder experimental formula of Eq. 27 he has found $\alpha = 1.0$)

η = 0.40

$\bar{u}_{max}, \bar{u}(z)$ = corresponding local mean values.

NUMERICAL ALGORITHMS
for the DISPERSION EQUATION

The numerical solution of the dispersion equation may be combined with the numerical solution of a nearly horizontal flow or a series of field data. Two examples based on the finite differences technique are given further on; the first concerns an explicit and the second an implicit numerical scheme. A numerical simulation based on the finite elements technique is given by Guymon et al, 1970 [45].

The main disadvantage of these numerical integration techniques is that the computational field usually includes open boundaries (say open bays), where boundary conditions are not easily and precisely specified. Moreover, large grid matrices require a considerably large computer time and storage capacity.

Koutitas, 1975 [46] has introduced a numerical integration technique for three-dimensional flows due to tangential stress on the free surface in a finite flow field of small depth. In order to investigate the general tendency of the mixing process (due to wind shear stress) of a constant waste-water discharge in a closed area (say the gulf of Thermaikos), he has solved numerically the dispersion equation

$$\frac{\partial \bar{c}}{\partial t} + \bar{u}\frac{\partial \bar{c}}{\partial x} + \bar{v}\frac{\partial \bar{c}}{\partial y} = D_H\ (\frac{\partial^2 \bar{c}}{\partial x^2} + \frac{\partial^2 \bar{c}}{\partial y^2}). \qquad \dots\dots (34)$$

The explicit numerical algorithm, on the two-dimensional grid of the *Fig. 5*, was

$$c_{i,j}^{n+1} = c_{i,j}^{n} - \frac{\Delta t}{2(\Delta x)} \left(\frac{u_{i-\frac{1}{2},j} + u_{i+\frac{1}{2},j}}{2}\right) (c_{i+1,j} - c_{i-1,j})$$

$$- \frac{v_{i,j-\frac{1}{2}} + v_{i,j+\frac{1}{2}}}{2} (c_{i,j+1} - c_{i,j-1}))$$

$$+ \frac{D_H \Delta t}{(\Delta t)^2} (c_{i+1,j} + c_{i-1,j} + c_{i,j+1} + c_{i,j-1} - 4c_{i,j}) \quad .. (35)$$

and the mean velocities over the depth were obtained by integration of the corresponding velocity diagrams.

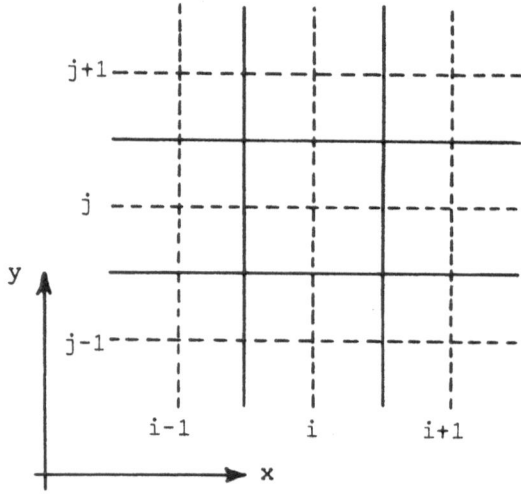

Fig. 5 : The Two-Dimensional Grid for the Algorithms of the Eqs. 35 and 37.

Oster et al, 1970 [47] have given an example of an implicit alternating direction numerical scheme, which had been introduced by Leendertse, 1970 [48]. Thus, they have solved the equation

$$\frac{\partial \bar{c}}{\partial t} + \bar{u} \frac{\partial \bar{c}}{\partial x} + \bar{v} \frac{\partial \bar{c}}{\partial y} = D_x \frac{\partial^2 \bar{c}}{\partial x^2} + D_y \frac{\partial^2 \bar{c}}{\partial y^2} \qquad \dots (36)$$

using the alternating direction, two-steps, implicit numerical scheme (*Fig. 6*).

1st Step

$$(a_{j-1} - r_x)\ c_{i,j-1}^{n+1} + (a_j + 2r_x + \tau)\ c_{i,j}^{n+1} + (a_{j+1} - r_x)\ c_{i,j+1}^{n+1}$$

$$= -(b_{i-1} - r_y)\ c_{i-1,j}^{n} - (b_i + 2r_y - \tau) c_{i,j}^{n} - (b_{i+1} - r_y) c_{i+1,j}^{n} \quad \ldots \ (37a)$$

2nd Step

$$(b_{i-1} - r_y)\ c_{i-1,j}^{n+2} + (b_i + 2r_n + \tau) c_{i,j}^{n+2} + (b_{i+1} - r_y) c_{i+1,j}^{n+2}$$

$$= -(a_{j-1} - r_x)\ c_{i,j-1}^{n+1} - (a_j + 2r_x - \tau) c_{i,j}^{n+1} - (a_{j+1} - r_x)\ c_{i,j+1}^{n+1} \ , \ \ldots \ (37b)$$

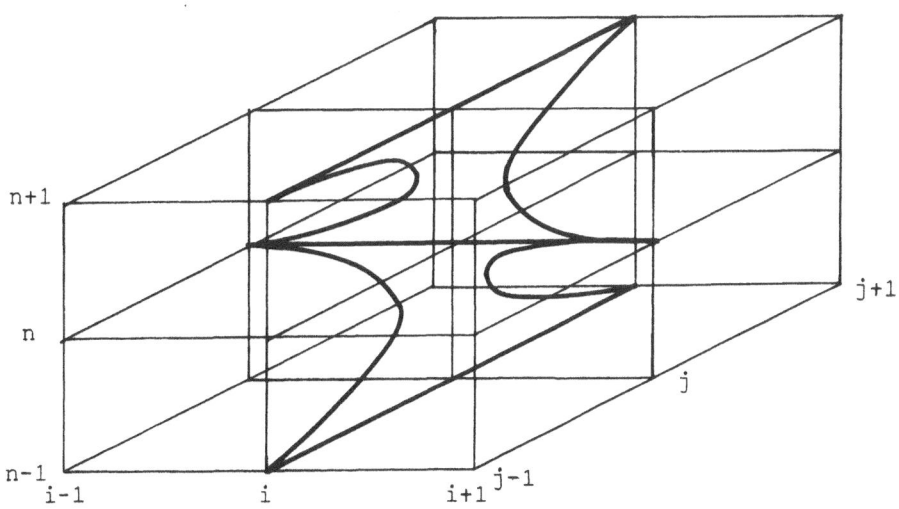

Fig. 6 : Computational Element of the
ADI Numerical Scheme of the
Eq. 37.

where;

$$r_x = \frac{D_x}{(\Delta x)^2} , \quad r_y = \frac{D_y}{(\Delta y)^2} , \quad \tau = \frac{1}{\Delta t} . \qquad \dots (38)$$

The coefficients a_{j-1}, a_j, a_{j+1}, b_{i-1}, b_i, b_{i+1} are calculated from the following table, in accordance with the direction of the corresponding mean velocity at the point (i,j).

$(u_{i,j-1/2}+u_{i,j+1/2})/2$	< 0	≥ 0
a_{j-1}	0	$-u_{i,j-1/2}/\Delta x$
a_{j+1}	$u_{i,j+1/2}/\Delta x$	0
a_j	$-a_{j-1}-a_{j+1}$	
$(v_{i-1/2,j}+v_{i+1/2,j})/2$	< 0	≥ 0
b_{i-1}	0	$-v_{i-1/2,j}/\Delta y$
b_{i+1}	$v_{i+1/2,j}/\Delta y$	0
b_i	$-b_{i-1}-b_{i+1}$	

ACKNOWLEDGEMENTS

The authors wish to express their gratitude to Dr. Eng. H. Hadjianghelou for reading the draft manuscripts and offering a number of suggestions for improvement.

NOTATION

$b =$		width of the diffused patch or field (m);
$c=\bar{c}+c' =$		concentration (ppm or $m\ell^{-1}$);

D_x, D_y, D_z =   dispersion coefficients (m^2/s)

Fr =   Froude number;

g =   acceleration of gravity (m/s^2)

g' = $g(\Delta\rho/\rho_0)$

h =   water depth (m)

K =   sink factor (s^{-1})

p =   pressure (N/m^2)

P =   factor of blocking effect

q =   discharge per unit length $(m^2/s$ or $(sm)^{-1})$

Re =   Reynolds number

S_t =   initial dilution

t =   time (s)

u, v, w =   local velocities (m/s^2)

$\bar{u}, \bar{v}, \bar{w}$ =   mean velocities (m/s^2)

u', v', w' =   fluctuating velocities (m/s^2)

u_* =   shear velocity (m/s^2)

x, y, z =   coordinates (m)

$\Delta\rho$ =   maximum density variation (kg/m^3)

$\varepsilon_x, \varepsilon_y, \varepsilon_z$ =   turbulent diffusion coefficients (m^2/s)

ν =   kinematic viscosity (m^2/s)

$\rho = \rho_0 + \rho'$ =   water density (kg/m^3)

τ_0 =   wall shear stress (N/m^2)

τ_w =   wind shear stress (N/m^2)

Ω =   Coriolis coefficient (s^{-1})

REFERENCES

1. Koh, R.C.Y., and Brooks, N.H., Fluid Mechanics of Waste-water Disposal in the Ocean, in *Annual Review of Fluid Mechanics,* Annual Reviews, Inc., Palo Alto, California, 187, 1975.

2. Rawn, A.M., Bowerman, F.R., and Brooks, N.H., Diffusers for Disposal of Sewage in Sea Water, *Trans.ASCE,* 126-III, 344, 1961.

3. *European Course on Heat Disposal from Power Generation in Water Environment,* Lecture notes in two volumes, Delft Hydraulics Laboratory, 1975.

4. Adams, E.E., Stolzenbach, K.D., and Harleman, D.R.H., *Near and Far Field Analysis of Buoyant Surface Discharges into Large Bodies of Water,* Ralf M. Parson Laboratory, MIT, Report no. 205, Cambridge Mass., 1975.

5. Koh, R.C.Y., Brooks, N.H., List, E.J., and Wolanski, E.J., *Hydraulic Modeling of Thermal Outfall Diffusers for the San Onofre*

Nuclear Power Plant, W.M. Keck Laboratory, California Institute
of Technology, Report no. KH-R-30, Pasadena, Cal., 1974.

6. Mihaljan, J.M., A rigorous exposition of the Boussinesq approximation applicable to a thin layer of fluid, *The Astrophysical
Journal,* 136, 1126, 1962.

7. Liseth, P., *Mixing of Merging Buoyant Jets from a Manifold in
Stagnant Receiving Water of Uniform Density,* University of California, Berkeley, Hydraulic Engineering Laboratory, Technical Rep. HEL 23-1, 1970.

8. Cederwall, K., *Buoyant Slot Jets into Stagnant or Flowing Environment,* W.M.Keck Laboratory, California Institute of Technology, Report no KH-R-25, 1971.

9. Abraham, G., *Jet Diffusion in Stagnant Ambient Fluid,* Delft
Hydraulics Laboratory, Publication no. 29, 1963.

10. Brooks, N.H., *Dispersion in Hydrologic and Coastal Environment,*
W.M. Keck Laboratory, California Institute of Technology,Report
no KH-R-29, 1972.

11. Carlucci, A.F., and Pramer, D., An evaluation of factors affecting the survival of escherichia coli in sea water, *Applied
Microbiology,* 8, 243, 1960.

12. Hadjianghelou, H., Ein Beitrag zu der Untersuchung des Verfalls
der Escheriscia-Coli im Meerwasser während der ersten 24 Stun-
den, *Gesundheits-Ingenieur,* 93, 330, 1972.

13. Hadjianghelou, H., Interaction of Physical and Biochemical Processes by Sewage Disposal in Sea Water, in *Proceedings of the
XVIth Congress of the IAHR,* vol. 3, 451, São Paulo, 1975.

14. Okubo, A., Oceanic Diffusion Diagrams, *Deep Sea Research,* 18,
789, 1971.

15. Einstein, A., Über die von der molecular-kinetischen Theorie
der Wärme geforderte Bewegung von in ruhenden Flüssigkeiten
suspendierten Teilchen, *Ann. Physik,* 17, 549, 1905.

16. Cannerson, C.G., Discussion on the Reference [43], *J. Hydr.
Div., Proceedings ASCE,* 86, 101, 1960.

17. de Vries, M., *Hydraulic Model Investigation,* Lecture notes,In-
ternational Courses in Hydraulic Engineering, Delft, 1973.

18. Xanthopoulos, Th., Contribution a une Syntèse Systématique de
quelques Notions et Méthodes de la Similitude Physique, *La
Houille Blanche,* 22, 633, Grenoble, 1967.

19. Vadot, L., La pollution atmosphérique, ses mecanismes et son
étude en laboratoire au moyen de modèles hydrauliques, *Het In-
genieurblad,* 13/14, 1, 1971.

20. Balafoutas, G.J., Chapon, Y., Catalan, J.P., Xanthopoulus, Th.,
Application des Modéles Hydrauliques à l'étude de la Pollution

de régions urbaines en bordure de mer, in *Proceedings of the XVIth Congress of the IAHR*, vol. 3, 141, São Paulo, 1975.

21. Aubert, M., et Désirotte, N., Etude en Bassin Hydraulique des Rejets d'eau Résiduaire en Mer, *Rev. Intern. Océanogr. Méd.*, 22-23, 5, 1971.

22. Partheniades, E., Beechley, B.C., and Jen, Y., Near Field Temperature Distribution in Shallow Waters due to Summerged Heated Water Jets, in *Proceedings of the XVth Congress of the IAHR*, vol. 2, 137, Istanbul, 1973.

23. Hadjianghelou, H., *Ein Beitrag zur Lösung des Problems der Oberflächendiffusion der Abwässer im Meer*, Hydraulics Department, School of Technology, Thessaloniki, 1975 (in greek with summary in German).

24. Fisher, H.B., and Holley, E.R., Analysis of the Use of Distorted Hydraulic Models for Dispersion Studies, *Water Resources Research.*, 7/1, 46. 1971.

25. Batchelor, G.K., *An Introduction to the Fluid Mechanics*, Cambridge University Press, 1970.

26. Long, R.R., Some Aspects of Turbulence in Stratified Flow, *Applied Mechanics Reviews*, 25, 1297, 1972.

27. Suquet, F., *Similitude des Phénomènes de Pollution Atmosphérique Utilisation de Modèles Hydrauliques*, SOGREAH, Rapport 8927, Grenoble, 1965.

28. Vadot, L., *Etude de la Diffusion des Panaches de Fumée dans l'atmosphère*, CITERA, Paris, 1965.

29. Balafoutas, G.J., Contribution des Modèles Hydrauliques à l' Etude de la Pollution Atmosphérique, Τεχνικά Χρονικά, 10/556, 911, Athens, 1972 (in Greek).

30. Sharp, J.J., Spread of Buoyant Jets at the Free Surface, *J.Hydr. Div.*, *Proceedings ASCE*, 95, 811 and 1771, 1969.

31. Daubert, A., et Graffe, O., Quelques Aspects des Ecoulements presque Horizontaux à Deux Dimensions en Plan et non Permanents Applications aux Estuaires, *La Houille Blanche*, 22, 8, 847, 1967.

32. Neumann, G., and Pierson, W.J., Jr., *Principles of Physical Oceanography*, Prentice-Hall, Inc., Englewood Cliffs, N.J.,1966.

33. Abbott, M.B., Damsgaard, Aa., and Rodenhuis, G.S., System 21' JUPITER' a Design System for Two-Dimensional Nearly Horizontal Flows, *Journ. of Hydr. Res.*, IAHR, 11, 1, 1, 1973.

34. Lai, C., Some Computational Aspects of One- and Two-Dimensional Unsteady Flow Simulation by the Method of Characteristics, *Int. Symposium on Unsteady Flow in Open Channels*, Paper D1, Newcastle -upon-Tyne, 1976.

35. Balafoutas, G.J., *Computation of the Two-Dimensional Two-Layer Flow by the Method of Characteristics,* Hydraulics Department, School of Technology, Thessaloniki, 1976 (in Greek in summary in English).

36. Monin, A.S., and Yaglom, A.M., *Statistical Fluid Mechanics: Mechanics of Turbulence,* vol. 1, the MIT Press, 1971.

37. Taylor, G., The Dispersion of Matter in Turbulent Flow through a Pipe, *Proceedings, Royal Society of London,* 223, A, 446,1954.

38. Elder, J.W., The Dispersion of Marked Fluid in Turbulent Shear Flow, J. of Fluid Mechanics, 5/4, 544, 1959.

39. Sullivan, P.J., *Dispersion in a Turbulent Shear Flow,* Thesis presented to the University of Cambridge, England, 1968.

40. Fisher, H.B., *Methods for Predicting Dispersion Coefficient in Natural Streams with Application to Gree-Duwarnisch River,* U.S. Geological Survey, Prof. Paper 582A, Washington, D.C., 1968.

41. Richardson, L.F., and Stommel, H., Note on Eddy Diffusion in the Sea, *J. Meteorol.,* 5, 238, 1948.

42. Batchelor, G.K., The Application of the Similarity Theory of Turbulence to the Atmospheric Diffusion, *Royal Meteor.Soc.Q.J.,* 76, 133, 1950.

43. Orlob, G.T., Eddy Diffusion in Homogeneous Turbulence, *J.Hydr. Div., Proceedings ASCE,* 85, 75, 1959.

44. Hadjianghelou, H., Über den Quer-Diffusionkoeffizienten in Micro-Scale Turbulenten Strömungen, *GWF Wasser-Abwasser,* 1976 (in Press).

45. Guymon, G.L., Scott, V.H., and Herrmann, L.R., A General Numerical Solution of the Two-Dimensional Diffusion-Convection Equation by the Finite Element Method, *Water Resources Research,* 6, 1611, 1970.

46. Koutitas, Ch. G., *Contribution to the Investigation of Wind generated Currents in Coastal Areas or Lakes,* Hydraulics Department, School of Technology, Thessaloniki, 1975 (in Greek with summary in English).

47. Oster, C.A., Sonnichsen, J.C., and Jaske, R.T., Numerical Solution of the Convective Diffusion Equation, *Water Resources Research,* 6, 1746, 1970.

48. Leendertse, J.J., A Water-Quality Simulation Model for Well-Mixed Estuaries and Coastal Seas, Vol. 1, in *Principles of Computation,* RM-6230-RC, The Rand Corporation, Santa Monica,1970.

BIBLIOGRAPHY

49. Abbott, M.B., *Numerical Methods,* International Courses in Hydraulic and Sanitary Engineering, Delft, 1974.

50. Abraham, G., *Reference Notes on Density Currents,* International Courses in Hydraulic Engineering, Delft, 1975.

51. Cederwall, K., *Hydraulics of Marine Waste-Water Disposal,* Chalmers Institute of Technology, Göteborg, Sweden, 1968.

52. Csanady, G.T., *Turbulent Diffusion in the Environment,* D.Reidel Publishing Co., 1973.

53. Hinze, J.O., *Turbulence,* 2nd ed., McGraw-Hill Book Co., 1975.

54. Raudkivi, A.J., abd Callander, R.A., Diffusion and Dispersion, in *Advanced Fluid Mechanics,* Chap. 6, Edward Arnold, London, 1975.

55. Turner, J.S., *Buoyancy Effects in Fluids,* Cambridge University Press, 1973.

56. Xanthopoulos, Th., *Contribution de l'Analyse Mathématique et des Applications en Laboratoire à la Généralisation de la Théorie et des Etudes Experimentales sur Modèles Physiques à l'Echelle Quelconque,* Hydraulics Department, School of Technology, Thessaloniki, 1965 (in Greek).

57. Yalin, M.S., *Theory of Hydraulic Models,* Macmillan Press Ltd., 1971.

58. *International Symposium on Stratified Flows,* Novosibirsk, 1972, Ed. ASCE, USA.

59. *Discharge of Sewage from Sea Outfalls,* London, 1974, Ed. Pergamon Press.

60. *Marine, Municipal and Industrial Waste-Water Disposal,* Sorrento, 1975, Ed. Pergamon Press.

RIVER BASIN MANAGEMENT IN THE RUHR DISTRICT

Klaus R. Imhoff

Director of Ruhrverband and Ruhrtalsperrenverein

A well known industrial zone has been developed between the Ruhr, the Rhine, and the Emscher, because of the important coal deposits and favorable traffic in this area. About 6 million people are living in that area, one tenth of West Germany's population. Approximately 80 million tons of coal are excavated and some 20 million tons of steel are smelted. In addition many other kinds of industry have settled there. Consequently, water consumption and wastewater flows are seven times higher than West Germany's areal average. By these figures it is readily understood that special measures had to be undertaken to guarantee water supply and wastewater disposal.

THE TASK of WATER SUPPLY

There are three main sources of water supply: the Ruhr River with a contribution of 600 million cubic meter per year, the Rhine River with 100 million cubic meter per year, and the deep sands around Haltern from which 120 million cubic meter per year can be pumped *(Fig. 1)*. The Ruhr contributes the greatest portion be-cause its water contains almost no iron and manganese and is of low hardness.

Approximately 6 m of gravel has been spread by the Ruhr River on a tight layer of rock *(Fig. 2)*. The surface is protected against pollution by 1 or 2 m of clay accumulation.

Before 1900, natural infiltration of river water was sufficient for water supply. After 1900, additional basins were constructed for artificial groundwater recharge.A layer of sand is distributed on the bottom of these basins to prevent clogging of the gravel body. The accumulated sludge must occasionally be removed. The filtered river water is further purified by aerobic biochemical processes in the gravel. After a filter path of 50 m and a retention time of 1 day, the water is of potable quality. For safety, 0.4 mg/ℓ of chlorine is added before the water is pumped to supply.

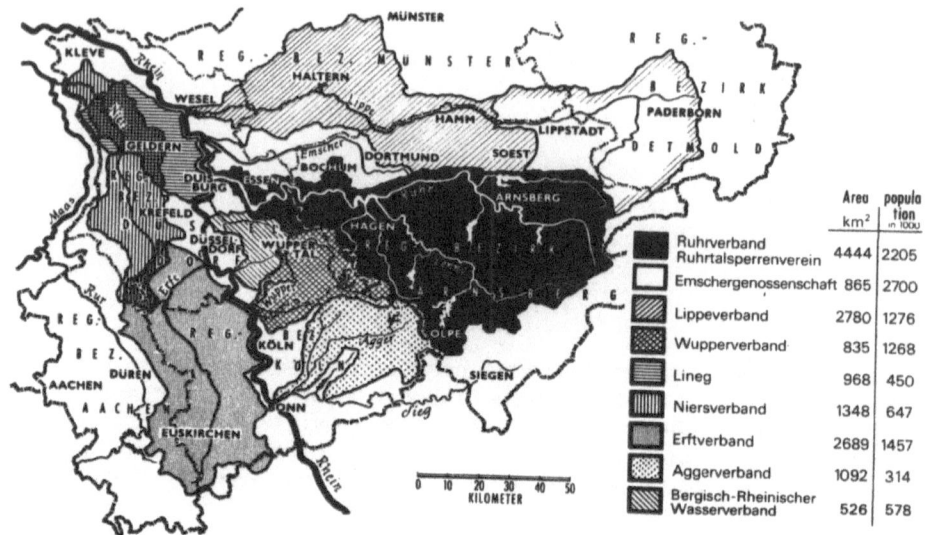

Fig. 1 : Water Associations in the State of North-Rhine-Westfalia.

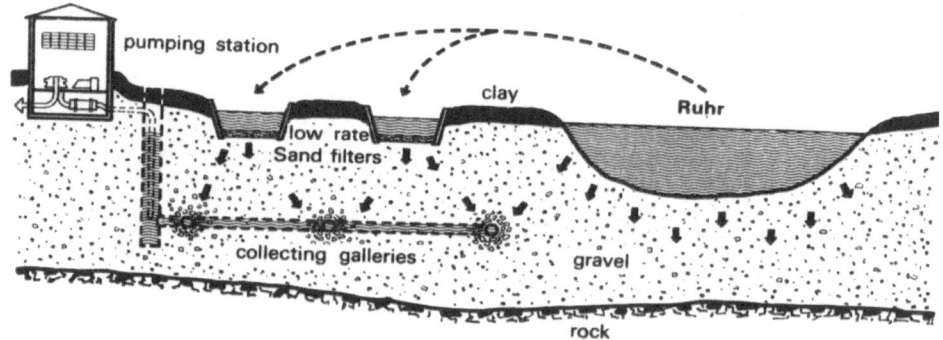

Fig. 2 : Groundwater Recharge in the Ruhr Valley

Occasionally, rapid sand filters (which may be mechanically back-washed) or chemical precipitation have been used to diminish clogging of the low-rate sand filters. In all cases, however, water is artificially drained underground to equalize temperatures and to gain time for pollution prevention.

WATER ASSOCIATIONS

While water treatment and distribution is performed by municipal or private water facilities, it has become the legal task of Ruhrtalsperrenverein to care for water quantity and the legal task of Ruhrverband to insure sufficient water quality in the Ruhr basin. Both associations were set up by law in 1913. By state law, owners of hydropower stations and water facilities must become paying members of Ruhrtalsperrenverein. On the other hand, Ruhrtalsperrenverein (for the water works), municipalities, and industries must contribute financially to Ruhrverband. Both associations are self-controlling bodies with a general assembly of all members, a board of directors, two managing directors, and a technical staff. The Minister of Food, Agriculture, and Forests of the State of North-Rhine-Westfalia controls that Ruhrverband and and Ruhrtalsperrenverein work according to their laws and statutes, as described by Fair [1].

Besides the Ruhrverband and the Ruhrtalsperrenverein, nine similar water associations have been set up in the state of North-Rhine-Westfalia *(Fig. 1)*. These associations are responsible for water pollution control within entire drainage basins. Not hampered by political borders, they design, construct, and operate the necessary plants. Thus, system-wide management has been enacted with the potential of equalizing and minimizing costs [2].

LOW-FLOW AUGMENTATION
by RESERVOIRS

The Ruhr is a relatively small river with an average flow of 75 m^3/sec and a lowest flow of 4 m^3/sec. Approximately 600 million cubic meters of water is drawn annually from the river for water supply. By evaporation and discharge to neighboring drainage basins, about 480 million cubic meters of that water is lost. This is equivalent to a flow of 15 m^3/sec that does not return to the river. During hot spells the loss may even reach 18 m^3/sec *(Fig. 3)*.

To make this water consumption possible and to compensate for water abstraction, the Ruhrtalsperrenverein has constructed a system of reservoirs with a total capacity of 470 million cubic meters [3]. After completion of the Bigge Dam, a low flow of 10 m^3/sec

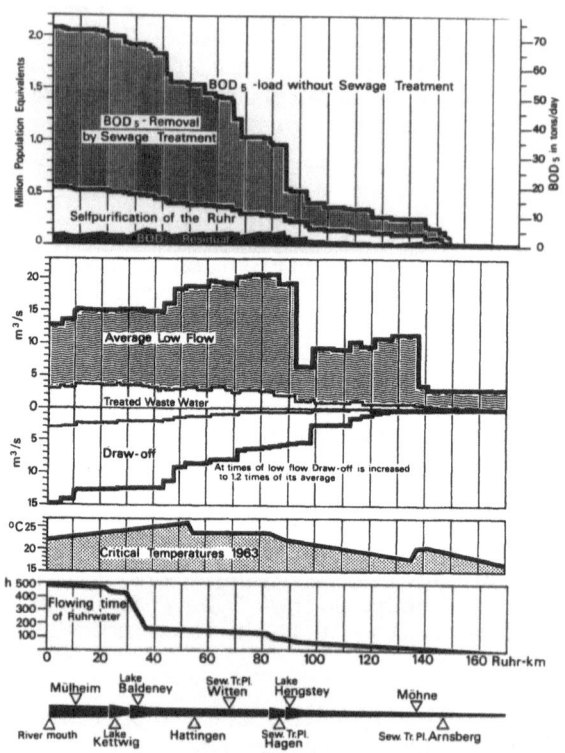

Fig. 3 : Water Pollution Control Plan
of the River Ruhr at Low Flow
Conditions.

can be guaranteed for the first time at the river mouth.

Today, about 25 percent of the drainage basin is controlled by reservoirs. During winter and spring, water is retained behind the dams because runoff from uncontrolled areas is sufficient to meet the demand. During summer and autumn, reservoirs discharge into the river system to augment low flow. It should be noted that a mixed system has been used instead of a direct reservoir supply by means of connecting pipelines. Direct reservoir supply would require not only water conduits but also additional reservoir capacity on the order of 300 million cubic meters.

WASTEWATER TREATMENT

Because water facilities infiltrate river water underground, biochemical oxygen demand (BOD) must not exceed 5 mg/ℓ in raw water, while on the other hand dissolved oxygen (DO) must be higher than 4 mg/ℓ. In total, 4 million BOD population equivalent load the Ruhr system, the portion for the main channel being on the order of 2.1 million *(Fig. 3)*. More than 100 waste treatment plants, most of which are small, retain approximately 75 percent of the gross pollution *(Figures 3 &4)*.

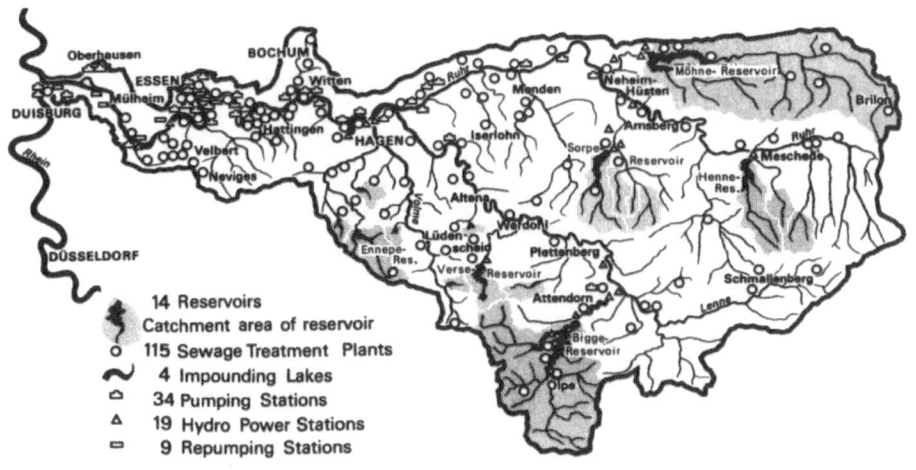

Fig. 4 : Plants of Ruhrverband and Ruhrtalsperrenverein

Tasks are divided between the communities and the Ruhrverband. The various communities are responsible for construction and operation of the sewerage system to the point where a waste treatment plant could be located. At this cross section, the responsibility of the Ruhrverband for wastewater treatment begins. Thus the Ruhrverband is able to realize the most economical system. In one case, a pumping station and connecting trunk sewer to a central treatment plant may be feasible; in another case, a local purification unit may be more suitable. Pretreatment of toxic industrial wastewater is demanded and controlled by local authorities and Ruhrverband personnel to insure that biological processes in the Ruhrverband plants have not deteriorated.

In many cases, trickling filters have been applied [4] because this system gives more reliable results in smaller conditions. For medium-sized and larger plants, the activated sludge process is currently preferred. Occasionally also, two-stage plants have been designed to protect neighboring water facilities. A combination of the activated sludge process and the trickling filter process *(Fig. 5)* has the advantage that toxic, non-iron metals may be precipitated by the addition of chemicals, thus, insuring the efficient operation of the trickling filter. On the other hand, a third clarifier, between aeration tank and filter, results in additional costs. Aeration time is frequently 1 hr and the BOD_5 loading of the filter not more than 400 g/day/m^3.

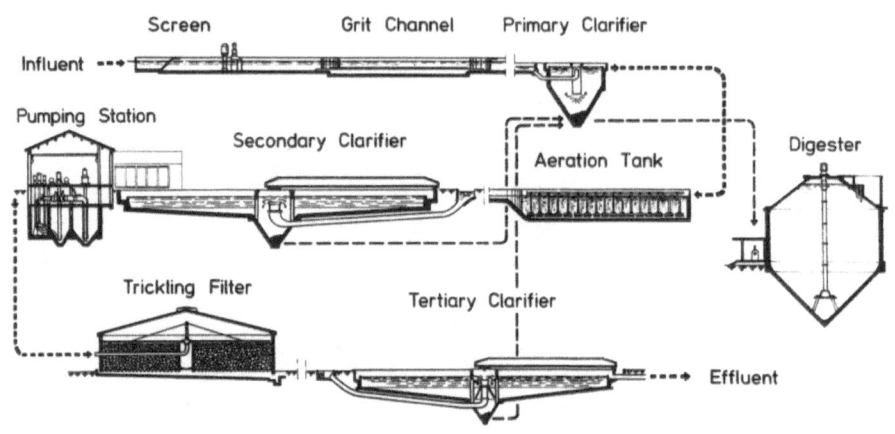

Fig. 5 : Schematic Diagram of a
Two-Stage Biological Plant

If the trickling filter becomes the first stage, it may be loaded with 800 g BOD_5/day/m^3. Then no intermediate sedimentation is required before wastewater flows into the activated sludge tank. Aeration time in this case may be restricted to one hour.

With normal BOD_5 concentrations of 250 to 300 mg/ℓ in raw wastewater, both two-stage plants produce effluent BOD's on the order of 15 mg/ℓ.

PICKLING and PLATING WASTES

Rincke [5] has described different processes used in West Germany. In the Ruhrverband area there are about 1,000 pickling and plating facilities ranging from one to 1,000 employees.

To avoid accumulation of toxic material in combined wastewater treatment plants, a committee suggested the following standards for discharges of pickling and plating works.

TABLE 1: Standards for Discharging of Pickling &
Plating Wastewater.

Settleable Solids	0.3	mg/ℓ
pH	6.5-9.0	
Fe	2	mg/ℓ
Cr	2	mg/ℓ
Cu	1	mg/ℓ
Ni	3	mg/ℓ
Zn	3	mg/ℓ
Cd	3	mg/ℓ
CN	0.1	mg/ℓ
Free Chlorine	0.5	mg/ℓ

Thus, pickling and plating plants are obliged to purify wastewater as effectively as possible.

To avoid sludge accumulation, the Ruhrverband promoted the development of sulfuric acid regeneration plants. Under normal conditions sulfuric acid regeneration is feasible if more than 40 tons of acid is consumed per year. Part of the warm pickling bath is cooled in one or several stages depending on the required iron content for the pickling process. For crystal separation, hoses that can also be installed outdoors are used. Accumulating iron sulphate as well as spent hydrochloric pickling solutions are shipped by the Ruhrverband to chemical industries and other consumers.

In the case of plating facilities it is most important that cyanide and chromium polluted wastewaters are separated. The destruction of cyanides is effected mainly by oxidation with chlorine. The reduction of chromium with iron sulfate at a pH value of 2.5 presents no problem. After decontamination the different wastes are mixed and neutralized, and sludges are separated and thickened. For further concentration, filter presses or drying beds may be applied. Then sludge cakes must be disposed of on controlled pits.

Adequate sludge treatment becomes a problem because the space in most facilities is inadequate. The Ruhrverband therefore constructed central dewatering plants for sludges from pickling and plating shops [6]. Sludge deposits have a clay sealing to prevent seeping underground. As far as possible, surface water is diverted. Water that has drained into the deposit is drawn off and treated in neighboring plants.

MEASURES AGAINST OIL POLLUTION

For modern civilization, water and oil are indispensable, but if they mix it becomes disasterous. According to a theoretical calculation, 1 liter of oil may spoil 1 million liters of water. Because about 70 major oil accidents occur annually in the Ruhr area, an emergency service has been organized. During weekends and holidays, experienced engineers stay at the phone so that no time is lost in the event of oil alarm.

Several devices have been developed by the Ruhrverband for oil pollution control [7]. An oil alarm apparatus functions by means of a rotating disk electrode. In case of oil pollution, the disk will be covered and its conductivity diminished. In addition, a floating oil skimming device may remove major oil spills from the water surface. The apparatus consists of rollers that move endless plastic strings. Oil attaches to this material and can be stripped off. Unfortunately both devices are no longer commercially available because not enough pieces could be sold.

To remove oil from the flowing water, a special truck has been equipped with flotable baffles, ropes, a skimming device, a boat, and oil containers. This truck is stationed in the center of the Ruhr drainage basin and can be directed to selected cross sections of the river.

Analyses of rain water from highways had shown that as much as 59 mg/ℓ of the oil may be contained. New highways in the neighborhoods of water facilities were therefore lined with tight drainage systems. The discharge is treated in oil separators. These tanks have been constructed by the street authority. The surface is calculated for a maximal hydraulic loading of 12 $m^3/hr/m^2$. Retention time equals 10 min [8].

IMPOUNDED LAKES

A total of four impounded lakes have been constructed to supplement the system of wastewater treatment plants and to degrade the residual pollution of effluents and storm overflows.

From the point of view of size, Lake Baldeney in the South of Essen is the largest with an average retention time of some 60 hours. One-third of the construction cost was paid by the city of Essen for its recreational value. The remainder has been financed by the income from hydropower. Numerous analyses from inflowing and outflowing water have demonstrated that the BOD_5 of river water is diminished from about 5.5 to 4.8 mg/ℓ. At the same time it should be pointed out that, until recently, between the control sections, the lake has been additionally loaded by 80,000 population equivalents. In terms of overall performance, Lake Baldeney therefore provided wastewater treatment to the value of 100,000 population equivalents [9].

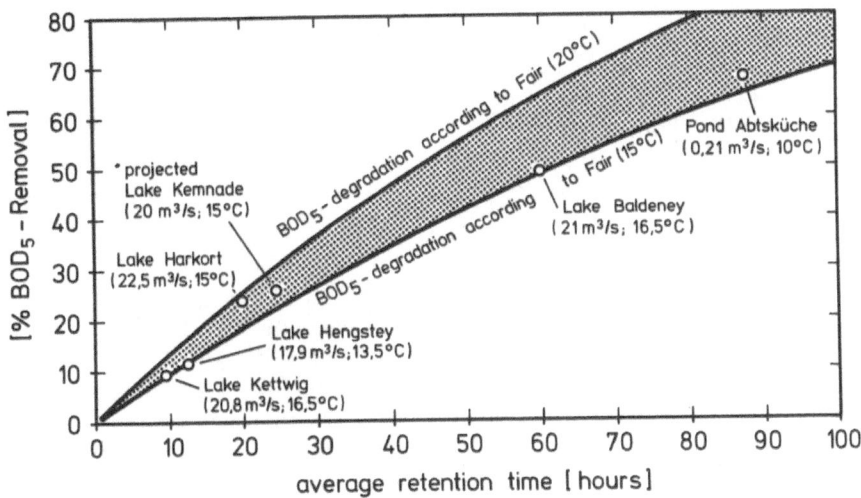

Fig. 6 : Percentage of BOD5
Removal in the Ruhr lakes

Figure 6 demonstrates that the percentage of BOD5 removal in purification lakes can be related to the effective or average retention time of river water. After tests with radio isotopes, it has been concluded that the average retention time is about 50 percent of the theoretical retention time. On this basis, average retention times have been calculated, and measured percentages of BOD5 removal have been plotted. The results follow the curves of Fair for aerobic decomposition.

In *Figure 6* the projected Lake Kemnade has also been plotted. Its purification value will be only some 20 percent of its calculated construction cost. Therefore, Lake Kemnade can only be realized if neighboring cities and the new university pay approximately 80 percent for its amenity value.

Measurements have shown that sludges are deposited in Ruhr lakes with river flows up to 10 times the average rate. Only floods on the order of 15 times average flow clean the lakes. Because a flood of this order has not occurred for 10 years, sludge deposits currently represent one-tenth of the capacity of Lake Baldeney. Sludge dredging would be expensive and cost 10 million DM but may become necessary.

INSTREAM AERATION

Due to the good self-purification process in Lake Baldeney, DO
decreases, for instance, from 6.5 to 4.0 mg/ℓ. Because phosphorus
on the order of 1.5 mg PO_4/ℓ and nitrates on the order of 15 mg/ℓ
are available, the number of algae may grow to 100,000 total cells/
ml of river water. After the algae bloom an oxygen depression is
often recorded. In such a case the oxygen profile may drop below
2 mg/ℓ (Figure 7). To compensate for this effect, three artificial aeration devices have been applied: a floating mechanical aerator, a compressed air aeration system, and a turbine aeration
installation [10]. Downstream from Lake Baldeney, additionally
two weirs may be put into operation by reducing the hydropower generation of the adjacent turbines.

Five continuously recording oxygen stations are operated at a
distance of approximately 10 river kilometers. In case of alarm,
additional longitudinal profiles are measured and instream aeration is put into operation as required by the situation. For the
lower Ruhr, instream aeration has proved to be a successful method
to maintain DO contents of 3 to 4 mg/ℓ [11].

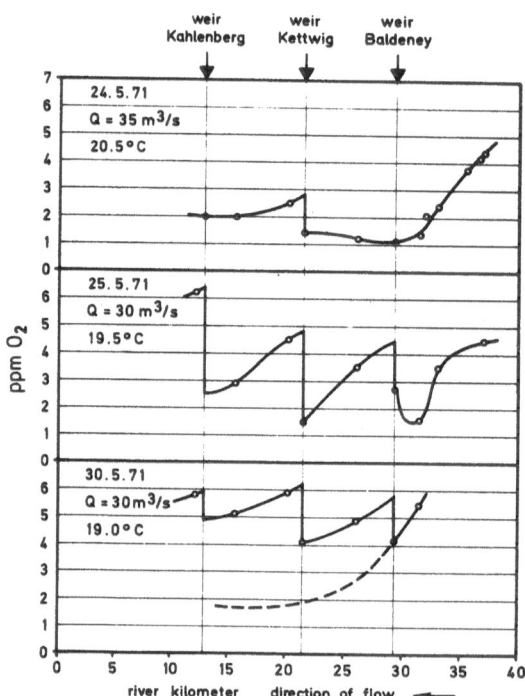

Fig. 7 :   Oxygen Profiles of the
           Lower Ruhr

WATER REUSE

Reservoirs and wastewater treatment plants discharge into the same river system from which water facilities obtain their raw water, thus providing water reuse. Its frequency depends on the 6.2 m^3/sec of treated wastewater effluent which is received in total and on river flow. At average flow conditions, reuse frequency equals 8 percent.

At average low flow conditions, draw-off has to be considered in the calculation. As may be seen from *Figure 3*, a total of 15 m^3/sec has been taken out. Reuse frequency therefore increases to 6.2:(13+15)= 0.22. By long experience it is known that this ratio can be permitted without jeopardizing drinking water supply.

The most critical situation for this system arose in 1959. That year, Germany experienced the driest summer in three decades. The Bigge Reservoir was still under construction, and unfortunately the Sorpe Reservoir had to be emptied for repair. Only 260 million cubic meters of reservoir volume was available, compared to today's 470 million cubic meters. By autumn the river flow ended at Essen, 35 km above the usual river mouth. For the supply of the lower water facilities, 100 million cubic meters of water was pumped upstream over the continuous row of weirs. At Essen, the Rhine and the Ruhr water met end conditions and were stagnant for two months. In this worst cross section, drinking water quality become very poor [12]. Chlorides increased from 100 mg/ℓ in May to almost 300 mg/ℓ in September.The situation continued until December 1959. Chlorides increased to 800 mg/ℓ in river water. The reuse factor reached almost 100 percent, at which time the tolerable limit was surpassed. About 7 percent of Essen's population suffered from nonbacterial gastroenteritis [13].

Additional reservoir volume of 100 million cubic meters would have been necessary to avoid pumping Rhine water. By this capacity, reuse frequency would have been lowered to 6.2/15= 0.42, which, from the water quality point of view, could be permitted [12]. Because an additional 210 million cubic meters is available, the system has a good safety factor.

COST ASPECTS

Since West Germany's currency reform in 1948, the Ruhrverband has invested approximately 400 million DM in new plants and the Ruhrtalsperrenverein some 600 million DM. In this connection it may be of interest to compare relative costs of the various possibilities of water quality management.

To compare river systems, a division of tasks has been developed between the Ruhr and the Emscher. Potable water is primarily obtained from the Ruhr, while wastewater is discharged to the Emscher. Thus the Ruhr River has to be kept clean by construction and operation of many small and expensive wastewater treatment plants.The Emscher River, on the other hand, has become an open sewer and is treated, before its confluence to the Rhine, by central facilities of a capacity of 5 million population equivalents. The treatment cost advantage of the Emscher population is equalized by the price of water that is received from the Ruhr valley.

Currently, 45 percent of the water pollution control expenses of the Ruhrverband is covered by the water facilities. The remaining 55 percent is collected from communities and industries by sophisticated cost assessment rules. The charges can be enforced like a tax.

The general concept of cost allocation may be explained by an example. If an industry discharges into a sewerage system at the end of which is a wastewater treatment plant operated by the Ruhrverband, the industry will be assessed according to the quantity of discharge and to its pollution load. The pollution load will be analyazed and rated on the basis of BOD population equivalents. In case of toxic wastewater, a bioassay test is used [14]. This procedure is reasonable because the treatment facilities of the association have been sized on the basis of flow and wastewater strength to be treated. If the same industry discharges to the river directly, it pays on the basis of pollution load. Wastewater quantity no longer is put into consideration.

The income from effluent charges is used to pay for additional water pollution control measures, such as wastewater treatment, impounded lakes, or instream aeration, to equalize the given pollution. It has to be pointed out that by receiving the effluent charge, the pollution has not been licensed. Such a license can only be granted by state authority after a formal procedure. In most cases, precise effluent standards are demanded.

In spite of the fact that Germany has good water laws, effluent standards are not met. Long experience of the water associations has shown that effluent charges are a practical way at least to collect money from a polluter and to finance equalizing measures. The federal government in Bonn strongly promotes the effluent charge principle and is preparing a law for the whole country. Moreover, the effluent charge may cause a polluter to pretreat or treat the wastewater if the relative costs are advantageous [15].

Some mention should be made of the relative costs of dilution water and wastewater treatment. To a certain degree, dilution water and wastewater treatment may be exchanged. With both methods available, their costs have been compared from the oxygen point of view. For the Ruhr system it was determined that dilution water is four times more expensive than wastewater treatment to maintain 4 mg/ℓ of oxygen in river water. It also has been shown that the oxygen balance can be affected much more cheaply by instream aeration than by increasing wastewater treatment plant efficiencies beyond 90 percent. From the oxygen point of view, relative costs of dilution water, wastewater treatment, and instream aeration are in the ratio of 16:4:1.

Reuse is another effective way to design a feasible system. If reuse were to be completely abolished, an additional 300 million cubic meters of reservoir volume would have to be constructed, and water conduits would have to be installed. According to an estimate, this would require for the Ruhr system an extra 2 billion DM, if the sites for additional reservoirs were available.

WATER POLLUTION CONTROL PLAN

Traditionally, two or three longitudinal profiles of the Ruhr River were determined per year. When the results of a certain cross section for a series of years were compared, it became evident that the available material was insufficient. Either flow or temperature or total cells were changing and influenced the picture to a degree that no conclusion could be drawn.

In 1965, therefore, the Ruhrverband began to sample once per week at seven cross sections. The analytical results are plotted by a computer over the water flow with temperature as the parameter. By this simple procedure average data can be obtained for all interesting flows.

The results become basic points in the longitudinal study (*Fig. 8*). Knowing the effluent load of the plants, industry and tributaries, the longitudinal profile can be calculated according to the mixing principle. To account for self-purification, the simplest assumption has been made: that it works in proportion to given residual pollution and to river length. More sophisticated concepts may be applied later. The obtained plan is a very useful tool to determine what has to be done and which plants have to be expanded first.

Figure 8 also gives a forecast of the possible effect of the water pollution control program. The degradation constants "k" for 1970 conditions have been used to calculate the 1977 BOD profile. Since the river water becomes cleaner the ecological conditions will change. That means that "k" will diminish and that the forecast will be too good.

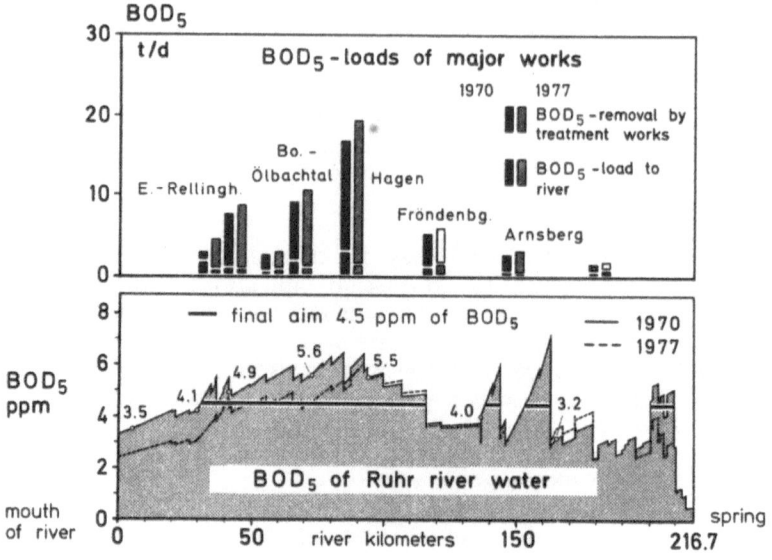

Fig. 8 : Water Pollution Control
Plan of the Ruhr River

RESEARCH

Only applied research may be conducted by the Ruhrverband. Some topics such as pickling and plating wastes, measures against oil pollution , impounded lakes, and instream aeration have already been mentioned. Additional experiments have been carried out with plastic media for trickling filters [16].

The pilot filters were 6 m high and 1 m wide. They were packed with four different media. Best results could be obtained with Cloisonyl pipes and hanging plastic sheets. The BOD volume loading ranged between 1,000 and 8,000 g/day/m^3. The BOD removal was between 40 and 80 percent for one passage of wastewater and no recirculation. Waste, silo seepage, and wastewater from a pulp mill and from a carcass utilization plant have been treated (Fig. 9).

Sludge disposal seems to be a perpetual problem. Digestion tanks and sludge lagoons used to be the solution. In Germany large digesters have often been prestressed in segments and have been pearshaped. Moreover, digesters have a bottom cone to facilitate sludge withdrawal and a top cone to restrict scum. This shape is much more expensive than the American cylindrical digester but perhaps more effective. It cannot be proven which is the better concept. Thus two processes have developed in the two countries. Digested sludge then was discharged into sludge lagoons. Despite good drainage possibilities, about 0.5 m^2 of area is required per capita.   In many cases this land is no longer available.   Consequently, mechanical sludge dewatering becomes a necessity.

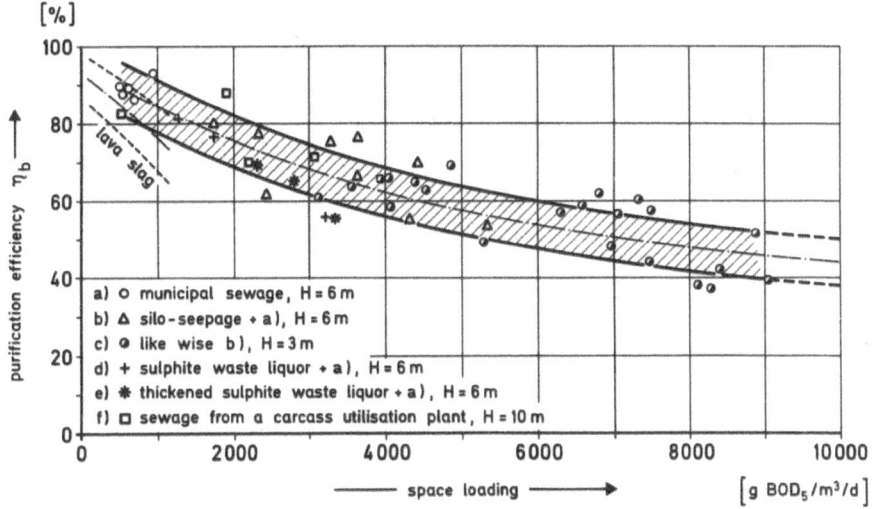

Fig. 9 :   Efficiency of Plastic
Medium Trickling Filters

A new machine, the belt press (Fig. 10), was tested and proved to be useful [17].   First, the sludge is flocculated by the addition of polyelectrolytes.   Some water may be released in the pre-dewatering zone.   Subsequently the cake is pressed between the two belts, the lower screen belt being cleaned by a water jet.   The solids content of the cake ranged for different sludges between 20 and 28 percent, the flocculant consumption between 50 and 150 g/cubic meter of sludge.

A sludge dewatering and incineration plant has been constructed according to the following concept (Fig. 11).   Thickened raw sludge can be stored in an aerated tank.   For conditioning, the sludge may be heated by the burned gases to 65°c.   In addition, lime, acid, or polyelectrolytes are required.   Water is removed by belt presses or by centrifuges in parallel because these machines should be compared in full scale.   Subsequently the dewatered sludge is burned in two fluidized bed furnaces.   Also screenings, grit, and spent oil can be added.   Costs on the order of two or three times the expenditure of the conventional process is anticipated.

Advanced treatment gradually becomes a research topic for the association, not in terms of "zero pollution" but in terms of 5 mg/ℓ suspended solids effluent quality.   The two-stage processes men-

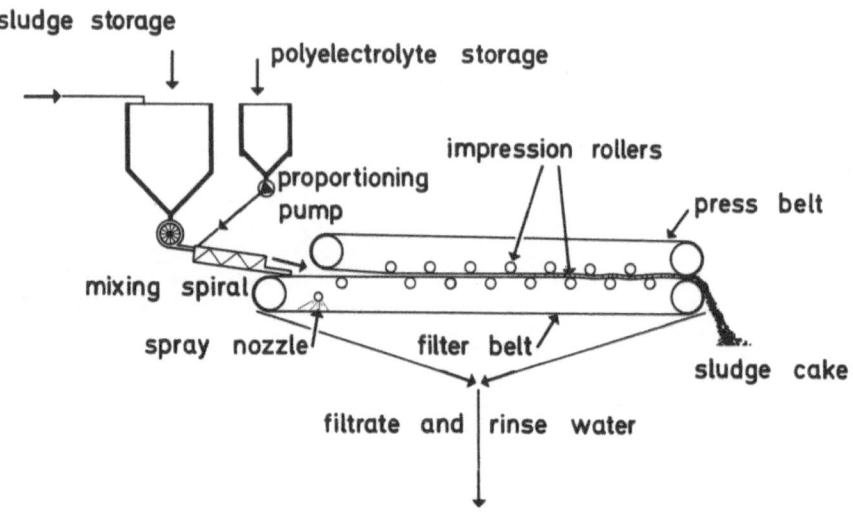

Fig. 10: Sludge Dewatering by a Belt Press

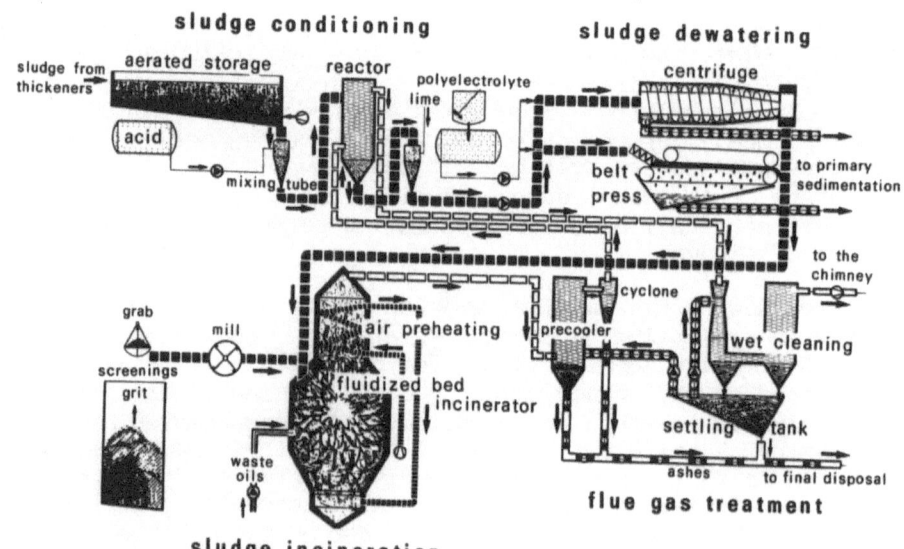

Fig. 11 : Sludge Treatment Process at
Bochum-Ölbachtal Wastewater
Facility

tioned and the impounded lakes are steps towards that aim. Last
year in three activated sludge plants phosphates have been simultaneously precipitated for the first time. During the next few
years, biological treatment efficiencies must be upgraded and the
hydraulic capacity of the plants must be enlarged. The rainwater
problem will also be carefully studied. More advanced treatment
systems will probably have to be installed after 1980.

CONCLUSIONS

Some aspects of water quantity and quality management at the
Ruhr have been described. It may be noticed that only known elements are applied, not secret processes or sophisticated third or
fourth treatment steps. The answer to successful water pollution
control lies elsewhere, in the creation of a powerful association
that is responsible for a drainage area. Let the state, water facilities, cities and industries control them. Make the water facilities strong enough. Employ devoted, hardworking engineers and
plant operators and pay them adequately. Have a good cost assessment system. With these prerequisites, the overall result will be
a good one.

The legal form of the Ruhr River association was determined a-
bout 60 years ago but the idea of a system-wide authority is nevertheless most promising. Pollution increases and water resources
cannot be augmented. Therefore, water pollution control must become more and more expensive. Increasing costs can only be minimized if a system-wide management is installed. Kneese and Cleary,
two experts from the U.S., very much promote this concept [15, 18].

REFERENCES

1. Fair, G.M., Pollution Abatement in the Ruhr District, *Jour.
   WPCF,* 34, 749 (1962).

2. Rincke, G., Economic Considerations on the Clarification Requirements in a Catchment Area, *Gas-u. Wasserfach,* 109, 44,
   1209 (1968).

3. Koenig, H.W., The Water Supply of the Ruhr District, *Aqua,* Qu-
   arterly Bull. Intl. Water Supply Assn., London, Eng., 3, (1966)

4. Rincke, G., Aspects for Wastewater Treatment with Trickling
   Filters, *Gas- u. Wasserfach,* 108, 24, 667 (1967).

5. Rinche, G., Pickling and Plating Wastes Treatment in Particular
   Plants in West Germany, Purdue Univ. W. Lafayette, Ind., *Eng.*

Bull., 72 (March 1966).

6. Imhoff, K. R., Sludge Removal in Plating Industry, *IWL-Forum 70/I-II,* Köln, Inst. für Gewerbliche Wasserwirtschaft und Luft-reinhaltung e. V., Germany.

7. Koenig, H.W. and Rinche, G., Measures of the Ruhr River Association against Oil Pollution of Waters, *Wasser u. Boden,* 18, 8, 257 (1966).

8. Imhoff, K. R., Oil Separation Tanks for Motor Roads in the Ruhr Valley, *Gas- u. Wasserfach,* 108, 2, 43.(1967).

9. Imhoff, K. R., On the Treatment of the Impounded Lakes of Ruhr River, *Gas- u. Wasserfach,* 106, 46, 1264 (1965).

10. Imhoff, K. R., Oxygen Balance and Artificial Reaeration of Lake Baldeney and the Lower Ruhr, Proc. 4th Int. Conf. Water Poll. Res., Prague, *Pergamon Press,* London, Eng. (1968).

11. Albrecht, D. and Imhoff, K.R., Experience with Artificial Aeration of Ruhr River, *Gas-u. Wasserfach,* 114, 3, 131 (1973).

12. Koenig, H. W., et al., Water Reuse in the Ruhr Valley with Particular Reference to 1959 Drought Period, Proc. 5th Int. Conf. Water Poll. Res., San Francisco, *Pergamon Press,* London, Eng. (1970).

13. Trüb, A., et. al., Archiv für Hygiene und Bakteriologie, Stutt-gart: Fischer, Ger., 5 (1961).

14. Bucksteeg. W., Problems of Assessment of Toxic Matter in Waste-waters and Possibilities for Creation of Reliable Assessment Principles, *Münchener Peitrage zur Abwasser-, Fischerei- und Flussbiologie,* München: Oldenbourg, 6, 70 (1959).

15. Kneese, A. V., The Economics of Regional Water Quality Management, *Johns Hopkins Univ. Press,* Baltimore, Md. (1964).

16. Rincke, G. and Wolters, N., Technology of Plastic Medium Trickl-ing Filters, Proc. 5th Int. Conf. on Water Poll. Res., San Francisco, *Pergamon Press,* London, Eng., II-15 (1970).

17. Imhoff, K. R., Sludge Dewatering Tests with a Belt Press, Water Res., *Pergamon Press,* London, Eng., 6, 515 (1972).

18. Cleary, E. J., ORSANCO Story, *Johns Hopkins Univ. Press,* Baltimore, Md. (1967).

SOLVED PROBLEMS

CHARACTERIZATION OF WASTEWATERS

1. Given the following BOD data:

Incubation time, days	BOD, mg/ℓ
1	14
2	26
3	38
4	46
5	56
6	62
7	65

Compute k and L_O by three (3) methods.

SOLUTION:

1. Compute k and L_O by:

    a) Thomas Graphical method: (Figure 1)

t	y	$(t/y)^{1/3}$
1	14	.415
2	26	.425
3	38	.429
4	46	.443
5	56	.447
6	62	.459
7	65	.476

$$L_O = \frac{1}{2.3ka^3}$$

$$k_{10} = 2.61(.012/.403) = .078/\text{day}$$

$$L_O = (.078(.403)^3(2.3))^{-1} = 85 \text{ mg}/\ell$$

Smooth the data: (Figure 2)

time, days	original BOD	corrected BOD
1	14	13
2	26	25
3	38	37
4	46	47
5	56	56
6	62	62
7	65	65

$\Sigma y = 305$ $\Sigma ty = 1470$

$\Sigma y / \Sigma ty = 2.08$

From Figure 3,

$$k_{10} = .07/\text{day}$$

$$\Sigma y / L_O = 2.35$$

$$L_O = 119 \text{ mg}/\ell$$

b) Log-difference method:

t	y(corrected)	daily difference
1	13	13
2	25	12
3	37	12
4	47	10
5	56	9
6	62	6
7	65	3

See **Figure 4**

$$k_e = 2.7 - 1.4/7 = .18$$

507

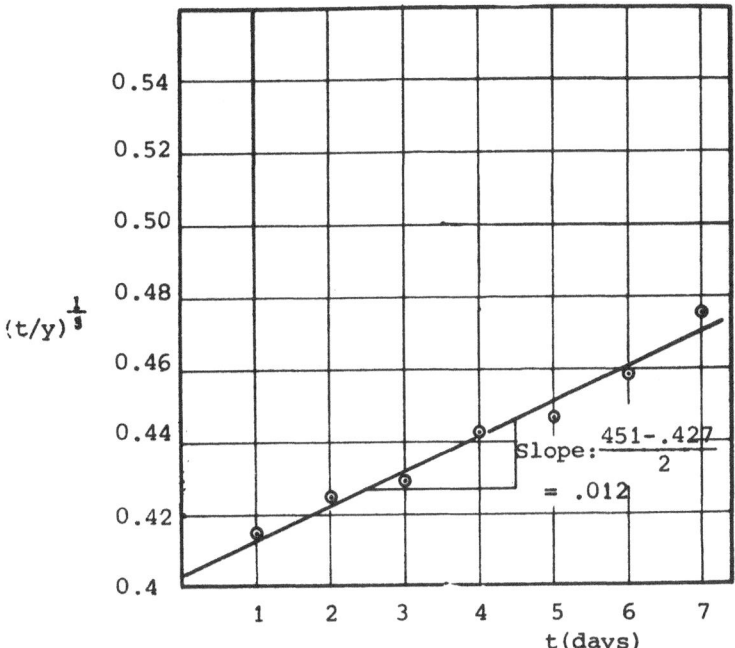

Fig. 1: Thomas Graphical Method

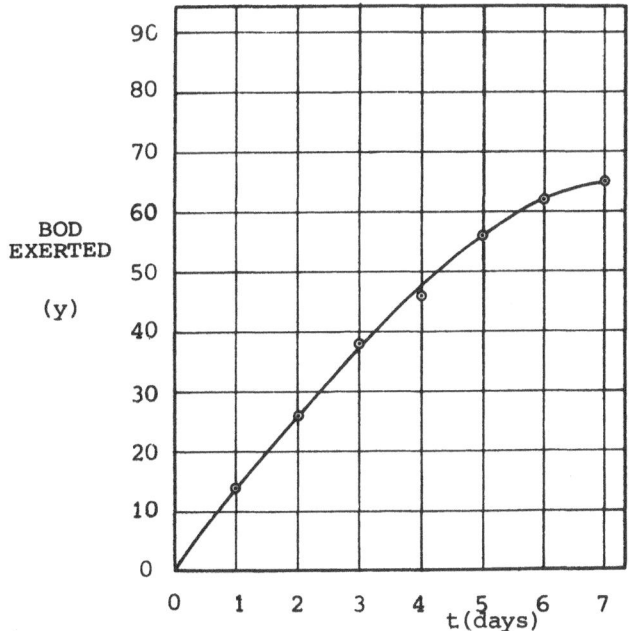

Fig. 2: Smoothed—Out Data

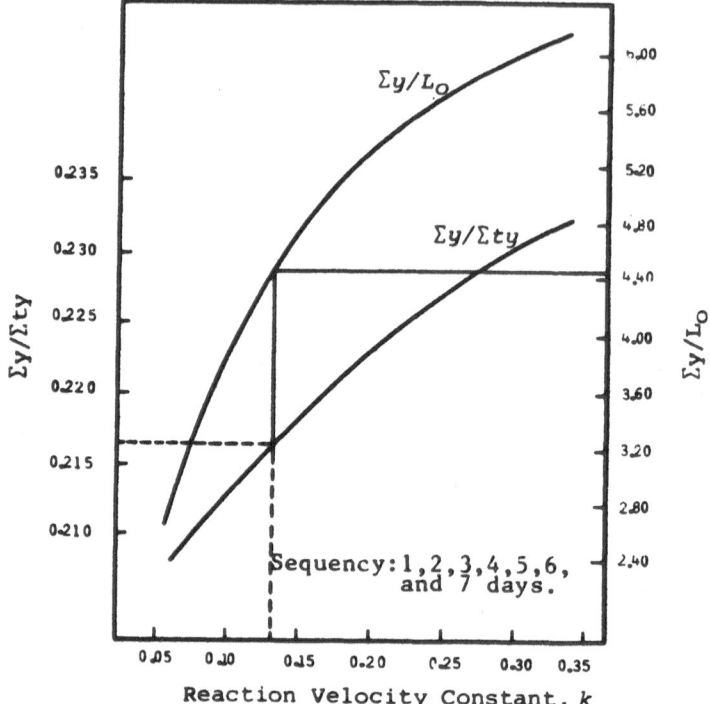

Fig. 3: Calculation of BOD Constants from the Method of Moments
(Moore, 1950)

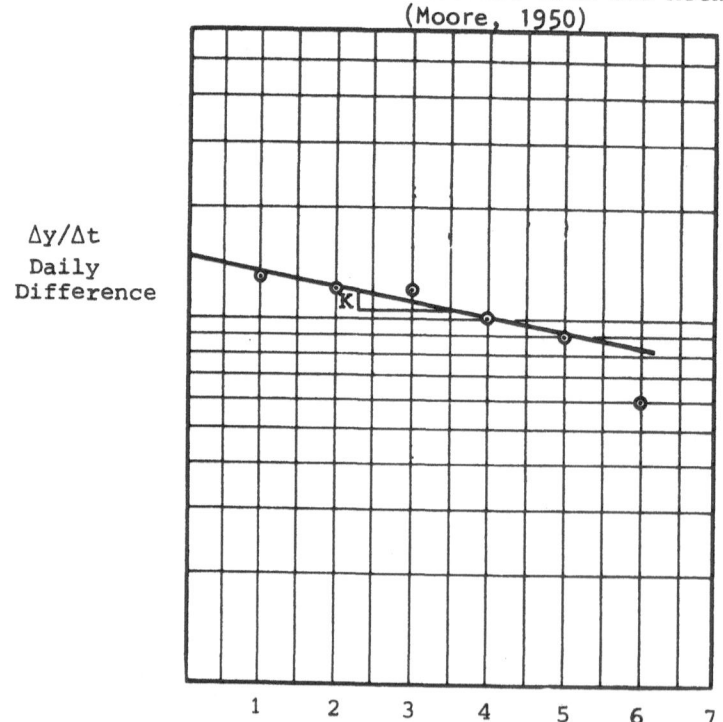

Fig. 4: Log-Difference Method

$k_{10} = .08$ day

intercept $= L_{o}k = 2.7$

$\exp (2.7) = 14.9 = .18L_{o}$

$L_{O} = 82.7$ mg/ℓ

PRINCIPLES OF BIOLOGICAL TREATMENT

1. A wastewater shows the following analysis:

    $COD = 2140$ mg/ℓ

    $BOD_{5} = 835$ mg/ℓ

    $k_{10} = 0.17$

    The treated effluent from an activated sludge plant with a sludge age of 12 days showed the following:

    $BOD_{5} = 17$ mg/ℓ

    $k_{10} = 0.08$

    Estimate the effluent BOD.

SOLUTION:

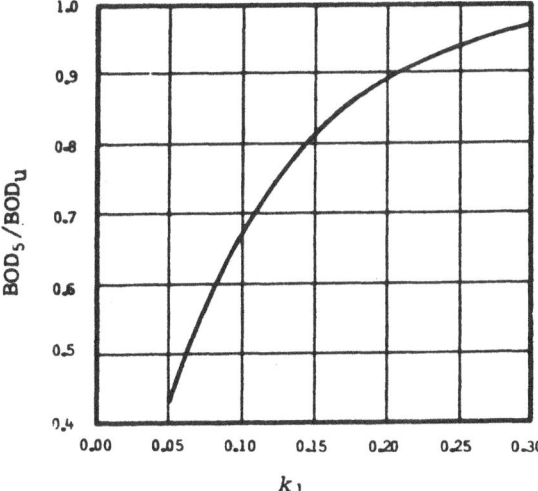

Fig. 5: Relationship between k_{1} and BOD_{5}/BOD_{u}.

From Figure 5:

k_{10}	BOD_5	BOD_5/BOD_u	BOD_u	$(BOD_u/0.92)$ x (degradable COD)
.17	835	.85	971	1055
.08	17	.6	28	31

$$\text{influent COD} = \frac{\text{influent ultimate BOD}}{0.92} + \text{nondegradable COD}$$

inorganics + nondegradable COD

= influent COD - influent ultimate BOD

= 2140 - 1055

= 1085

$$\text{effluent COD} = \frac{\text{effluent ultimate BOD}}{0.92} + \text{nondegradable COD} +$$

nondegradable residue (assumed 2% of influent COD)

given: nondegradable residue = 1085

effluent COD = 31 + 1085 + (.02 x 2140)

= 1159 mg/ℓ

2. The following data was obtained on the treatment of a waste-water. Correlate the data and develop a kinetic model.

Test 1:

Detention time, days	0.82	0.84	0.97	1.02
MLVSS, mg/ℓ	2380	2630	2710	3370
Influent TOC, mg/ℓ	87.8	215	401	730
Effluent TOC, mg/ℓ	11.0	14.4	27.8	41.0

Test 2:

Detention time, days	0.18	0.38	0.66	1.26
MLVSS, mg/ℓ	2530	2800	2790	2750
Influent TOC, mg/ℓ	75.3	186	370	760
Effluent TOC, mg/ℓ	12.8	18.8	29.5	43.7

SOLUTION:

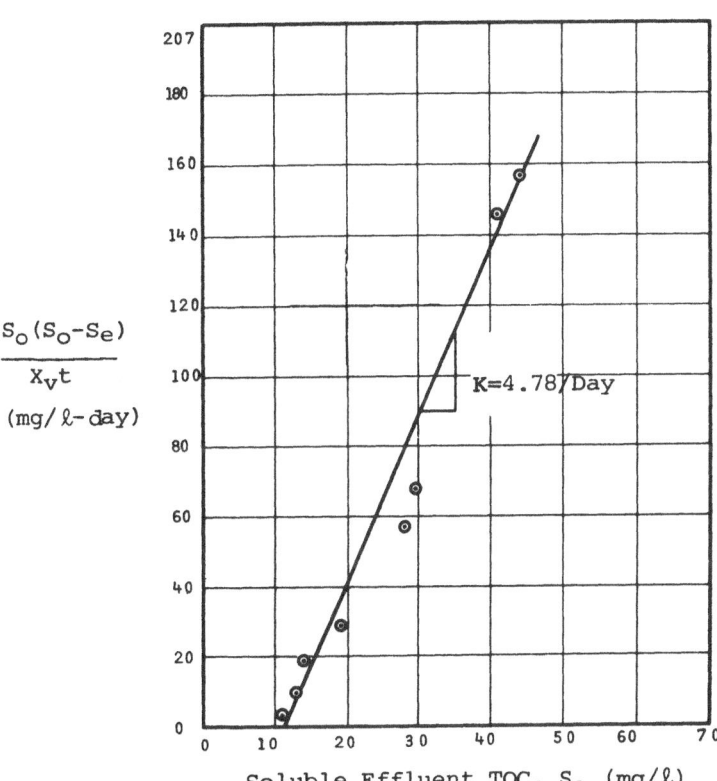

$$\frac{S_o(S_o-S_e)}{X_v t}$$

$(mg/\ell\text{-}day)$

K=4.78/Day

Soluble Effluent TOC, S_e (mg/ℓ)

Fig. 6: Correlation of Data

S/S_o	$(S_o-S)/X_v t$	$(S_o(S_o-S))/X_v t$	S
0.1253	0.0394	3.46	11.0
0.0670	0.0908	19.52	14.4
0.0693	0.1420	56.94	27.8
0.0562	0.2004	146.29	41.0
0.1700	0.1372	10.33	12.8
0.1011	0.1571	29.22	18.8
0.0797	0.1849	68.41	29.5
0.0575	0.2067	157.09	43.7

See Figure 6;

$$\frac{S_o - S}{X_v t} = k_1 \frac{S}{S_o}$$

$$k_1 = 4.78/day$$

S_o = influence TOC, mg/ℓ S = effluent TOC, mg/ℓ

X_v = MLVSS, mg/ℓ t = time, days

3. Table 1 shows a summary of activated sludge results treating an organic chemicals wastewater. Determine a and b from a plot of these data.

TABLE 1

UNIT	1	2	3
F/M	0.113	0.298	0.410
t, days	0.690	0.240	0.167
V, ℓ	20.0	20.0.	20.0
MLVSS, mg/ℓ	2020.0	2210.0	2310.0
BOD, S_o, mg/ℓ	158.0	158.0	158.0
BOD, S_e, mg/ℓ	7.0	5.0	5.0
x	0.66	0.73	0.74
ΔX_v, g/d	1.09	3.84	6.16

SOLUTION:

$$\Delta X_v = aS_r - bxX_v$$

UNIT	1	2	3
ΔX_v, g/d	1.09	3.84	6.16
X_v, g	40.4	44.2	46.2
S_r, g/d	4.38	12.8	18.3
$\Delta X_v / xX_v$	0.041	0.118	0.179
S_r / xX_v	0.164	0.393	0.533

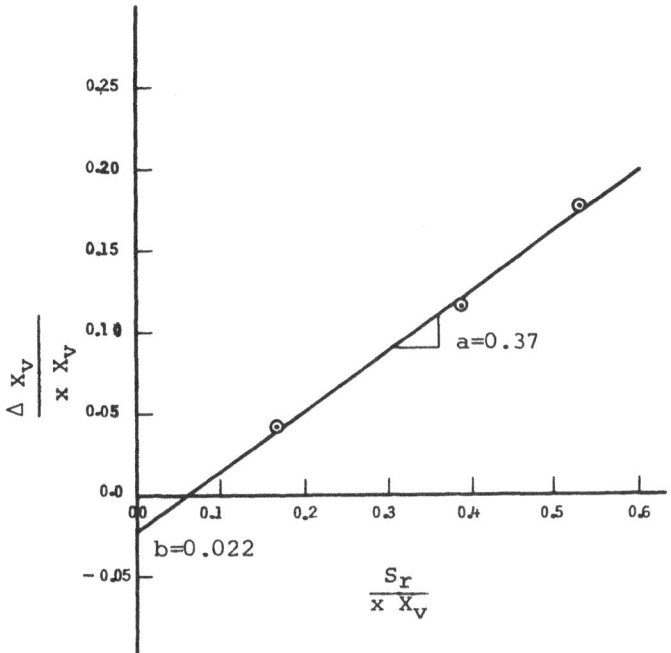

Fig. 7: Sludge Production Relationship

4. An organic chemicals plant is designing an activated sludge plant. The following data was generated from a pilot plant study:

Flow	=	11356.2 m³/day
BOD₅	=	935 mg/ℓ
NH₃-N	=	5 mg/ℓ
a	=	0.62
b	=	0.15/day
MLVSS	=	3000 mg/ℓ
BOD₅ eff	=	25 mg/ℓ
F/M	=	0.2

Compute the nitrogen requirements.

SOLUTION:

$$\text{lb/day N} = \frac{0.123 \times \Delta x_v}{0.77} + 0.07 \frac{(0.77 - x)\Delta x_v}{0.77}$$

$$\Delta x_v = aS_r \cdot Q - bxX_v \cdot Qt$$

$$S_r = BOD_5 - BOD_{eff} = 935 - 25 = 910 \text{ mg/}\ell$$

$$F/M = \frac{S_o}{X_v t} \qquad t = \frac{S_o}{X_v (F/M)} = \frac{935}{3000(0.2)} = 1.56 \text{ days}$$

$$S_r = 910 \text{ mg/}\ell \times 11356.2 \text{ m}^3/\text{day} \times \frac{1 \text{ kg}}{1000 \text{ g}} = 10334 \text{ kg/day}$$

$$X_v = 3000 \text{ mg/}\ell \times 11356.2 \text{ m}^3/\text{day} \times \frac{1 \text{ kg}}{1000 \text{ g}} \; 1.56 \text{ days} = 53147 \text{ kg}$$

$$x = \frac{aS_r + bX_v - \sqrt{(aS_r + bX_v)^2 - 4bX_v(0.77aS_r)}}{2bX_v}$$

$$= \frac{0.62 \times 10334 + 0.15 \times 53147 - \sqrt{(0.62 \times 10334 + 0.15 \times 53147)^2 - 4 \times 0.15 \times 53147 \times 0.77 \times 0.62 \times 10334}}{2(0.15)(53147)}$$

$$= 0.46$$

$$\Delta X_v = 0.62(10334) - 0.15(0.46)(53147) = 2746 \text{ kg/day}$$

$$\text{kg/day N req'd} = \frac{0.123(0.46)(2746)}{0.77} + 0.07\frac{(0.77 - 0.46)(2746)}{0.77}$$

$$= 278.6 \text{ kg/day}$$

$$\text{kg/day N available} = 5 \text{ mg/}\ell \cdot 11356 \text{ m}^3/\text{day} \cdot 10^3 \text{lt/m}^3 \, 1 \text{ kg/}10^6 \text{mg}$$

$$= 56.78 \text{ kg/day}$$

$$\begin{aligned}
\text{kg/day N to be supplied} &= \text{kg/day N req'd} - \text{kg/day N available} \\
&= 278.6 - 56.78 \\
&= 221.82 \text{ kg/day}
\end{aligned}$$

OXYGEN TRANSFER AND AERATION EQUIPMENT SELECTION

1. An aeration basin has a volume of 9463.5 m^3 with an oxygen requirement of 454.54 kg/hr, for the following conditions:

$$
\begin{array}{ll}
\text{temperature} & = 25°C \\
\text{CL} & = 1.5 \text{ mg/}\ell \\
\alpha \text{ diffused air} & = 0.75 \\
\alpha \text{ turbine air} & = 0.85 \\
\alpha \text{ surface air} & = 0.85 \\
\beta & = 0.90
\end{array}
$$

a) Design an aeration system using:

    1) air diffusers
    2) turbine aerators
    3) high speed surface aerators
    4) static aerators

    Adjust the basin geometry to the equipment selected.

b) Estimate the Kw-hr for each case.

SOLUTION:

$$O_{2 \text{ req}} = R_r = 454.54 \text{ kg/hr}$$

1) Diffused Air System:

    a) Assume

$$D = 4.57 \text{ m} \qquad W = 7.62 \text{ m} \quad \rightarrow \quad L = 271.64 \text{ m}$$

      Assume seran wrapped tubes with: (see following page)

$$C = 0.0032 \qquad m = 1.283 \qquad p = 0.571$$

$$\ell - m = -1.64 \qquad G_s = 0.2832 \text{ m}^3/\text{min}$$

$$N = C \, G_s^{(\ell - m)} \, \frac{H^m}{W^p} \, (C_{sw} - C_L) \, 1.02^{(T - 20)} \, \alpha \text{ diffused}$$

$$C_{sw} = \beta C_{sm} = \beta C_s [\frac{1}{2}(P_b + \frac{O_t}{21})]$$

$$= (0.9)(8.4)[\frac{1}{2}(1.44 + 0.92)] = 8.92$$

$$P_b = P_o + \gamma h = 1 \text{ atm} + (\frac{1 \text{ g}}{cm^3} \times \frac{980 \text{ cm}}{s^2} \times 457 \text{ cm})/(1.014 \times 10^6 \frac{dynes}{cm^2-atm})$$

$$= 1.44 \text{ atm}$$

Assume 10% efficiency of O_2 transfer

$$O_t = \frac{(21 - 0.1(21))}{(21 - 0.1(21) + 79)} \times 100 = 19.3\%$$

$$N = 0.00032 (0.2832 \text{ m}^3/\text{min})^{-1.64} \frac{(4.57 \text{ m})^{1.283}}{(7.62 \text{ m})^{0.571}} \times (8.92 - 1.5) \times$$

$$1.024^{(25-20)} \times 0.75$$

$$= 0.358 \text{ kg } O_2/ \text{ hr-unit}$$

$$\text{\# Tubes} = \frac{R_r}{N} = \frac{454.54 \text{ kg } O_2/\text{hr}}{0.358 \text{ kg } O_2/\text{hr-unit}} = 1267 \text{ units}$$

$$\text{Spacing} = \frac{271.34 \text{ m} \times 100 \text{ cm/m}}{1267 \text{ units}} = 21.44 \text{ cm/units}$$

b) Actual Air Flow Requirements:

$$G_s = \frac{454.54 \text{ kg } O_2/\text{hr}}{0.45 \text{ kg } O_2/0.34 \text{ sm}} \times \frac{100\%}{21\%} \times \frac{1 \text{ hr}}{60 \text{ min}} \times \frac{1}{0.10 \text{ eff}} \times \frac{1}{0.75 \text{ eff}}$$

$$= 363 \text{ s.m}^3/\text{min} \times 1 \text{ min}/60 \text{ sec}$$

$$= 6.05 \text{ m}^3/\text{s}$$

Now:

Assume: $P_c = 5.513 \times 10^4 \text{ N/m}$ or   0.544 atm

$$HP = \frac{G_s P_c}{746 \ E_c} \qquad E = 70\% \rightarrow HP = \frac{6.05 \text{ m}^3/\text{s} \times 5.513 \times 10^4 \text{ N/m}^2}{746 (\text{Nm/s-HP}) \times 0.70}$$

$$= 639$$

2) Turbine Aeration:

Assume $D = 4.57$ m $V = 9465$ m^3 $A = 2071.2$ m^2

Assume   204.5 m^2/aerator

Therefore # Aerators = A/204.5 = 10 aerators

Total O_2 transfer   = 454.54 kg O_2/hr (given)

Therefore O_2 transfer/aerator = 454.54/10 = 45.45 kg O_2/hr-unit

Compute Air Flow:

$$O_2 \text{ transfer eff.} = \frac{\text{kg } O_2/\text{hr-unit} \times 1 \text{ hr/60 min}}{G_s \times 0.232 \text{ kg } O_2/\text{kg air} \times 1.2}$$

Assume

O_2 transfer eff. = 0.2

Therefore $0.2 = \dfrac{45.45}{G_s \times 0.232 \times 60 \times 1.2}$ →

$$G_s = \frac{45.45}{0.2 \times 0.232 \times 60 \times 1.2}$$

$$= 13.6 \text{ m}^3/\text{min/unit} \times 1 \text{ min/60 s} = 0.227 \text{ m}^3/\text{s-unit}$$

therefore compressor HP

$$HP_c = \frac{G_s P_c \times 144}{746 \text{ E}}$$ where $P_c = 41365$ N/m^3 (assumed)

E = compressor efficiency
= 0.7 (assumed)

$$HP_c = \frac{0.227 \times 41365}{746 \times 0.7} = 18 \text{ HP required}$$

For Actual Oxygen Transfer Conditions:

$$HP_{req} = \frac{HP_c}{\dfrac{\beta C_{sw} - C_L}{C_s} \; 1.024^{(T-20)} \alpha \text{ turbine}}$$

$$= \frac{18}{\dfrac{0.9(9.92) - 1.5}{9.1} \times 1.024^5 \times 0.85} = 23 \text{ HP}$$

Assume:

$$HP_r / HP_c = 1 \quad \rightarrow \quad HP_r = HP_c = 23 \text{ HP} \quad \rightarrow \quad HP_{total} = HP_c + HP_r$$

$$= 46 \text{ HP/unit}$$

Therefore,

$$\text{Transfer eff.} = \frac{\text{kg } O_2/\text{hr-unit}}{HP_{total}/\text{unit}} = \frac{45.45 \text{ kg } O_2/\text{hr-unit}}{46 \text{ HP}}$$

$$= 0.99 \text{ kg } O_2/\text{HP}$$

$$\text{Total HP} = HP_{total}/\text{unit} \times \text{no. units} = 46 \text{ HP/unit} \times 10 \text{ units}$$

$$= 460 \text{ HP}$$

3) High Speed Surface Aerators:

Assume: $D = 3.05 \text{ m}$ $W = 39.63 \text{ m}$ $\rightarrow$ $L = 7.83 \text{ m}$

$$N = N_o \left(\frac{0.9(9.92) - 1.5}{9.1} \right) 1.024^{(25-20)} (0.85)$$

$$= 0.78 N_o$$

Select a Power Level:

0.066 HP/m^3 $N_o = 1.28 \text{ kg } O_2/\text{HP-hr}$

Therefore, $N = 0.78(1.28) = 1.00 \text{ kg } O_2/\text{HP-hr}$

$$HP = R/N = \frac{454.54 \text{ kg } O_2/\text{hr}}{1 \text{ kg } O_2/\text{HP-hr}} = 455 \text{ HP}$$

Power Level = 455/9463.5 = 0.048 (checks)

Therefore, HP = 455 HP and N = 1.00 kg O_2/HP-hr

Use 9 aerators at 50 HP each;

4) Static Aerators:

Assume: D = 4.57 m   W = 30.5 m   L = 67.89 m

O_2 transfer efficiency = 10% G_s = 0.283 m^3/min/unit

= 0.00472 m^3/s

$$O_2 \text{ transfer eff.} = \frac{N_o}{G_s \times 0.232 \times 3600 \times 1.2} \quad \rightarrow \quad N_o$$

N_o = O_2 trans. eff. x G_s x 0.232 x 3600 x 1.2

= 0.1 x 0.00472 x 0.232 x 3600 x 1.2

= 0.473 kg O_2/hr-unit

$$N = N_o (\frac{\beta C_{sw} - C_L}{C_s}) \alpha \text{ diff } x \ 1.024^{(T-20)}$$

$$= 0.473 (\frac{0.9(9.92) - 1.5}{9.1}) 0.75 \times 1.024^5$$

= 0.326 kg O_2/hr-unit

units = R/N = 454.54 kg/0.326 = 1394 units

Spacing: E = 0.7 ,   P = 55154 N/m^2 (assumed)

$$HP = \frac{G_{total} \times P \times 144}{746(E)} \quad \rightarrow G_{total} = 1395 \text{ units x } 0.00472 \ m^3/s$$

$$= \frac{6.58 \times 55154}{746(0.7)} \qquad = 6.58 \ m^3/s$$

= 695 HP   or

Power Level = 695/9463.5 = 73.44 HP/1000 m^3

Summary

D m	W m	L m	O$_2$ trans. eff.	Spacing cm	Kw	Power Level $\frac{Kw}{1000 \ m^3}$	N $\frac{kg \ O_2}{hr-unit}$	Units
Diffuser Air:								
4.6	4.6	272	7.5%	21.3	474	50.0	0.36	1267
Turbine Aeration:								
4.6	15.2	136	20.0%	-	343	36.3	-	10
High Speed Surface Aerators:								
3.0	39.6	79.3	-	-	340	35.9	0.99	9
Static Aerators:								
4.6	30.5	68.6	10.0%	-	526	55.2	0.33	1389

ACTIVATED SLUDGE

1. An activated sludge plant is to be designed to meet the following requirements:

    Flow = 3785.4 m^3/day
    BOD$_5$ = 720.0 mg/ℓ (Avg.)
    BOD$_5$ = 1000.0 mg/ℓ (Max.)

    The summer effluent soluble BOD$_5$ (20°C) is to be 25 mg/ℓ.
    The effluent suspended solids is estimated as 40 mg/ℓ.
    The following data was generated from pilot plant studies:

BOD removal rate coefficient	= 16/day @ 20°C
Maximum F/M	= 0.7
a	= 0.58
a´	= 0.35
b	= 0.15/day @ 25°C
MLVSS	= 3200 mg/ℓ
θ_s	= 1.08

% volatile   = 80
f   = 0.3 mg BOD/mg SS

Design:

1) the aeration basin volume
2) the excess sludge, ΔX_v
3) the oxygen requirements

Compute the soluble and total effluent BOD during winter operation with an air temperature of $-12.2°C$ assuming an influent wastewater temperature of $26.67°C$.

SOLUTION:

Estimate the underflow solids from the final clarifier as 20,000 mg/ℓ (from experimental data)

Using an average recycle rate of 25%, compute the MLSS and MLVSS

$$X_a = \frac{RX_R}{(Q + R)} = \frac{0.25 \ (20,000)}{1.25} = 4000 \ mg/\ell$$

$$X_v = 0.8 \times 4000 = 3200 \ mg/\ell$$

For a solids flux of 232 kg/m²/day the clarifier area is:

$$\frac{(Q + R)X_a}{flux(1000)} = \frac{1.25(3785)(4000)}{232 \times 1000} = 81.6 \ m^2$$

The aeration basin volume for an effluent soluble BOD of 25 mg/ℓ is computed:

$$k \ \frac{S_e}{S_o} = \frac{S_r}{X_v t}$$

$$16 \ \frac{25}{720} = \frac{695}{3200t}$$

$$t = 0.39 \ days = 9.4 \ hrs$$

$$V = 0.39 \times 3785 \ m^3 = 1476 \ m^3$$

$$F/M = \frac{720}{3200 \times 0.39} = 0.58$$

Therefore, K controls; design for $F/M = 0.58$

The degradable fraction is computed:

$$X_v, \text{ kg} = \frac{3200 \times 1476}{1000} = 4723 \text{ kg}$$

$$S_r = \frac{695 \times 3785}{1000} = 2630 \text{ kg}$$

$$x = \frac{sS_r - bX_v - \sqrt{(aS_r - bX_v)^2 - 4bX_v(0.77aS_r)}}{2bX_v}$$

$$= 0.68$$

The excess sludge, ΔX_v is:

$$\Delta X_v = aS_r - bxX_v$$

$$= 0.58(5830) - 0.15(0.68)(10,400)$$

$$= 1054 \text{ kg/day}$$

The effluent suspended solids is:

$$\frac{40 \text{ mg/}\ell \times 3785 \text{ m}^3/\text{day}}{1000} = 151 \text{ kg/day}$$

Sludge for disposal is 904 kg/day

The oxygen requirements under maximum conditions are:

$$O_2/\text{day} = a'S_r + 1.4bxX_vV \qquad b' = 1.4b$$

$$= \frac{0.35 \ (980 \times 3785)}{1000} + \frac{(1.4 \times 0.15) \times 0.68 \times 10400 \times 1476}{1000}$$

$$= 3490 \text{ kg/day}$$

The estimated Kw in the aeration basins is 131.2 Kw. The aeration basin temperature using low speed surface aerators is estimated:

$$T_w = \frac{T_i Q + 1864 T_a \, Kw}{Q + 1864 \, Kw}$$

$Q = 158,000$ kg/hr

$$= \frac{26.7 \times 158,000 + 1864 \times 131.2 \, (-12.2)}{158,000 + 1864 \times 131.2}$$

$= 3.1°C$

$$K_{5°} = K_{20°} \, 1.08^{(5-20)}$$

$$= \frac{16}{3.2} = 5.0$$

$$S_e = \frac{S_o^2}{KX_v t + S_o}$$

$$= 75 \text{ mg}/\ell$$

The total effluent BOD is 75 mg/ℓ + 0.30 mg BOD/mg SS x 40

$= 87$ mg/ℓ

2. The following data was obtained on the zone settling velocity of an activated sludge:

MLSS, mg/ℓ	ZSV, m/hr
3000	3.96
6000	1.22
10000	0.49
20000	0.15

For an underflow concentration of 20,000 mg/ℓ, size a final clarifier for a MLSS of 4000 mg/ℓ. Compute the recirculation rate. The wastewater flow is 18927 m^3/day.

SOLUTION:

MLSS, mg/ℓ	MLSS, kg/m^3	ZSV, m/hr	ZSV, m/day	G kg/m^2/day
5000	3.00	3.96	95.1	285.3
6000	6.00	1.22	29.3	126.7
10000	10.02	0.49	11.7	117.2
20000	3.98	0.15	3.7	73.3

Assume X_R = 20000 mg/ℓ $\qquad$ X_A = 4000 mg/ℓ

Given Q = 3785 m^3/day

Compute recirculation rate and size the final clarifier

$$R/Q = \frac{X_A}{X_R - X_A} = \frac{4000 \text{ mg}/\ell}{(20000 - 4000)\text{mg}/\ell} = 0.25$$

Q = 3785 m^3/day $\qquad$ R = 946 m^3/day

Size clarifier for overflow rate:

ZSV = 2.44 m/hr $\quad$ for $\quad$ C = 4000 mg/ℓ = X_A (see Figure 8)

O.R.= 24 x ZSV x F_C $\qquad$ where $\qquad$ F_C = clarifier scale-up factor, assume 0.7

$\qquad$ = 24 x 2.44 x 0.7

$\qquad$ = 41 m^3/m^2/day

$$A_1 = \frac{Q \text{ m}^3/\text{day}}{\text{O.R. m}^3/\text{m}^2\text{-day}} = \frac{18925 \text{ m}^3/\text{day}}{41 \text{ m}^3/\text{m}^2\text{-day}} = 461 \text{ m}^2$$

Size clarifier for thickening:

$$A_2 = \frac{(Q + R)X_V}{\text{Solids flux}} \qquad \begin{array}{l} \text{For } X_V = 4000 \\ \text{Solids flux} = \mathbf{232.2 \text{ kg/m}^2/\text{day}} \\ \text{(See Figure 9)} \end{array}$$

$$= \frac{(946 + 3785)(4000)}{232.2 \ (1000)} = 81.5 \text{ m}^2$$

Therefore, use A_1 = 461 m^2 for design;

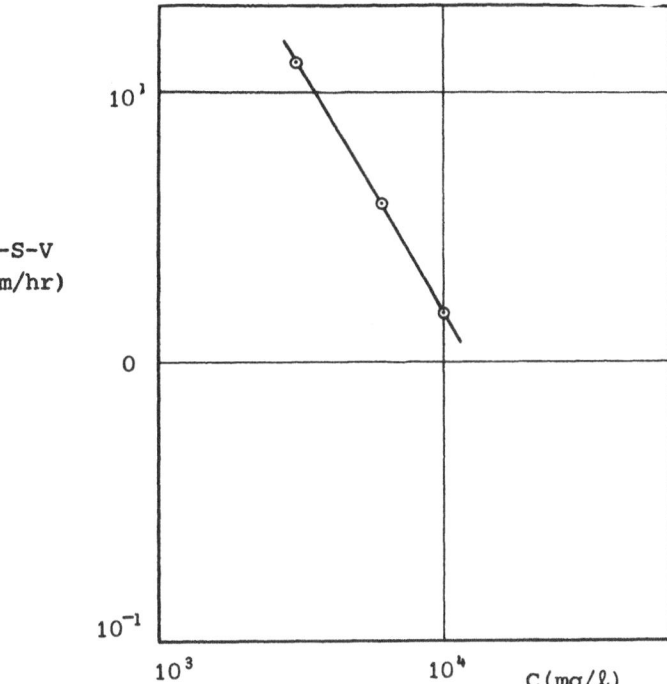

Fig. 8 : Relation of ZSV to C.

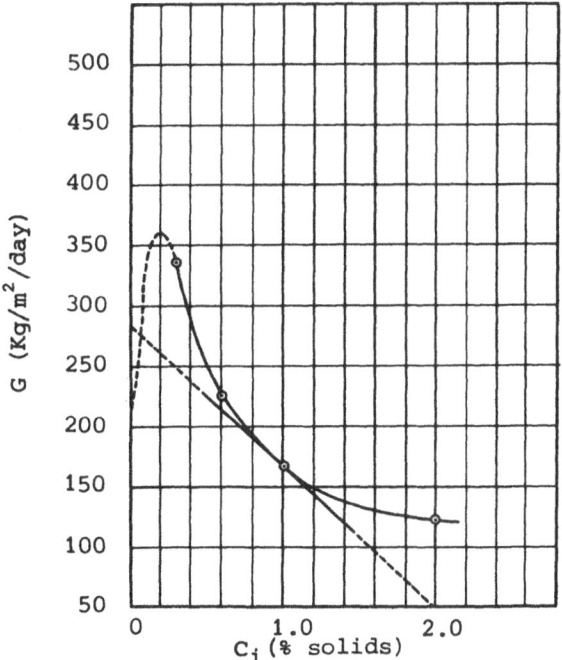

Fig. 9 : Relation of G to C_i

AERATED LAGOON

1. A brewery wastewater is to be pre-treated in an aerobic lagoon with a retention period of one day. Estimate the effluent BOD and suspended solids.

$$BOD_5 = 2500 \text{ mg/}\ell$$
$$k = 40/\text{day}$$
$$a = 0.5$$
$$\text{Inf SS} = 500 \text{ mg/}\ell$$
$$b = 0.1/\text{day}$$

SOLUTION:

$$\frac{S_e}{S_o} = \frac{1 + bt}{akt}$$

$$S_e = \frac{1 + 0.1/\text{day}(1 \text{ day})}{0.5(40/\text{day})(1 \text{ day})} \; (2500 \text{ mg/}\ell)$$

$$= 137 \text{ mg/}\ell \quad \text{as } BOD_5$$

$$X_v = \frac{aS_r}{1 + bt} = \frac{(0.5)(2500 - 137.5) \text{ mg/}\ell}{1 + (0.1/\text{day})(1 \text{ day})} = 1074 \text{ mg/}\ell$$
as biological VSS

$$\text{Effluent BOD} = 137 + 0.3 \; \frac{\text{mg BOD}}{\text{mg VSS}} \; (1074 \text{ mg/}\ell \text{ VSS}) = 459 \text{ mg/}\ell \text{ BOD}$$

$$\text{Effluent VSS} = 1074 + 500 = 1574 \text{ mg/}\ell$$

ANAEROBIC TREATMENT

(See article by Downing and Kell, "General Concepts of Anaerobic Treatment)

1. A wastewater contains anaerobically digestible organic matter composed mainly of carbohydrates having an empirical formula $C_{10}H_{22}O_{10}$; calculate

   a) the approximate composition of the gas produced by weight and volume.

b) its calorific value.

c) the concentration of organic matter that would have to be in the waste in order for the digester to be maintained at $35°C$ using the digester gas as the only fuel (assuming no heat losses) when the temperature of the incoming wastewater is $15°C$.

Take the calorific value of pure methane at STP to be 37.1 J/ml.

SOLUTION:

From Buswell's equation, the number of molecules of carbon dioxide and methane produced from 1 molecule of the carbohydrate are respectively 4.75 and 5.25. The proportion of the two gases by volume are thus 47.5 and 52.5%. The proportions by weight are calculated by multiplying the number of molecules of each by the molecular weights and expressing the weights as percentages of the total. This leads to 28.6 percent methane and 71.4 percent carbon dioxide (using round figures of 16 and 44 for the molecular weights). Since the mixture of gases contains 52.5 percent methane by volume the calorific value will be the same percentage of that of the pure gas, that is 19.4 J/ml; If the organic content of the wastewater is x percent then there are x kilo in 100 . Since from Buswell's equation 304 kg carbohydrate yield 84 kg methane x kilo yield 84x/304 kg methane; some 16 g (i.e. the gram molecular weight) methane occupy 22.4 ℓ at STP.

Thus 84 x/304 kg methane would occupy

$$\frac{84 \cdot 10^3x \cdot 22.4 \cdot 10^3ml}{16 \cdot 304} = 0.39 \cdot 10^6x \text{ ml}$$

$$= 37.1 \cdot 0.39 \cdot 10^6x \text{ J}$$

(Since 4.2 J = 1 Cal)

$$= 8.8 \cdot 0.39 \cdot 10^6x \text{ Cals}$$

The number of calories required to heat 100 ℓ through 20°C would be

$$100 \cdot 10^3 \cdot 20$$

Thus   x = 0.6 percent.

2. An anaerobic digester is operating at a pH value of 7.3 and producing gas containing 40 percent CO_2. Estimate the concentration to which the concentration of acetic acid could rise before the pH fell below the minimum level permitting satisfactory digestion (say 6.4).

SOLUTION:

As can be seen in the text given as reference

$$HCO_3^- + HAc \rightleftharpoons H_2O + CO_2 + Ac^-$$

thus, 60 g acetic acid would neutralise 61 g bicarbonate alkalinity. From Fig. 3 of the reference given above, a digester having a pH value of 7.3 could contain about 5000 mg/ℓ alkalinity when the carbon dioxide content of the gas was about 40 percent. Alkalinity could thus be reduced according to Fig.3 by just over 4000 mg/ℓ before pH value fell to 6.4, so that tolerable concentration of fatty acids (expressed as acetic acid) would be about 4000 mg/ℓ;

3. A wastewater containing an organic substrate on which methano-bacteria grow with a minimum doubling time of 2 days and have a Michaelis constant of 50 mg/ℓ is to be treated in a uniformly-mixed continuous digester without recycle; calculate

    a) the minimum retention time necessary to maintain digestion.

    b) the concentration of substrate in the effluent when the retention time is twice the minimum.

SOLUTION:

Doubling time, t_D, is related to growth constant k_m by the equation

$$\log_e 2 = 0.693 = k_m t_D$$

Thus if t_D =2 days

$$k_m = 0.35/d$$

Using a slight simpler approach than that leading to Equation 9 (see reference) one can say that to retain a microbial population in the reactor the minimum sludge-age must just equal the reciprocal of the maximum growth-constant of the bacteria. Thus

$$T_{min} = 1/0.35 \text{ d}$$

$$= 2.86 \text{ d}$$

Again using an approach which is slightly simpler than that leading to equation 8 (see reference) in that the separate fractions of the substrate used for growth and maintenance are not distinguished leads to

$$1/T = k_m S/(K_s + S)$$

as the equation from which the concentration of substrate issuing from the reactor can be calculated.

When

$$T = 2T_{min} = 5.72 \text{ d}$$

$$S = 50 \text{ mg}/\ell$$

WASTE STABILIZATION PONDS

1. Given the following information, design a pond system to produce a 90 percent reduction in BOD.

Topic	Value
BOD, Influent 5-day, 20°C	250 mg/ℓ
BOD, Influent, Ultimate, 20°C	305 mg/ℓ
BOD, Effluent 5-day 20°C (90% removal)	25 mg/ℓ
BOD, Effluent, Ultimate, 15°C	27 mg/ℓ
Photosynthetic Efficiency (f)	4 percent
Quantity	7570 m^3/day
Reaction Coefficient (Base e) k_1 (5°C)	0.102/day
Reaction Coefficient K_1 (15°C)	0.24/day
Reaction Coefficient K_1 (20°C)	0.35/day
Reaction Coefficient K_1 (30°C)	0.80/day
Reaction Coefficient K_1 (35°C)	1.2/day
Temperature (water, coldest month)	5°C
Temperature Coefficient θ	1.072 to 1.085

SOLUTION:

Design Based on Surface Loading

In the southwestern USA, many ponds have been built on
the basis of 5.6 gm of 5-day BOD, 20°C per sq.m per day.
For the northern areas in the USA, these loadings have
been reduced to as little as 1.12 gm per sq.m per day.
The organic load to the example pond is

250 mg/ℓ x 7570 m^3/day x kg/10^{6}mg x 10^{3}ℓ/m^3 = 1880 kg/day

Using a loading of 0.0056 kg/sq.m per day and a depth of
1.80 m , the surface area and volume are

$$A = \frac{1880 \text{ kg/day}}{0.0056 \text{ kg/m}^2\text{-day}} = 335714 \text{ m}^2 = 33.6 \text{ Ha}$$

V = 335714 m^2 x 1.80 m = 604285 m^3

The retention time t is

$$t = \frac{604285 \text{ m}^3}{7570 \text{ m}^3/\text{day}} = 80 \text{ days}$$

The pond layout should be one rectangular pond or two
equally-sized rectangular ponds, with the flexibility
of introducing wastes into either pond. Ponds should
always be as close to rectangular as possible to reduce
dead spaces.

Design Based on Empirical Equation (GLOYNA-HERMAN RELATIONSHIP)

The use of this equation is particularly applicable for
rapid estimates involving the effects of temperature. Note
in this problem it is suggested that the design temperature
be based on the coldest month and on the 5-day BOD, 20°C.

It is necessary to correct for temperature and convert to
ultimate BOD. Also, to compensate for a possible sludge
effect, estimates should be based on ultimate BOD values.
However, it should be recalled that retention time will be
increased slightly as a result of evaporation in some areas
of the country where evaporation exceeds rainfall.

$$V = 0.0349 \ Q \ S_o \ \Theta^{(35-T)} ff' \qquad (1')$$

where

V	=	Pond Vol. (cu.m)
S_o	=	Influent ultimated BOD (mg/ℓ)
Q	=	Flow (cu.m/day)
T	=	Average temperature of coldest month (°C)
f	=	Algal toxicity factor (1.0 for domestic wastes)
f'	=	Sulfide correction factor (1.0 if SO_4 is less than 500 mg/ℓ)

$$V = 0.0349 \times 7570 \ m^3/day \times 305 \ mg/ℓ(1.072^{(35-5)}(1.0)(1.0))$$
$$= 648,730 \ m^3$$

With a depth of 1.80 m, the surface area, as based on a facultative zone of 1.50 m is 648,730/1.50 = 432490 m^2, and the retention time is 86 days; The surface loading is

$$\frac{1880 \ kg/day}{432490 \ m^2} = 0.0043 \ kg/m^2\text{-}day$$

Design Based on Basic Kinetic Equation (MARAIS RELATIONSHIP)

The BOD removal relatiohship from 2 single, completely mixed ponds can be expressed as:

$$\frac{S_e}{S_o} = \frac{1}{1 + K_1 t} \qquad (2)$$

where

S_o	=	Influent BOD (mg/ℓ)
S_e	=	Effluend BOD
K_1	=	Organic removal coefficient (days^{-1})
t	=	Detention time (days)

532

For a series of ponds, the relationship is

$$\frac{S_e}{S_o} = \frac{1}{(1 + K_1 t) \ldots \ldots (1 + K_n t)} \tag{3}$$

The organic removal coefficient is a function of water temperature:

$$K_{35} = K_T \ \theta^{(35-T)}$$

For a single pond, and assuming winter conditions control, then K_T can be calculated

$$1.2/\text{day} = K_T \ 1.085^{(35-5)}$$

$$K_T = 0.104$$

Assuming a 25 mg/ℓ effluent requirement, then

$$25 = \frac{250}{1 + (0.104)t}$$

$$t = 86 \text{ days}$$

The surface area, assuming an aerobic layer depth of 1.0 m

Volume = (86 days)(7570 m³/day) = 651,020 m³

651,020 m³/1.0 m = 651,020 m²

The surface loading $= \frac{(250 \text{ mg/ℓ}) (7570 \text{ m}^3/\text{day}) (\text{kg}/10^6\text{mg}) (10^3 ℓ/\text{m}^3)}{651,020 \text{ m}^2}$

$$= 0.0029 \text{ kg/m}^2\text{-day}$$

It may be desirable to use several ponds in series. Assume a series of three ponds with the following organic removal velocities (5°C).

Estimated Value *	Predicted S_e^*
$K_1 = 0.114$	$S_{e_1} = 75$ mg/ℓ
$K_2 = 0.075$	$S_{e_2} = 40$ mg/ℓ
$K_3 = 0.063$	$S_{e_3} = 25$ mg/ℓ

First Pond Volume Requirement:

$$\frac{75}{250} = \frac{1}{1 + (0.114)t}$$

$8.55t = 175$

$t \quad = 20.5$ days

Volume $= (20.5 \text{ days})(7570 \text{ m}^3/\text{day}) = 155,190 \text{ m}^3$

Second Pond Volume Requirement:

$$\frac{40}{75} = \frac{1}{1 + 0.075t}$$

$3.0t = 35$

$t \quad = 11.7$ days

Volume $= (11.7 \text{ days})(7570 \text{ m}^3/\text{day}) = 88,570 \text{ m}^3$

Third Pond Volume Requirement:

$$\frac{25}{40} = \frac{1}{1 + 0.063t}$$

$1.57t = 15$

$t \quad = 9.6$ days

volume $= (9.6 \text{ days})(7570 \text{ m}^3/\text{day}) = 72,670 \text{ m}^3$

---

* From treatibility studies.

The total retention time requirement and organic surface loading, assuming 1.0 m effective depth, to each pond is calculated as follows:

Pond #1

$$\text{Loading} = \frac{1880 \text{ kg/day}}{155,190 \text{ m}^3/1.0\text{m}} = \frac{0.0121 \text{ kg BOD}}{\text{m}^2\text{-day}} \qquad t = 20.5 \text{ days}$$

Pond #2

$$\text{Loading} = \frac{568 \text{ kg/day}}{88,570 \text{ m}^3/1.0 \text{ m}} = \frac{0.0064 \text{ kg BOD}}{\text{m}^2\text{-day}} \qquad t = 11.7 \text{ days}$$

Pond #3

$$\text{Loading} = \frac{303 \text{ kg/day}}{72,670 \text{ m}^3/1.0 \text{ m}} = \frac{0.0042 \text{ kg BOD}}{\text{m}^2\text{-day}} \qquad t = 9.6 \text{ days}$$

$$\text{Total Retention time} = 41.8 \text{ days}$$

2. Accelerated Photosynthetic Design System

    Design Criteria

Hydraulic Loading	$1123 - 1872 \text{ m}^2/\text{ha/day}$
Organic Loading	$0.0392\text{--}0.0784 \text{ kg BOD}_5/\text{m}^2/\text{day}$
Detention Time	$2 - 4$ days, depending on climatic conditions

    Design Basis

Average wastewater flow	$6056 \text{ m}^3/\text{day}$
Average wastewater BOD_5	$200 \text{ mg/}\ell$
Length of daily effluent drawoff	8 hours
Freeboard	0.30 m minimum
Detention time (winter)	4 days
Hydraulic loading	$1872 \text{ m}^3/\text{ha/day}$[a]
Organic loading	$0.0392 \text{ kg BOD}_5/\text{m}^2/\text{day}$

Use two equal size basins (3028 m^3 daily flow each) to provide flexibility during initial operation to allow better control of process parameters.

SOLUTION:

Basin Design

Hydraulic loading = 1872 m^3/ha/day

Surface area required = 6056/1872 = 3.2 ha

Organic loading = 392 kg BOD$_5$/ha/day

$$\text{Surface area required} = \frac{(6056)\,(200)\,(10^3\,\ell/m^3)\,(Kg/10^6 mg)}{392}$$

$$= 3.1 \text{ ha}$$

Hydraulic loading controls ÷ use 2 basins with 1.60 ha (16,000 m^2) each

Volume required = 6056 m^3/day (3 days) = 18168 m^3 total

$$= 9084 \text{ m}^3/\text{basin}$$

$$\text{Basin depth} = \frac{9084 \text{ m}^3}{16000 \text{ m}^2} = 0.57 \text{ m} \qquad = 0.57 \text{ m}$$

$$\text{add } 0.30 \text{ m freeboard} = \underline{0.30 \text{ m}}$$

$$\text{Total Channel Depth} = 0.87 \text{ m}$$

Basin outer dimensions 97.5 m x 166.2 m x 0.86 m deep (2 basins) 6.10 m channel width *(Fig. 10)*.

Aerator Design

Assume night time conditions control.
Use power level = 1.97 Kw/1000 m^3
Volume per basin = 9084 m^3
Horsepower per basin = 9084(1.97/1000) = 17.9 Kw per basin
minimum

Use standard brush aerators,
Use 2 units/basin
4 units total;

536

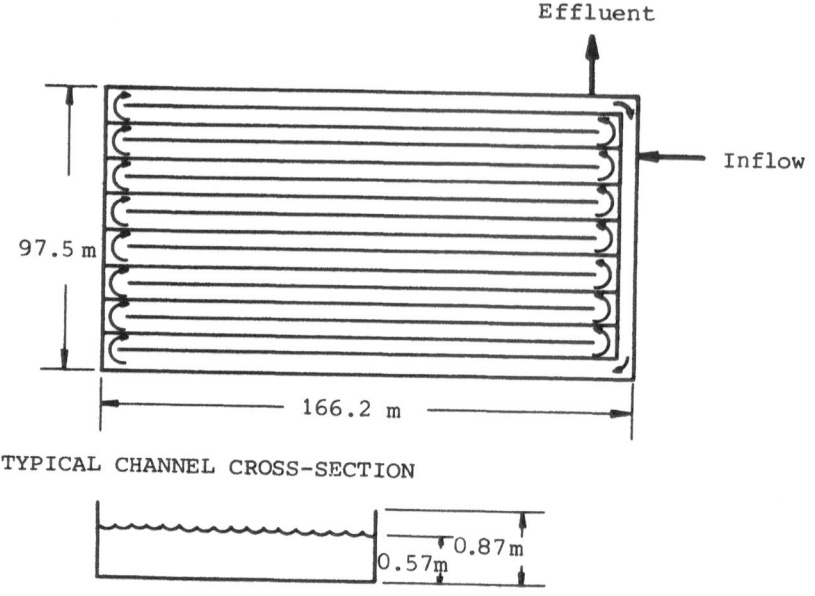

Effluent

Inflow

97.5 m

166.2 m

TYPICAL CHANNEL CROSS-SECTION

0.87 m

0.57 m

Fig. 10 : Dimensions of the Basin

ADVANCED BIOLOGICAL TREATMENT PROCESSES

(See Article by Downing & Kell,
"Advanced Biological Treatment Processes").

1. Settled sewage with a BOD of 250 mg/ℓ is to be treated in an
    activated sludge plant carrying 4000 mg/ℓ activated sludge
    in the aeration units. If the pH value of the sewage is 7,
    the temperature is 16°C and the ratio of the returned sludge
    flow to sewage flow is 1:1 estimate the minimum period of re-
    tention of sewage in the aeration units to maintain the con-
    sistent nitrification.

SOLUTION:

All equations mentioned in this problem are related to the
article by Dowing and Kell "Advanced Biological Treatment
Processes".

Using the symbols adopted in the text

L = 250

$X = 4000$

$P = 7.0$

$p = 1$

$\theta = 16$

It then follows from Equation 8 that

$k_n = 0.18(1 - 0.83(0.2))e^{0.12}$

$= 16/\text{day}$

From Equation 9, 10 and 11 we have

$X = (0.96)(1)(250)(0.2 + 0.25T_S^{-1/2})/2$

$= 23 + 29T_S^{1/2}$

and

$T_h = T_S/2$

The minimum period of retention which must just be exceeded to achieve consistent nitrification can then be calculated from Equation 7 (obtained as the limit of the inequality 6) so that

$T_S/2 = (23 + 29T_S^{-1/2})/(4000)(0.16)$

$T_S = 0.07 + 0.99/T_S^{1/2}$

This equation is most conveniently solved by trial and error. Thus as an initial attempt one might try $T_S = 0.2$ which leads to

$0.2 \neq * 0.07 + 0.09/0.45$

$\neq 0.27$

As a second attempt try $T_S = 0.3$ leading to

$0.3 \neq 0.07 \text{ T } 0.09/0.55$

$\neq 0.23$

A third attempt reveals the correct value of T_S as $0.25d$.

2. A nitrified mixed liquor containing 30 mg/ℓ nitrate-N, 0.2 mg/ℓ nitrite-N, 8 mg/ℓ dissolved oxygen and 3000 mg/ℓ suspended sludge passes through a settling tank to an anaerobic denitrifying unit. Calculate the concentration of methanol (as COD) needed to ensure complete denitrification assuming that in passing through the settling tank the supernatant to be denitrified is in contact with an average concentration of 2000 mg/ℓ sludge for 0.5 h and the rate of respiration of the sludge is 20 mg/ℓ h in the mixed liquor as it leaves the aeration unit (and the rate of consumption per unit mass remains constant.)

SOLUTION:

The change in content of dissolved oxygen in the effluent after passing through the sedimentation tank is 6.7 mg/ℓ. The concentration remaining is thus 1.3 mg/ℓ. Substitution of this value, plus those of the two forms of oxidised nitrogen in Equation 15 leads to M_c = 75.5 mg/ℓ.

INDEX

AUTHOR'S INDEX